Remote Sensing for Land Administration

Remote Sensing for Land Administration

Editors

Rohan Bennett
Peter van Oosterom
Christiaan Lemmen
Mila Koeva

MDPI • Basel • Beijing • Wuhan • Barcelona • Belgrade • Manchester • Tokyo • Cluj • Tianjin

Editors
Rohan Bennett
Swinburne University
of Technology
Australia

Peter van Oosterom
TU Delft
The Netherlands

Christiaan Lemmen
University of Twente
The Netherlands

Mila Koeva
University of Twente
The Netherlands

Editorial Office
MDPI
St. Alban-Anlage 66
4052 Basel, Switzerland

This is a reprint of articles from the Special Issue published online in the open access journal *Remote Sensing* (ISSN 2072-4292) (available at: https://www.mdpi.com/journal/remotesensing/special_issues/Land_Administration).

For citation purposes, cite each article independently as indicated on the article page online and as indicated below:

LastName, A.A.; LastName, B.B.; LastName, C.C. Article Title. *Journal Name* **Year**, *Article Number*, Page Range.

ISBN 978-3-03943-054-3 (Hbk)
ISBN 978-3-03943-055-0 (PDF)

Cover image courtesy of Paula Dijkstra.

Contents

About the Editors . . . . **vii**

Rohan Bennett, Peter van Oosterom, Christiaan Lemmen and Mila Koeva
Remote Sensing for Land Administration
Reprinted from: *Remote Sens.* **2020**, *12*, 2497, doi:10.3390/rs12152497 . . . . **1**

Seula Park and Ahram Song
Discrepancy Analysis for Detecting Candidate Parcels Requiring Update of Land Category in Cadastral Map Using Hyperspectral UAV Images: A Case Study in Jeonju, South Korea
Reprinted from: *Remote Sens.* **2020**, *12*, 354, doi:10.3390/rs12030354 . . . . **9**

Mila Koeva, Claudia Stöcker, Sophie Crommelinck, Serene Ho, Malumbo Chipofya, Jan Sahib, Rohan Bennett, Jaap Zevenbergen, George Vosselman, Christiaan Lemmen, Joep Crompvoets, Ine Buntinx, Gordon Wayumba, Robert Wayumba, Peter Ochieng Odwe, George Ted Osewe, Beatrice Chika and Valerie Pattyn
Innovative Remote Sensing Methodologies for Kenyan Land Tenure Mapping
Reprinted from: *Remote Sens.* **2020**, *12*, 273, doi:10.3390/rs12020273 . . . . **29**

Cheonjae Lee and Walter Timo de Vries
Bridging the Semantic Gap between Land Tenure and EO Data: Conceptual and Methodological Underpinnings for a Geospatially Informed Analysis
Reprinted from: *Remote Sens.* **2020**, *12*, 255, doi:10.3390/rs12020255 . . . . **57**

Sophie Crommelinck, Mila Koeva, Michael Ying Yang and George Vosselman
Application of Deep Learning for Delineation of Visible Cadastral Boundaries from Remote Sensing Imagery
Reprinted from: *Remote Sens.* **2019**, *11*, 2505, doi:10.3390/rs11212505 . . . . **83**

Mila Koeva, Shayan Nikoohemat, Sander Oude Elberink, Javier Morales, Christiaan Lemmen and Jaap Zevenbergen
Towards 3D Indoor Cadastre Based on Change Detection from Point Clouds
Reprinted from: *Remote Sens.* **2019**, *11*, 1972, doi:10.3390/rs11171972 . . . . **105**

Jingya Yan, Siow Wei Jaw, Kean Huat Soon, Andreas Wieser and Gerhard Schrotter
Towards an Underground Utilities 3D Data Model for Land Administration
Reprinted from: *Remote Sens.* **2019**, *11*, 1957, doi:10.3390/rs11171957 . . . . **127**

Xue Xia, Claudio Persello and Mila Koeva
Deep Fully Convolutional Networks for Cadastral Boundary Detection from UAV Images
Reprinted from: *Remote Sens.* **2019**, *11*, 1725, doi:10.3390/rs11141725 . . . . **149**

Bujar Fetai, Krištof Oštir, Mojca Kosmatin Fras and Anka Lisec
Extraction of Visible Boundaries for Cadastral Mapping Based on UAV Imagery
Reprinted from: *Remote Sens.* **2019**, *11*, 1510, doi:10.3390/rs11131510 . . . . **163**

Claudia Stöcker, Serene Ho, Placide Nkerabigwi, Cornelia Schmidt, Mila Koeva, Rohan Bennett and Jaap Zevenbergen
Unmanned Aerial System Imagery, Land Data and User Needs: A Socio-Technical Assessment in Rwanda
Reprinted from: *Remote Sens.* **2019**, *11*, 1035, doi:10.3390/rs11091035 . . . . **183**

About the Editors

Rohan Bennett is an Associate Professor in Information Systems with the Swinburne Business School, Melbourne, Australia; and a Geodetic Advisor with Netherlands Kadaster. He specializes in spatial information systems and land information management. He has previously held posts with the University of Twente (NL), Kadaster International (NL), and University of Melbourne (AU). He holds degrees in Science, Engineering, and a Ph.D. in Land Administration from the University of Melbourne. He led and worked on the Euro Commission H2020 project 'its4land' from 2016–17 and is currently involved with project work on the application of blockchain, UAVs, and auto feature extraction in land rights management.

Peter van Oosterom is a Professor of GIS Technology at the Faculty of Architecture and the Built Environment. The Chair concentrates on spatial information infrastructure. A central component is durable geoinformation that can be shared and re-used, based on joint definitions of data sets and services. Proceeding from technology and technology development, the Chair contributes to the realization of a spatial data infrastructure based on vario scale spatiotemporal 2D–5D models and processes for application areas like cadastre and real-time GIS. Van Oosterom has contributed to developments in 3D geometry and data modeling, the results of which are now part of the Oracle spatial Database Management System. Together with the Netherlands Cadastre, van Oosterom introduced a worldwide standard for the land administration domain model (ISO 19152).

Christiaan Lemmen is a Full Professor at ITC, the Faculty of Geo-information Science and Earth Observation, University of Twente, the Netherlands (part-time). He is also a Geodetic Advisor with Kadaster International, the international branch of the Netherlands Cadastre, Land Registry and Mapping Agency. Christiaan is an Editor of the International Standard for the Land Administration Domain Model (LADM), published as ISO – 19152. The second edition of LADM is under development since 2018. He is a Co-Chair of the Land Administration Domain Working Group of the Open GeoSpatial Consortium. Christiaan is the Director of the OICRF, which is the Office International du Cadastre et du Régime Foncier (the international Office for Cadastre and Land Records), a documentation center for land administration, and a permanent institution of the International Federation of Surveyors (FIG). He holds a Ph.D. in Land Administration from Delft University of Technology, The Netherlands.

Mila Koeva is an Assistant Professor working in 3D Land Information. She holds a Ph.D. in 3D modeling in architectural photogrammetry from the University of Architecture, Civil Engineering and Geodesy in Sofia. She also holds MSc. degrees in Engineering (Geodesy) from the same institution, obtained in 2001. Her main areas of expertise include 3D modeling and visualization, 3D Cadastre, 3D Land Information, UAV, digital photogrammetry, image processing, producing large-scale topographic, and cadastral maps, GIS, application of satellite imagery for updating cadastral information among others. Currently, she is also a Co-chair of ISPRS WG IV/10 "Advanced Geospatial Applications for Smart Cities and Regions" and a Contributing Editor for GIM International. In 2017, she was ranked as one of the TOP 3 teachers of ITC 2016–2017. From 2017–2020, she was the Project Coordinator of its4land, a multidisciplinary European Commission Horizon 2020 project, involving 8 academic and private-sector partners and 6 countries in Europe and Africa.

remote sensing

MDPI

Editorial

Remote Sensing for Land Administration

Rohan Bennett [1,2,*], Peter van Oosterom [3], Christiaan Lemmen [2,4] and Mila Koeva [4]

1 Swinburne Business School, Swinburne University of Technology, Hawthorn 3122, Victoria, Australia
2 Kadaster International, Land Registry and Mapping Agency of the Netherlands, 7311 KZ Apeldoorn, The Netherlands; chrit.lemmen@kadaster.nl
3 Faculty of Architecture and the Built Environment, Delft University of Technology, 2600 AA Delft, The Netherlands; p.j.m.vanoosterom@tudelft.nl
4 ITC Faculty, University of Twente, 7500 AE Enschede, The Netherlands; m.n.koeva@utwente.nl
* Correspondence: rohanbennett@swin.edu.au

Received: 29 July 2020; Accepted: 30 July 2020; Published: 4 August 2020

Abstract: Land administration constitutes the socio-technical systems that govern land tenure, use, value and development within a jurisdiction. The land parcel is the fundamental unit of analysis. Each parcel has identifiable boundaries, associated rights, and linked parties. Spatial information is fundamental. It represents the boundaries between land parcels and is embedded in cadastral sketches, plans, maps and databases. The boundaries are expressed in these records using mathematical or graphical descriptions. They are also expressed physically with monuments or natural features. Ideally, the recorded and physical expressions should align, however, in practice, this may not occur. This means some boundaries may be physically invisible, lacking accurate documentation, or potentially both. Emerging remote sensing tools and techniques offers great potential. Historically, the measurements used to produce recorded boundary representations were generated from ground-based surveying techniques. The approach was, and remains, entirely appropriate in many circumstances, although it can be timely, costly, and may only capture very limited contextual boundary information. Meanwhile, advances in remote sensing and photogrammetry offer improved measurement speeds, reduced costs, higher image resolutions, and enhanced sampling granularity. Applications of unmanned aerial vehicles (UAV), laser scanning, both airborne and terrestrial (LiDAR), radar interferometry, machine learning, and artificial intelligence techniques, all provide examples. Coupled with emergent societal challenges relating to poverty reduction, rapid urbanisation, vertical development, and complex infrastructure management, the contemporary motivation to use these new techniques is high. Fundamentally, they enable more rapid, cost-effective, and tailored approaches to 2D and 3D land data creation, analysis, and maintenance. This Special Issue hosts papers focusing on this intersection of emergent remote sensing tools and techniques, applied to the domain of land administration.

Keywords: UAV; LiDAR; automated feature extraction; cadaster; land registration; land use planning; SDGs

1. Introduction

Land administration is the process of recording, securing, and disseminating information about land tenure, value, use, and development, within a jurisdiction [1]. As a study area, soft systems methodology, socio-technical systems, and geo-information science provide contemporary theoretical foundations [2]. Conventionally, the primary unit of concern is the land parcel-a multi-dimensional spatial extent, with associated rights, and a party who is said to hold those rights [3]. In practice, land administration aggregates all the land parcel information into a system, enabled through the mandate of prescribed administrative roles, processes, and supportive technologies [4]. The functioning system enables land dispute resolution, transaction controls, credit access, land transfer, land valuation, land taxation, changes in land use, and development decisions [5]. Modern land administration systems utilize information technology: existing paper records are digitized, manual recording and dissemination processes digitalized, with all aspects being constituted into a publicly accessible land information system [6].

In more developed contexts, land administration systems have generally developed, over the centuries, from narrowly purposed land registries and cadastres, into whole-of-jurisdiction multi-purpose systems, to support the contemporary policy objectives of good governance and sustainable development [7,8]. In developing contexts, much effort is afforded in establishing or renewing systems. Challenges stemming from anachronous colonial-era systems; limited techniques for recording customary forms of tenure; the lack land rights records of women; the want of legal documentation for the legitmate rights of the poor and vulnerable; and scarce technical skills and capacity—all conspire to impede responsible land administration development [9–11]. Meanwhile, all systems, regardless of the country context, are challenged by the emergent demands of rapid urbanization, climate-change response, digital transformation (and e-services), gender equality, and demands for openness and transparency [12,13]. In practical terms, this results in research, development and implementations relating to 3D (and 4D) cadastres [14,15], domain standardization and interoperability [3,16,17] (e.g., ISO 19152 Land Administration Domain Model LADM and ISO TC211), land rights fractionalization and new database structures [18] (e.g., big data analytics, NoSQL, and blockchain), fit-for-purpose land administration [19], and novel land data collection approaches [20–22].

The abovementioned developments confirm spatial information is a central concern of land administration systems. Spatial information delineates the boundaries between lands parcels. These boundaries can be 'general' or 'fixed' in nature [1]. To be recognized, the boundaries require 'perception' of existence by a community; an explicitly defined 'purpose'; an actual or physical field 'presence', a defined 'period-in-time' of application, and a 'presentation' or documented representation (which may be digital) [23]. More recently, clarification of the interrelationship between the physical, documented, and digital aspects is provided [24,25].

Conventionally, ground-based survey methods, often underpinned by local or national geodetic networks, support making observations, demarcating boundaries, and surveying boundary information. This in turn enables the derivation of boundary coordinates [24]. These coordinates enable a parcel to be recorded and represented on survey sketches, cadastral plans, cadastral maps, and in the contemporary era, cadastral databases. Over time, surveying tools and techniques have improved: rudimentary rope surveys, plane table usage, theodolite application, and electronic distance measurement (EDM), have given way to total stations and high-precision global navigation satellite systems (GNSS) (Note: whilst the survey and receiver are ground-based, the space segment (i.e. GNSS satellites) are not) [1,8]. These approaches have been argued to deliver the prescribed coordinate accuracies, often set, it must be said, overly bluntly, at centimeter-level in surveying laws and regulations [19].

More recently, increasing attention is afforded to applying advances in photogrammetry and remote sensing to the domain land administration. In the context of land administration, these approaches are considered 'indirect' surveying techniques, in contrast to 'direct' or ground-based methods [26]. Photogrammetry is the techniques and tools for extracting multi-dimensional geospatial information from images, needed for mapping activities [27]. Remote sensing is the process of scanning or monitoring the physical characteristics of the surface of the earth, measuring the emitted radiation at a distance. For this purpose, cameras, sensors or scanners with RGB (red, green, blue), hyperspectral, LiDAR (light detection and ranging), or RADAR (radio detection and ranging) capabilities are used. These sensors can be mounted on conventional aerial vehicles (i.e., airplanes), space-borne satellites (i.e., enabling high and very-high resolution satellite imagery (HRSI and VHRSI), unmanned aerial vehicles (UAVs), amongst others [28]. It is worth noting that whilst application of photogrammetry and remote sensing to land administration has a lengthy history [29], it is the novel tools and techniques emerging in those fields that drive renewed and increased interest. These are specifically focused on the transfer of 'cadastral intelligence' from human actors to machines [30], and make use of enhanced image processing tools, artificial intelligence, and machine learning techniques [31].

The abovementioned developments in remote sensing and photogrammetry, applied to land administration, are the focus of this special issue. The aim is to provide a snapshot of contemporary experimentations, demonstrations, implementations, and impacts from a diverse range of case contexts. From a broad perspective, the special issue presents: (i) comparisons of alternate remote sensing techniques for 2D and 3D data capture relevant to land administration (including UAV imagery, VHRSI, RADAR, LiDAR, and multi-spectral approaches); (ii) design and testing of techniques for 2D and 3D cadastral feature extraction from remotely-sensed data sources (including machine learning, pattern recognition, neural networks, semi-automated methods, algorithm design, and object-based approaches); (iii) modelling of data production workflows for scaled 2D and 3D cadastral production (including segmentation techniques, line extraction, contour generation, and pre/post-processing requirements); and (iv) observations from illustrative cases highlighting leading practices in data integration and utilization for 2D and 3D land administration (including both city, provincial, and national level examples). More specifically, the special issue consists of nine (9) individual works; developed by multi-disciplined research and practitioner teams; across Europe, Asia, and Africa; variously using qualitative and quantitative research methods with a diversity of daya sources; all with reference back to land administration applications (Table 1). The next section outlines each work and synthesizes the overarching contribution of the special issue.

Table 1. Remote sensing applications for land administration presented in this issue.

Source	Title	Country	Applications	Techniques	Data
Park and Song	**Discrepancy Analysis for Detecting Candidate Parcels Requiring Update of Land Category in Cadastral Map Using Hyperspectral UAV Images: A Case Study in Jeonju, South Korea**	South Korea	Automated land use classification; land tenure and cadastral updating; land value	Convolutional neural network (CNN); Inconsistency comparison	UAV hyperspectral; cadastral map
Koeva et al.	**Innovative Remote Sensing Methodologies for Kenyan Land Tenure Mapping**	Kenya	Land tenure and cadastral mapping	UAV survey; Machine learning; Nominal Group Technique (NGT); Semi-Structured Interviews; Questionnaires; Group Discussion	UAV RGB; sketch maps; human perceptions
Lee and De Vries	**Bridging the Semantic Gap between Land Tenure and EO Data: Conceptual and Methodological Underpinnings for a Geospatially Informed Analysis**	North Korea	Land tenure and cadastral mapping; land use change	Research synthesis; Manual image interpretation	HRSI and aerial RGB (Google Earth); Landsat7; academic literature
Crommelinck et al.	**Application of Deep Learning for Delineation of Visible Cadastral Boundaries from Remote Sensing Imagery**	Ethiopia, Kenya, Rwanda	Land tenure and cadastral mapping	Image segmentation (MCG); machine learning (RF and CNN)	Aerial RGB UAV RGB; cadastral map
Koeva et al.	**Towards 3D Indoor Cadastre Based on Change Detection from Point Clouds**	The Netherlands	Land tenure and cadastral updating	Point cloud change detection techniques	Point clouds (mobile mapping system, Zeb-Revo); Riegl; architectural plans
Yan et al.	**Towards an Underground Utilities 3D Data Model for Land Administration**	Singapore	Land tenure and cadastral mapping; land use and development	3D modelling and visualisation	Stream EM GPR and Leica Pegasus Two photo and laser scanning data; cadastral data
Xia et al.	**Deep Fully Convolutional Networks for Cadastral Boundary Detection from UAV Images**	Rwanda	Land tenure and cadastral mapping and updating	Machine learning; Fully Convolutional Network (FCN); Multi-Resolution Segementation (MRS); Globalized Probability of Boundary (gPb)	UAV RGB; cadastral map

Table 1. *Cont.*

Source	Title	Country	Applications	Techniques	Data
Fetai et al.	**Extraction of Visible Boundaries for Cadastral Mapping Based on UAV Imagery**	Slovenia	Land tenure and cadastral updating and mapping	Exelis Visual Information Solutions (ENVI) feature extraction	UAV RGB; GNSS; cadastral map
Claudia Stöcker et al.	**Unmanned Aerial System Imagery, Land Data and User Needs: A Socio-Technical Assessment in Rwanda**	Rwanda	Land Use; Land Tenure; Land Development Cadastral Updating	NGT; interviews; workshop; UAV survey	UAV RGB; GNSS; cadastral map

2. Overview of Contributions

The works making up this Special Issue are presented in reverse chronological order of acceptance. This was considered the most straight forward approach to administer and was also deemed to be the most unbiased. Further justification lies in the fact that whilst there are demonstratable clusters of like-contributions in terms of geographic focus, applications, techniques, and data used, the number of contributions (9) is not considered onerous to comprehend, and each individual work brings its own unique contribution.

First, inspired to deliver an alternative to ground-based surveying for the collection of non-boundary cadastral information, Park and Song [32] present a study aimed at remote identification of the discrepancy between existing cadastral maps (which include use information), and current on-ground land uses. The proposed method involves updating the existing land cover attributes of a land parcel maps using UAV hyperspectral imagery classified using CNN, and then composing a discrepancy map showing land use differences. The experimental results out of South Korea demonstrate performance relies heavily on the classification. An advantage of the approach is suggested to be its flexibility regarding modification of the matching criteria between land use categories and land coverage. The authors argue the method could be applied in other contexts and could significantly reduce the time and effort for land use monitoring and field surveying.

Second, Koeva et al. [33] introduce a suite remote sensing approaches, developed, adapted, applied, and tested for the case of land tenure recording in Kenya. These include a unique ontological analysis approach using smart sketch maps (SmartSkeMa); UAV application; and automatic boundary extraction techniques, based based on the acquired UAV images. To ensure applicability of the proposed methodologies, local community needs and the broader governance implications are examined. For the case location of Kajiado, the results show that SmartSkeMa requires little expertise for immediate use, UAVs have high potential for creating up-to-date base maps, and automatic boundary extraction appears an effective method for demarcation of visible tenure boundaries, even compared to traditional methodologies and manual delineation.

Third, Lee and de Vries [34] seek to bridge the semantic gap between land tenure concepts and remotely sensed or earth observation (EO) data. Specifically, they investigate the circumstances where it is possible to rely on EO data to inform land administration applications, such as cadastral mapping and land use change monitoring. Based on reviews of available EO data sources and techniques, they hypothesize that it is possible to both qualify and quantify specific types of land tenure. Furthermore, Lee and de Vries aim to standardize the identification and categorization of certain objects, environments, and semantics visible in EO data that can support (re-)interpretation of land tenure relations. Using North Korean as a case location, they illustrate that land tenure information, in conjunction with EO data and image interpretation, support land administration practice.

Fourth, in a continuation of earlier work, Crommelinck et al. [35] present results from a semi-automatic boundary delineation workflow, comprising image segmentation, boundary classification and an interactive delineation based on UAV data. The application of CNN for boundary line classification is shown to eliminate the previous need for Random Forest (RF) feature generation, and delivers a 71% accuracy result. For the interactive delineation component, more intuitive delineation functionalities, covering more application cases, are presented for large data sets in Kenya, Rwanda, and Ethiopia.

The results show that the new approach is more effective in terms of minimizing clicks and time requirements, compared to manual delineation of visible parcels boundaries. The most significant advantages are observed for rural areas, where the delineation effort per parcel requires 38% less time and 80% fewer clicks, compared to manual delineation.

Fifth, Koeva et al. [36] investigate the emerging areas of 3D cadastres and indoor boundary recording. They demonstrate that most 3D cadastre research has focused on interrelations at the level of buildings and infrastructures: analysis of interrelations in terms of indoor spaces has yet to be adequately explored. The promising research illustrates the opportunity to use 3D point clouds to establish 3D cadastral boundary data in indoor environments. The internal geometry changes of a building, temporally speaking, can be automatically detected from point clouds. The geometry changes can be linked with a data model such as LADM and included in a 3D spatial database, to support updating a 3D cadastre. The permanent changes (e.g., to walls and rooms) are automatically distinguished from dynamic changes (e.g., human, furniture) and are able to be linked to the space subdivisions.

Sixth, Yan et al. [37] head below the surface to tackle the issue of underground land tenure mapping and land use recordation. Increasing urban density and activity is resulting in public infrastructures moving underground. However, precise and detailed spatial information of underground infrastructure, the ownership of those underground objects, and knowledge on the interdependence of infrastructures below and above the ground is still often missing. The research explores how to create reliable 3D underground utility network maps for use in land administration. They investigate current issues pertaining to existing underground utility databases, using Singapore as a case study. A framework for underground utility data governance is proposed to manage the work process from data capture to data usage. An initial design of the 3D underground utility data model is introduced. It describes 3D geometric and spatial information about underground utilities data, and connects it to the cadastral parcel for land administration. Additionally, data from mobile Ground Penetrating Radar is integrated with the existing utility data in a 3D model in support of land administration of underground utilities.

Seventh, Xia et al. [38] explore the potential of deep FCNs for the novel application of cadastral boundary detection in urban and semi-urban areas, based on UAV images. They test the performance of FCNs against other state-of-the-art machine learning techniques, including MRS and gPb, in two case study sites in Rwanda. Experimental results show that FCNs outperformed MRS and gPb in both study areas and achieved an average accuracy of 0.79 in precision, 0.37 in recall and 0.50 in F-score. Therefore, they are able to effectively extract cadastral boundaries, especially when a large proportion of cadastral boundaries are visible. This automated method could minimize manual digitization and reduce fieldwork, thus facilitating the current cadastral mapping and updating practices.

Eigth, Fetai et al. [39] seek to further developments on automated boundary extraction, based on the use of high-resolution optical sensors mounted on UAV platforms. They investigate the potential of the ENVI feature extraction (FX) module for data processing using Slovenia as a case location. The results of the accuracy assessment showed that almost 80% of the extracted boundaries are correct, when compared to the reference data. Fetai et al. argue the proposed workflow has the potential to accelerate and facilitate the creation of cadastral maps, especially for developing countries. However, they caution that the boundaries must be validated by landowners and other beneficiaries after being extracted.

Finally, in a detailed socio-technical study for Rwanda, Stöcker et al. [40] determine the overarching alignment between land administration stakeholder perceptions, the characteristics of the UAV data acquisition workflows, and the final spatial data products obtainable. Additionally, three different UAVs are tested for the quality of data obtainable and the possibilities for using of the technology within the current institutional setting. A priority list of needs for cadastral and non-cadastral information, as well as insights into operational challenges, and data quality measures, of UAV-based data products is presented. It is concluded from the results that UAVs match most of the prioritized needs in Rwanda. However, it is also revealed that institutional and capacity conditions undermine the potential.

Overall, the selected works in this Special Issue demonstrate the growing opportunities for applying innovations in remote sensing and photogrammetry (including laser scanning) to the domain of land administration. This applies to outdoor, indoor, underground, and above ground environments. The opportunities highlighted cover a wide range of land administration applications, from 2D land tenure data capture, cadastral updating, land use classification in developing contexts, through to underground utilities administration, 3D cadastral development, and indoor boundary identification in more developed contexts. The range of available and applicable remotely sensed data types and techniques is shown to be highly relevant to the land sector, and this should encourage further fusion of the disciplines into the future.

Author Contributions: Conceptualization and writing—original draft preparation, R.B.; writing—review and editing, M.K., P.v.O., C.L. All authors have read and agreed to the published version of the manuscript.

Funding: This research received no external funding.

Acknowledgments: The authors would like the acknowledge the support of colleagues at academic and governmental organisations that supported this work, in particular Kadaster Netherlands, Swinburne University of Technology, ITC Faculty University of Twente, and TU Delft.

Conflicts of Interest: The authors declare no conflict of interest.

References

1. Dale, P.; McLaughlin, J. *Land Administration*; Oxford University Press: Oxford, UK, 2000.
2. Çağdaş, V.; Stubkjær, E. Doctoral research on cadastral development. *Land Use Policy* **2009**, *26*, 869–889. [CrossRef]
3. Lemmen, C.; Van Oosterom, P.; Bennett, R. The land administration domain model. *Land Use Policy* **2015**, *49*, 535–545. [CrossRef]
4. Enemark, S.; Williamson, I.; Wallace, J. Building modern land administration systems in developed economies. *J. Spat. Sci.* **2005**, *50*, 51–68. [CrossRef]
5. Henssen, J. Land Registration and Cadastre Systems: Principles and Related Issues. In *Lecture Notes of Technische Universität München*; Technische Universität München: München, Germany, 2010.
6. Bennett, R.; Wallace, J.; Williamson, I. Organising land information for sustainable land administration. *Land Use Policy* **2008**, *25*, 126–138. [CrossRef]
7. Williamson, I.; Ting, L. Land administration and cadastral trends—A framework for re-engineering. *Comput. Environ. Urban Syst.* **2001**, *25*, 339–366. [CrossRef]
8. Williamson, I.; Enemark, S.; Wallace, J.; Rajabifard, A. *Land Administration for Sustainable Development*; ESRI Press Academic: Redlands, CA, USA, 2010.
9. Zevenbergen, J.; De Vries, W.; Bennett, R.M. (Eds.) *Advances in Responsible Land Administration*; CRC Press: Boca Raton, FL, USA, 2015.
10. Zevenbergen, J.; Augustinus, C.; Antonio, D.; Bennett, R. Pro-poor land administration: Principles for recording the land rights of the underrepresented. *Land Use Policy* **2013**, *31*, 595–604. [CrossRef]
11. Hendriks, B.; Zevenbergen, J.; Bennett, R.; Antonio, D. Pro-poor land administration: Towards practical, coordinated, and scalable recording systems for all. *Land Use Policy* **2019**, *81*, 21–38. [CrossRef]
12. Bennett, R.; Rajabifard, A.; Williamson, I.; Wallace, J. On the need for national land administration infrastructures. *Land Use Policy* **2012**, *29*, 208–219. [CrossRef]
13. Bennett, R.; Rajabifard, A.; Kalantari, M.; Wallace, J.; Williamson, I. Cadastral futures: Building a new vision for the nature and role of cadastres. In Proceedings of the FIG Congress, Sydney, Australia, 11–16 April 2010; pp. 11–16.
14. van Oosterom, P.; Bennett, R.; Koeva, M.; Lemmen, C. 3D Land Administration for 3D Land Uses. *Land Use Policy* **2020**, 104665. [CrossRef]
15. Stoter, J.E.; van Oosterom, P. *3D Cadastre in an International Context: Legal, Organizational, and Technological Aspects*; CRC Press: Boca Raton, FL, USA, 2006.
16. ISO. *Geographic Information—Land Administration Domain Model (LADM)*; International Standard ISO 19152; ISO: Geneva, Switzerland, 2012.

17. Shnaidman, A.; van Oosterom, P.; Lemmen, C. LADM Refined Survey Model. In Proceedings of the 8th Land Administration Domain Model Workshop, Kuala Lumpur, Malaysia, 1–3 October 2019; van Oosterom, P., Lemmen, C., Rahman, A.A., Eds.; FIG International Federation of Surveyors: Copenhagen, Denmark, 2019; pp. 73–82.
18. Bennett, R.M.; Pickering, M.; Sargent, J. Transformations, transitions, or tall tales? A global review of the uptake and impact of NoSQL, blockchain, and big data analytics on the land administration sector. *Land Use Policy* **2019**, *83*, 435–448. [CrossRef]
19. Enemark, S.; Bell, K.C.; Lemmen, C.H.; McLaren, R. *Fit-for-Purpose Land Administration*; International Federation of Surveyors (FIG): Copenhagen, Denmark, 2014.
20. Koeva, M.; Bennett, R.; Gerke, M.; Crommelinck, S.; Stöcker, C.; Crompvoets, J.; Ho, S.; Schwering, A.; Chipofya, M.; Schultz, C.; et al. Towards innovative geospatial tools for fit-for-purpose land rights mapping. In *International Archives of the Photogrammetry, Remote Sensing and Spatial Information Sciences-ISPRS Archives, Proceedings of the ISPRS Geospatial Week 2017, Wuhan, China, 18–22 September 2017*; ISPRS: Hannover, Germany, 2017; Volume 42, pp. 37–43.
21. Asiama, K.; Bennett, R.; Zevenbergen, J. Participatory land administration on customary lands: A practical VGI experiment in Nanton, Ghana. *ISPRS Int. J. Geo Inf.* **2017**, *6*, 186. [CrossRef]
22. Lengoiboni, M.; Richter, C.; Zevenbergen, J. Cross-cutting challenges to innovation in land tenure documentation. *Land Use Policy* **2019**, *85*, 21–32. [CrossRef]
23. Bennett, R.; Kitchingman, A.; Leach, J. On the nature and utility of natural boundaries for land and marine administration. *Land Use Policy* **2010**, *27*, 772–779. [CrossRef]
24. Bennett, R.M.; Molen, P.V.; Zevenbergen, J.A. Fitted, Green, and Volunteered: Legal and Survey Complexities of Future Boundary Systems. *Geomatica* **2012**, *66*, 181–193. [CrossRef]
25. Grant, D.; Enemark, S.; Zevenbergen, J.; Mitchell, D.; McCamley, G. The Cadastral triangular model. *Land Use Policy* **2020**, *97*, 104758. [CrossRef]
26. Crommelinck, S.; Bennett, R.; Gerke, M.; Nex, F.; Yang, M.Y.; Vosselman, G. Review of automatic feature extraction from high-resolution optical sensor data for UAV-based cadastral mapping. *Remote Sens.* **2016**, *8*, 689. [CrossRef]
27. Konecny, G. *Geoinformation: Remote Sensing, Photogrammetry and Geographic Information Systems*; CRC Press: Boca Raton, FL, USA, 2014.
28. Toth, C.; Jóźków, G. Remote sensing platforms and sensors: A survey. *ISPRS J. Photogramm. Remote Sens.* **2016**, *115*, 22–36. [CrossRef]
29. Dale, P.F.; McLaughlin, J.D. *Land Information Management: An Introduction with Special Reference to Cadastral Problems in Third World Countries*; Oxford University Press: Oxford, UK, 1988.
30. Bennett, R.; Asiama, K.; Zevenbergen, J.; Juliens, S. The Intelligent Cadastre. In Proceedings of the FIG Commission 7 AGM, St Juliens, Malta, 16–19 November 2015.
31. Bennett, R.M.; Gerke, M.; Crompvoets, J.; Alemie, B.K.; Crommelinck, S.C.; Stöcker, E.C. Building Third Generation Land Tools: Its4land, Smart Sketchmaps, UAVs, Automatic Feature Extraction, and the GeoCloud. In Proceedings of the Annual World Bank Conference on Land and Poverty, Washington, DC, USA, 20–24 March 2017.
32. Park, S.; Song, A. Discrepancy analysis for detecting candidate parcels requiring update of land category in cadastral map using hyperspectral UAV Images: A case study in Jeonju, South Korea. *Remote Sens.* **2020**, *12*, 354. [CrossRef]
33. Koeva, M.; Stöcker, C.; Crommelinck, S.; Ho, S.; Chipofya, M.; Sahib, J.; Bennett, R.; Zevenbergen, J.; Vosselman, G.; Lemmen, C.; et al. Innovative remote sensing methodologies for Kenyan land tenure mapping. *Remote Sens.* **2020**, *12*, 273. [CrossRef]
34. Lee, C.; de Vries, W.T. Bridging the Semantic Gap between Land Tenure and EO Data: Conceptual and Methodological Underpinnings for a Geospatially Informed Analysis. *Remote Sens.* **2020**, *12*, 255. [CrossRef]
35. Crommelinck, S.; Koeva, M.; Yang, M.Y.; Vosselman, G. Application of deep learning for delineation of visible cadastral boundaries from remote sensing imagery. *Remote Sens.* **2019**, *11*, 2505. [CrossRef]
36. Koeva, M.; Nikoohemat, S.; Oude Elberink, S.; Morales, J.; Lemmen, C.; Zevenbergen, J. Towards 3D Indoor Cadastre Based on Change Detection from Point Clouds. *Remote Sens.* **2019**, *11*, 1972. [CrossRef]
37. Yan, J.; Jaw, S.W.; Soon, K.H.; Wieser, A.; Schrotter, G. Towards an underground utilities 3D data model for land administration. *Remote Sens.* **2019**, *11*, 1957. [CrossRef]

38. Xia, X.; Persello, C.; Koeva, M. Deep fully convolutional networks for cadastral boundary detection from UAV images. *Remote Sens.* **2019**, *11*, 1725. [CrossRef]
39. Fetai, B.; Oštir, K.; Kosmatin Fras, M.; Lisec, A. Extraction of visible boundaries for cadastral mapping based on UAV imagery. *Remote Sens.* **2019**, *11*, 1510. [CrossRef]
40. Stöcker, C.; Ho, S.; Nkerabigwi, P.; Schmidt, C.; Koeva, M.; Bennett, R.; Zevenbergen, J. Unmanned Aerial System imagery, land data and user needs: A socio-technical assessment in Rwanda. *Remote Sens.* **2019**, *11*, 1035. [CrossRef]

remote sensing

MDPI

Article

Discrepancy Analysis for Detecting Candidate Parcels Requiring Update of Land Category in Cadastral Map Using Hyperspectral UAV Images: A Case Study in Jeonju, South Korea

Seula Park and Ahram Song *

Department of Civil and Environmental Engineering, Seoul National University, 1 Gwanak-ro, Gwanak-gu, Seoul 08826, Korea; seula90@snu.ac.kr
* Correspondence: aram200@snu.ac.kr

Received: 6 December 2019; Accepted: 20 January 2020; Published: 21 January 2020

Abstract: The non-spatial information of cadastral maps must be repeatedly updated to monitor recent changes in land property and to detect illegal land registrations by tax evaders. Since non-spatial information, such as land category, is usually updated by field-based surveys, it is time-consuming and only a limited area can be updated at a time. Although land categories can be updated by remote sensing techniques, the update is typically performed through manual analysis, namely through a visually interpreted comparison between the newly generated land information and the existing cadastral maps. A cost-effective, fast alternative to the current surveying methods would improve the efficiency of land management. For this purpose, the present study analyzes the discrepancy between the existing cadastral map and the actual land use. Our proposed method operates in two steps. First, an up-to-date land cover map is generated from hyperspectral unmanned aerial vehicle (UAV) images. These images are effectively classified by a hybrid two- and three-dimensional convolutional neural network. Second, a discrepancy map, which contains the ratio of the area that is being used differently from the registered land use in each parcel, is constructed through a three-stage inconsistency comparison. As a case study, the proposed method was evaluated using hyperspectral UAV images acquired at two sites of Jeonju in South Korea. The overall classification accuracies of six land classes at Sites 1 and 2 were 99.93% and 99.75% and those at Sites 1 and 2 are 39.4% and 34.4%, respectively, which had discrepancy ratios of 50% or higher. Finally, discrepancy maps between the land cover maps and existing cadastral maps were generated and visualized. The method automatically reveals the inconsistent parcels requiring updates of their land category. Although the performance of the proposed method depends on the classification results obtained from UAV imagery, the method allows a flexible modification of the matching criteria between the land categories and land coverage. Therefore, it is generalizable to various cadastral systems and the discrepancy ratios will provide practical information and significantly reduce the time and effort for land monitoring and field surveying.

Keywords: Cadastral map; hyperspectral UAV; land category; land cover; land use; deep learning

---

1. Introduction

Cadastral maps show the boundaries and ownership of land parcels that separate adjacent land plots. These maps contain spatial information, such as shape, size, boundary, and location, as well as non-spatial information, such as land use, value, and tenure, which are uniquely encoded in textural or attribute files [1]. Moreover, cadastral maps are available as large-scale base maps with micro-level mapping [2]. As cadastral maps are related to personal properties, accurate cadastral mapping can improve agricultural productivity and support the national development policy [3]. Moreover, a well-structured cadastral map is a prerequisite for improved land management services [4].

A cadastral map is updated by modifying the spatial and non-spatial data of the existing cadastral maps to reflect the latest land information. High-quality cadastral mapping requires updating the changes in land use information and the spatial division of property units [5]. The land use type, which indicates the purpose of use, is registered and managed as an attribute of "land category" in a cadastral system. Therefore, the items of land category can be assigned according to their land use type, such as "Building site," "Parking lot," and "Road." Cadastral map updates are essential for not only recording the most recent land ownership and property division changes in a timely manner but also effectively managing the land information. For example, updating is necessary when the land is suddenly changed by new sub-divisions, transfer of land use, and natural disasters [6]. Furthermore, from the aspect of tax imposition, which is a main purpose of land use management by cadastral mapping [7], updating cadastral maps is crucial because the tax imposed on land owners depends on their land use type. Frequent updates of cadastral information can better manage illegal land use, whereby landowners register false land uses to reduce their taxes.

The procedure of updating cadastral maps can be divided into three steps: (1) extracting meaningful features and generating new data, (2) comparing new data with the existing base map and detecting changes, and (3) updating the base map with those changes and verifying the consistency of the updated map and actual information [8–10]. As the step of extracting relevant features, both up-to-date spatial and non-spatial information, such as parcel boundaries and land category, can be generated. Traditionally, cadastral surveying is performed by field work, aerial monitoring, and satellite data acquisition [5]. Although field surveying acquires accurate land information, it is extremely time-consuming and requires well-trained manpower for wide-area implementation. Remote sensing (RS) can be an effective alternative to field work because it is cheaper and faster compared to conventional cadastral surveys [3], and it is a useful data source for many base map-updating activities [8]. Cadastral boundaries set by roads, building, and water are visible in RS images and can be mapped from them.

To consider both generation of cadastral information and the further step of updating, which include a comparison between the generated information and existing cadastral maps, an integrated method that improves the efficiency of cadastral mapping and updating was proposed [5]. Using three bands of QuickBird satellite data, a digital and elevation model, and global navigation satellite system (GNSS) data, this method registers fused images to the existing cadastral map. After superimposing the boundaries of the cadastral map on the fused images, the map is updated by visual interpretation using a participatory geographic information system. Furthermore, the cadastral image was updated using CARTOSAT-2 panchromatic satellite images with 1.0 m resolution and Geo-eye multispectral images with GNSS data and 0.5 m resolution [6]. In this study, cadastral maps were updated by extracting the parcels from those images, along with three parameters (area, perimeter, and position) related to spatial elements, while non-spatial elements were not considered. Wassie et al. [11] proposed a procedure for extracting cadastral boundary information by semi-automatically using the WorldView-2 satellite data. In this study, the procedure of comparing the extracted information with the existing one, which is reference digital parcel boundaries, was performed in two ways: visual interpretation and quantitative analysis. The recentness of the information was not considered during the comparison, because this study aimed at verifying the accuracy of the extracted information rather than change detection for updating. Furthermore, the proposed procedure only focused on the parcel boundary, which comprises spatial elements.

Several previous studies on cadastral mapping and updating dealt with non-spatial updates, but focused only on extracting up-to-date information from the source date. The comparing and detecting changes step of updates were dependent on visual interpretation or performed in a limited range [2,9,10,12–14]. Specifically, Khadanga et al. [14] classified land use in cadastral parcels extracted from high-resolution satellite imagery through object-based image analysis (OBIA). The result layer of

OBIA was written into a shapefile and compared with a digitized map of the cadastral parcels. The digitizing was manually performed and the comparisons were visually analyzed. Avramović et al. [12] updated the status of rural land use only from digital cadastral maps. Although they compared the land category items in the cadastral map with those in the real estate cadaster, they did not provide the details of the comparison.

To automate the comparison between the newly generated and existing cadastral information, the authors of [15] suggested a map-query-based comparison between the cadastral map and the land-cover map from satellite images. They generated the land cover map from Landsat TM satellite images and matched the land cover classes with the land category items in South Korea [15]. However, the spatial resolution of satellite images is relatively high (30 m) and obtaining images at the desired time is hindered by the time resolution and noise, such as cloud. After generating the land cover map, the authors performed a binary analysis of the pixel-level inconsistency between the land cover and cadastral maps. Although they analyzed heterogeneous data at the pixel-level, they calculated only the ratio of inconsistent area to the entire test area without considering the inconsistency by parcels or land category items. In conclusion, to improve the efficiency of the overall update process, it is vital to automate the comparison of up-to-date information with existing cadastral maps and the detection of parcel discrepancies.

When improving the efficiency of updating cadastral maps, one must consider the elaborateness of the latest cadastral information generated through various cadastral surveys. Although aerial and satellite surveying techniques acquire data over large-scale regions with superior spatial resolution, they are influenced by weather conditions, old acquisition time, and military security problems [3,5]. Therefore, unmanned aerial vehicles (UAVs) have recently been deployed for extracting up-to-date cadastral information. UAVs are cost-effective, especially in local applications, and acquire real-time data at high spatial resolution [10,16,17]. Manyoky et al. [18] noted that UAVs collect detailed information. Moreover, UAV-based methods enable an efficient documentation of the non-spatial information in cadastral maps, such as land use and vegetation. Relevant features are often extracted from orthophotos generated from UAV images using various feature-extraction methods such as image classification, segmentations, and line extraction [9]. As an example of cadastral mapping and updating with UAV imagery, areas subjected to landslides, which manifest as a sudden change in land use, were automatically detected from UAV images [19]. The detected changes in land use provided the basis information for synchronizing the cadastral information. However, the target area was manually extracted through an overlay analysis between the information extracted from the UAV and the cadastral map. Moreover, the updates were performed on limited target parcels (landslide areas) rather than the whole area. Manyoky et al. [18] compared the use of UAVs with the tachymeter–GNSS combined method in cadastral mapping and updating. The acquired points for generating and updating the cadastral maps were classified by land cover, such as vegetation types, buildings, and streets. However, the authors did not thoroughly describe the data processing steps for generating and updating the cadastral information.

As mentioned earlier, the land category as non-spatial data directly affects land value estimation and thus needs to be up-to-date in a short cycle. The inconsistency between the registered land category and actual land use when updating cadastral maps is a well-reported problem in cadastral mapping [12,13,17,20]. In South Korea, items of land category are determined based on the primary use of each parcel [21], which is directly related to the assessment value of the land [15]. Therefore, it is a legal obligation to correct the registration if the registered land information differs from the actual land use information [21]. New information can be updated by the land owner's registration. Fines or imprisonment may be imposed for those who do not inform about a change in land category or a false notification [21,22].

As the accuracy of an update is associated with ownership of property, the update must be verified through a field survey. This is especially important for updates of cadastral systems (including maps). Detecting the areas requiring update is crucial for reducing the target area of the field survey and improving the effectiveness of the field work. An automatic process would facilitate cadastral mapping and updating. To this end, the present study proposes a new discrepancy analysis method that automatically detects candidate parcels requiring an update of their land category information. The proposed method is implemented in two stages: generating up-to-date land category information and comparing the new information with the existing cadastral information. To effectively extract the land cover, we use hyperspectral UAV images and a deep learning approach. Hyperspectral UAV images can spectrally discriminate similar materials that cannot be identified in RGB or multi-spectral images captured at specific times. In the latter stage, the land category information generated from a UAV is compared with the existing cadastral map managed by the government. The comparison process generates a discrepancy map representing the parcels requiring update. The major contributions of the proposed method are stated below.

- For generating up-to-date land category information, we combine two-dimensional (2D) and three-dimensional (3D) convolutional neural networks (CNNs) to classify hyperspectral UAV images, and hence, generate the latest land category information at specific times and intervals. Furthermore, the environmental settings for learning are demonstrated and the classification results are analyzed to understand when the proposed network was applied to hyperspectral UAV images.
- For detecting discrepancies between the new information and the existing cadastral information, the efficiency of updating the registered land category is improved by a new technique that automatically compares two sets of non-spatial information under different criteria and structures.

The remainder of this paper is organized as follows. Section 2 proposes our discrepancy analysis method, and Section 3 describes the datasets, environmental conditions of the experiments, and the results of a case study in South Korea. Finally, the conclusions are provided in Section 4.

2. Methods

As shown in Figure 1, the proposed method of discrepancy analysis comprises two main parts: (i) classifying hyperspectral UAV images using the hybrid CNN for generating the land cover map and (ii) comparing the consistency between the cadastral map and land cover map for detecting inconsistent parcels through a query-based approach.

(1) The hybrid CNN with 2D and 3D kernels extracts the spatial–spectral features from hyperspectral UAV images and obtains a land cover map depicting the regions covered by forests, crops, bare soils, water, roads, and buildings. The images input to the hybrid CNN are pre-processed by principle component analysis (PCA) to reduce the number of redundant spectral bands and the computational cost. The hybrid CNN then classifies images by extracting various meaningful feature maps. The resulting land cover map provides the latest land information on sites.
(2) To the procedure that automatically detects inconsistent parcels, two maps are input: the existing cadastral map, which is managed by the government, and the land cover map, which is generated from hyperspectral UAV image classification. To compare the heterogeneous datasets with vector and raster structured data, the procedure adopts an encoding–decoding approach. The final output is a discrepancy map generated through query-based comparison of the mapping information in the land category items and the land cover classes in different frameworks.

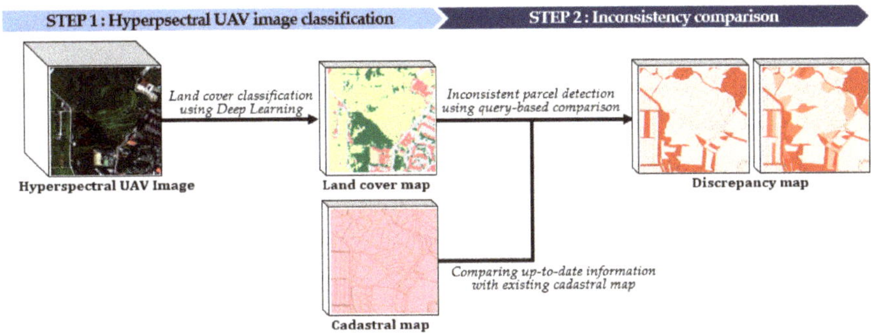

Figure 1. Proposed process of discrepancy analysis. UAV: unmanned aerial vehicle.

2.1. Step 1: Hyperpsectral UAV Image Classification For Generating the Land Cover Map

Hyperspectral images (HSIs) contain hundreds of narrow continuous bands over a wide range of the electromagnetic spectrum. Therefore, they provide more detailed spectral information than multispectral images and can spectrally discriminate similar materials. A land cover map derived from HSIs distinguishes distinct classes, such as forest and crop land, which are included in the land category framework. Level-I classes of land cover can be regarded as the usage information. In this sense, the land cover information from HSIs contains not only the land surface materials but also the land use. Furthermore, the latest land surface information of the target area can be extracted from UAV images taken at the desired time point in the desired interval.

HSI classification methods should consider the high dimensionality of the dataset. Traditionally, HSIs have been classified by pattern recognition algorithms, such as nearest neighbor, decision trees, and linear functions [23]. k-nearest neighbor (k-NN) clustering is a representative simple method that measures the similarities between the training and test data by using their Euclidean distances. Support vector machines remove the curse of dimensionality by determining the boundaries in a high-dimensional space, using the kernel method [23].

More recently, HSI classification has been performed by deep learning approaches. Deep learning replaces the hand-crafted feature-engineering process, which requires expert experience and careful parameter settings, with automatic extraction of the meaningful features contained in high-dimensional bands [24]. CNNs have been widely applied to HSI classification tasks [25–28]. Many studies have successively classified the items in hyperspectral images using 2D-CNNs, which extract features from spatial domains [25,26]. Efficient feature extraction by 2D-CNNs requires a data transformation process, such as data reduction, to convolute all bands of the input image. As HSIs include hundreds of spectral bands, the convolutions require several kernels, which introduces the over-fitting problem and increases the computational cost. 2D convolution is computed as follows:

$$v_{l,j}^{x,y} = \phi\left(\sum_{n}\sum_{h=0}^{H-1}\sum_{w=0}^{W-1} w_{ljn}^{hw} o_{(l-1)n}^{(x+h)(y+w)} + b\right), \tag{1}$$

where $v_{l,j}^{x,y}$ is the pixel value of position $(x,\ y)$ on the jth feature map in layer l (the layer of the current operation); ϕ is the activation function; b is a bias parameter; and w_{ljn}^{hw} is the weight value at position (h, w) in the nth shared $H \times W$ kernel, where n is the number of feature maps in the $(l-1)$th layer. $o_{(l-1)n}^{(x+h)(y+w)}$ is the input at position $(x+h)(y+w)$ and (h, w) denotes its offset to (x, y).

3D-CNNs simultaneously extract the spatial and spectral features [27,28]. A 3D-CNN preserves the original input data by avoiding complex data reconstruction and considers the relationships among channels; however, 3D-CNNs are more computationally complex than 2D-CNNs. In classes with similar textures over many spectral bands, they can perform worse than 2D-CNNs [29]. The pixel value at position (x, y, z) in the jth 3D feature cube of the lth layer is given as follows:

$$v_{l,j}^{x,y,z} = \phi\left(\sum_{n}\sum_{h=0}^{H-1}\sum_{w=0}^{W-1}\sum_{r=0}^{R-1} w_{ljn}^{hwr} o_{(l-1)n}^{(x+h)(y+w)(z+r)} + b\right), \tag{2}$$

where R is the spectral dimension of the 3D kernel and w_{ljn}^{hwr} is the weight value at position (h, w, r), connected to the nth feature in the $(l-1)$th layer. $o_{(l-1)n}^{(x+h)(y+w)(z+r)}$ represents the input at position $(x+h)(y+w)(z+r)$ and (h, w, r) denotes its offset to (x, y, z).

The abovementioned limitations can be resolved by hybridizing 2D- and 3D-CNNs [29]. In the hybrid spectral CNN (HybridSN), the output of a 3D-CNN is input to a 2D-CNN. This configuration learns the spatial representation at a more abstract level, with lower model complexity, compared to the 3D-CNN alone. The present study proposes a new hybrid 2D-CNN and 3D-CNN for effectively classifying hyperspectral UAV images (Figure 2). The network comprises 2D- and 3D-CNN branches in convolutional layers, which generate various meaningful feature maps from the input. First, spectral redundancy is removed by PCA along the spectral bands of the original HSI. The PCA image is then processed through the convolutional layers with 2D and 3D kernels. The first convolutional layers of both branches have eight filters, and the subsequent convolutional layers of the 2D and 3D branches have 16 and 32 kernels, respectively. The outputs of the 3D convolutional layers are converted to a 2D shape and the feature maps obtained from both branches are combined to form the spectral and spatial feature maps. These maps are input to the fully connected layers. Finally, the pixels are classified into land cover classes. In the next process, these land cover classes are mapped to the land category items in the cadastral map. To reduce the complexity of the mapping and to generalize the model, we adopt level-I types of land cover, namely forests, crop lands, roads, buildings, bare soil, and water bodies.

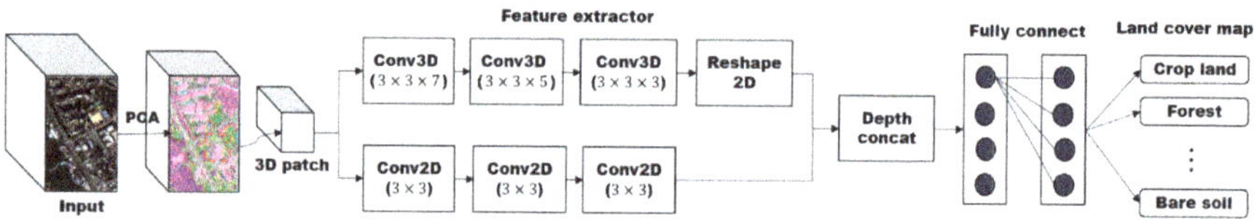

Figure 2. Process of step1 in discrepancy analysis: hyperspectral UAV image classification for generating the land cover map. PCA: principle component analysis.

2.2. *Step 2: Inonsistency Comparison Between the Cadastral Map and Land Cover Map*

Our proposed pixel-level inconsistency comparison automatically detects the areas of inconsistent land use between the registered and actual land information. In a previous study [30], a restructured land use map was generated in vector format, which assigned the actual land cover classes from the imagery as attributes and the cadastral boundary as the geometry. Although this map compares the registered land categories in cadastral maps with the actual land use, it is limited to the primary land use, which occupies the maximum area in each parcel. An elaborate comparison must consider all land uses in each parcel. Figure 3 shows the process of comparing the actual land cover and cadastral map at the pixel-level, which considers both minor and primary uses.

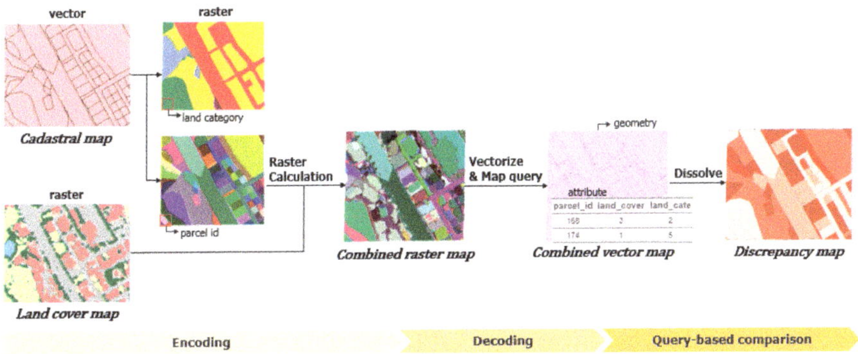

Figure 3. Process of step 2 in discrepancy analysis: inconsistency comparison between cadastral and land cover maps.

The proposed automatic comparison technique is then divided into three stages: "Encoding," "Decoding," and "Query-based comparison" (Figure 3). Because the cadastral map and land cover map are constructed in vector and raster formats, respectively, the automatic inconsistency comparison must convert the heterogeneous datasets into the same structure prior to the overlay analysis [15]. The first encoding step performs raster conversion using the cadastral map attributes. For this purpose, the land category and parcel ID are assigned to each pixel of the rasterized cadastral maps, which have the same pixel size as that of the land cover map. A combined raster map is then generated with coded values V_{ij} combining the land cover C_{ij}, land category U_{ij}, and parcel ID P_{ij} values. An encoding query is expressed as follows:

$$V_{ij} = P_{ij} \times 10^4 + U_{ij} \times 10^2 + C_{ij} \ \forall\ (i,\ j). \quad (3)$$

The second decoding stage vectorizes the combined raster map, which results in a vector map combining the land cover and land category information. The attributes of the vector map include the parcel ID, land cover, and land category, and their values are assigned by decoding the pixel values.

The combined vector map includes both the land category values and land cover values in a unit area. Therefore, the inconsistent area can be automatically extracted through a query-based comparison between the corresponding values in the previous stage. The land category items are defined in terms of land use, and each item can contain multiple usages. For example, a "building site" may include buildings and bare land, and a "school site" may include buildings, bare land, trees, and grass. However, when extracting the land cover information from the imagery, the materials and/or objects covering the land are extracted from the spectral characteristics of the image. When constructing a query for comparing these two maps, we must define mapping rules that determine the discrepancy between the land category items and land cover classes, which are classified under different criteria. However, establishing an absolute standard for mapping land category items to land cover classes is restricted because the land category items differ among country-specific cadastral systems and the number of available classification classes depends on the quality of the imagery. In the case study (Section 4.1), the mapping between land cover classes and land category items is performed under the Korean Cadastral System as a guideline. An automatic comparison can be queried based on the corresponding mapping information; the query result can automatically determine the discrepancy between the land category and land cover. The discrepancy map can be generated by dissolving the area based on parcel IDs. From the discrepancy map, we can calculate the portions of inconsistent areas where the registered land category is different from the actual land cover in each parcel. Because the discrepancy map is generated by comparing both the primary and minor land uses, it provides reference data for the automatic detection of parcels that must be divided. Table 1 shows the proposed algorithm of a pixel-level comparison for detecting inconsistent areas; moreover, this algorithm can be automated in the model builder of ArcGIS 10.1 [31] (Figure 4).

Table 1. Proposed algorithm of pixel-level inconsistency comparison.

```
Input:  Land Cover Map (LC, Rraster)
        Cadastral Map (CM, Vector)
--------------------------------------------------------------------
# Encoding
R_Cate = rasterized CM by assigning values with "land category"
R_PI = rasterized CM by assigning values with "Parcel ID"
w, h = width, height (image extent) of LC
CRM (Combined Raster Map) = empty raster layer with w × h
For each pixel (i,j) on CRM:
    P(i, j) = assigned value of pixel (i,j) on R_PI
    U(i, j) = assigned value of pixel (i,j) on R_Cate
    C(i, j) = assigned value of pixel (i,j) on LC
  Combined(i,j) = P(i, j) × 10^4 + U(i, j) × 10^2 + C(i, j)
  assign value of Combined(i,j) on pixel (i,j) to generate CRM
end
--------------------------------------------------------------------
Intermediate Output: Combined Raster Map (CRM, Raster)
--------------------------------------------------------------------
# Decoding
CVM (Combined Vector Map) = Raster to Polygon (CRM)
For each polygon i on CVM:
    PV(i) = pixel value of polygon i
    CVM(i).p_id = PV(i)/10^4
    CVM(i).category = (PV(i)%10^4 − PV(i)%10^2)/10^2
    CVM(i).cover = PV(i)%10^2
    CVM(i).area_ia, CVM(i).area_ca = 0
end
--------------------------------------------------------------------
Intermediate Output: Combined Vector Map (CVM, Vector)
--------------------------------------------------------------------
# Query-based comparison
Make query Q using mapping information between land category and land cover
TF = Execute query Q on CVM
If TF == true:
    CVM(i).area_ia = calculate area of polygon i
else:
    CVM(i).area_ca = calculate area of polygon i
end
DM (Discrepancy Map) = Dissolve (CVM) based on p_id with summation of area_ia and area_ca
For each polygon i on MDA:
    DM (i).ic_ratio= area_ia(i)/area(i)
end
--------------------------------------------------------------------
Output: Discrepancy Map (DM, vector)
```

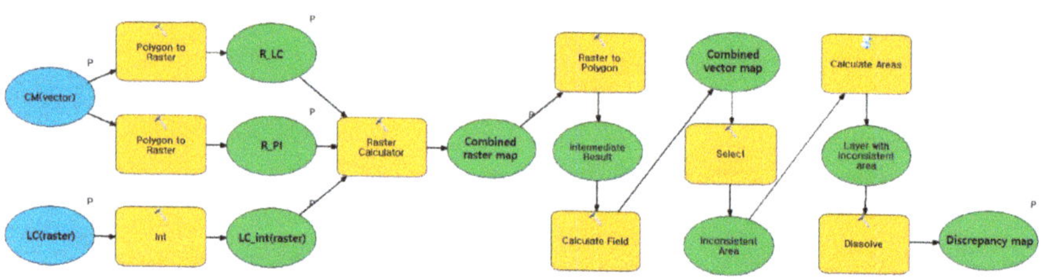

Figure 4. Automated model for detecting inconsistent area.

The generated discrepancy map can be utilized for another purpose: detecting parcels requiring division. Specifically, because the cadastral map was created by assigning one land category value (based on the primary use) per parcel, parcels with a high ratio of minor-use area must be divided for efficient land management [21]. The proposed process reflects all uses of the land. Therefore, it automatically detects inconsistent areas while detecting the parcels that must be divided into different land use statuses.

3. Dataset

The dataset contained the hyperspectral UAV images acquired at two sites of Jeonju City in South Korea. The hyperspectral UAV images were acquired on September 19, 2019, by a DJI Matrice 200 UAV equipped with hyperspectral sensors (Corning microHSI SHARK 410). This platform had accurate flight controls and inherent stability. Its spatial resolution was 15 cm and spectral resolution was 4 nm over 150 bands ranging from 398.78 to 996.74 nm. The flight path of the UAV was selected to follow the waypoint at a flight height of 200 m. The whole study area (890 m × 730 m) was covered in 15 courses. Study sites of area 600 m × 600 m, where the errors associated with camera shaking and gematric problems were few, were selected from the whole area. The images were registered using the geographic map projection WGS-84. The center coordinates of Sites 1 and 2 were (35°48′19″ N, 127°05′45″ E) and (35°47′16″ N, 127°07′14″ E), respectively (Figure 5). These sites included crop lands, forests, and building areas. Owing to the high spatial resolution of the hyperspectral UAV images, objects such as vehicles, the centerlines of roads, and shadows, besides buildings and trees, could be identified. As such information was unnecessary for updating the cadastral map, the spatial resolution of the images was reduced to 60 cm to limit the number of classification classes and reduce the memory requirements of deep learning. Prior to the classification, the images were pre-processed by geometric and radiometric corrections based on GNSS and field spectrometry data.

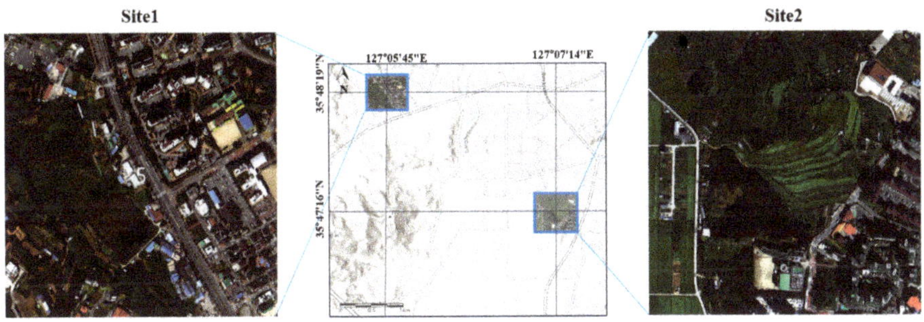

Figure 5. Locations of the two study sites in South Korea, along with their UAV hyperspectral images. The background map was obtained from ArcGIS (a geographic information system (GIS) for working with maps and geographic information maintained by Esri) world map [32]. The hyperspectral UAV images were obtained on 19 September 2019.

Figure 6 shows the cadastral maps of the study sites. There were 284 and 250 parcels in the cadastral maps of Sites 1 and 2, respectively. We obtained the most recently updated serial cadastral map taken in January 2018. In Korea, land categories of cadastral maps can be divided into 28 items, and a cadastral map can be divided into 28 main land categories. The study sites included 17 land category items: building sites, paddy fields, fields, park sites, school sites, roads, forests, reservoirs, miscellaneous land, sites for religious use, parking lots, ditches, factory sites, cemeteries, gas station sites, sports areas, and ranches.

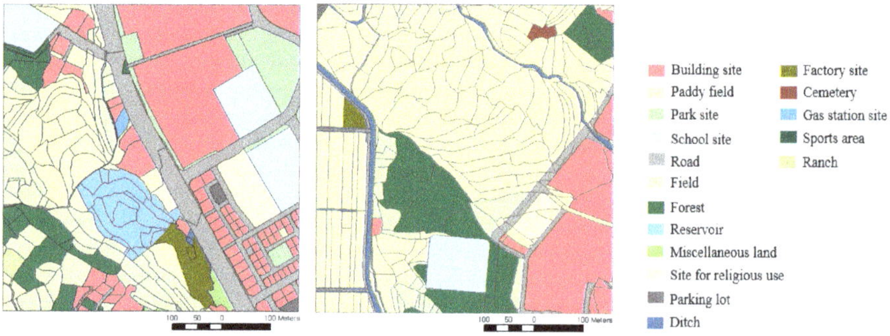

(a) (b)

Figure 6. Cadastral map of the study sites in South Korea: (**a**) Site 1, (**b**) Site 2.

4. Results

4.1. Classification Results

The hyperspectral UAV images were classified by the proposed hybrid CNN. The network was optimized in 30 epochs of Adam with a learning rate of 10^{-3} and a batch size of 256. The Adam optimizer is a combination of the stochastic gradient descent with momentum and RMSprop, and has relatively low memory requirements and is quite computationally efficient [33]. At the start of each iteration, the network was randomly initialized. The ground-truth data were manually defined from the field data. The field work acquired the spectral libraries and types of surface materials. The ground truth was composed of 88,567 pixels and contained six classes: crops, forests, roads, buildings, bare soil, and water. The classes that could be mapped to the land category items were then defined. The various crop lands and grass covers were combined into "crop land," and relatively high trees were classified as "forest." Colored roofs, such as blue, brown, and white, were all classified as "buildings." "Bare soil" represented ground without buildings and vegetation, and "road" encompassed asphalt roadways. The ground-truth data were randomly divided into training, validation, and test samples. Sixty percent of the ground-truth data (53,140 pixels) were used as training samples, which were subdivided into validation and training data at a ratio of 7:3 to avoid overfitting problems. The remaining 40% of the ground-truth data (35,427 pixels) were reserved as the test samples. The performance of the proposed network was estimated from the classification accuracy of the test data.

To confirm the effectiveness of the hybrid network, the classification accuracies of the 2D and 3D-CNNs were compared. Both networks were composed of three convolutional layers and used the same variables as those used by the hybrid CNN. In each experiment, the performance of the network was evaluated by the F1 scores of the six classes and the overall accuracy (OA). The F1 score measured the classification accuracy in terms of the precision and recall scores (Equation (4)). Precision defines the fraction of correctly retrieved instances among all instances, and recall is the fraction of correctly retrieved instances among all correct instances.

$$\text{F1 score} = 2 \times \frac{Precision \times Recall}{Precision + Recall}. \quad (4)$$

Figure 7 shows the classification losses and accuracies in each epoch for the training and validation samples of Sites 1 and 2. The hybrid CNN achieved lower classification loss and higher accuracy compared to 2D-CNN and 3D-CNN. Relative to 2D-CNN, the loss reduction and accuracy improvement in the hybrid CNN became more noticeable with increasing epoch number. Although 3D-CNN also achieved higher accuracy than 2D-CNN, it was less accurate and incurred higher losses than the hybrid

CNN at Site 1. According to the results, 3D-CNN was more useful for classifying hyperspectral images than 2D-CNN but combining the 2D and 3D CNNs improved the classification performance.

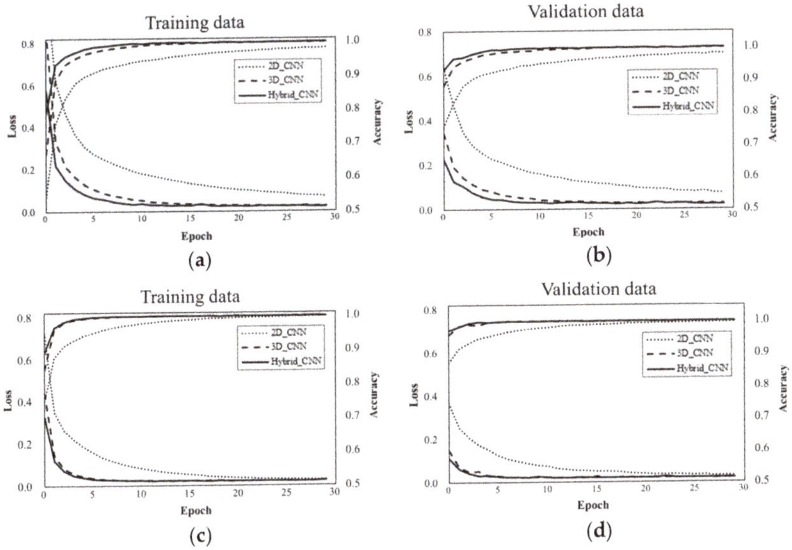

Figure 7. Classification losses and accuracies in each epoch on (**a**) training samples, (**b**) validation samples at Site 1, (**c**) training samples, (**d**) validation samples at Site 2.

Figure 8 shows the classification results of the hyperspectral UAV images using the hybrid CNN. The F1 scores and overall accuracies of the six classes are listed in Table 2. The OAs of the land cover classifications at Sites 1 and 2 were 99.93% and 99.75%, respectively. Because the ground-truth data did not cover the entire study area, it was not the classification accuracy of the entire image but rather that of a randomly selected test sample location. According to Table 2, all six classes were well classified. As there was no water at Site 2, the results of this site were divided into five classes. Forests and roads obtained a lower F1 score than the other classes, because the spectral characteristics of crop land and forest were very similar. Furthermore, roads, parking lots, and car were classified into the "road" class and various colored rooves were classified into the "building" class. Moreover, areas that appeared to be farmland with low vegetation were classified as "bare soil." Pixel-level classification errors in the results can be considered as insignificant because an inconsistency comparison will be conducted at the parcel level.

Table 2. Classification results of the South Korean sites: F1 score and overall accuracy.

	F1 Score						OA (%)
	Crop Land	**Forest**	**Road**	**Building**	**Water**	**Bare Soil**	
Site1	0.9998	0.9990	0.9993	0.9994	1.0000	0.9984	99.93 ± 0.1
Site2	0.9999	0.9935	1.0000	0.9984	-	0.9705	99.75 ± 0.1

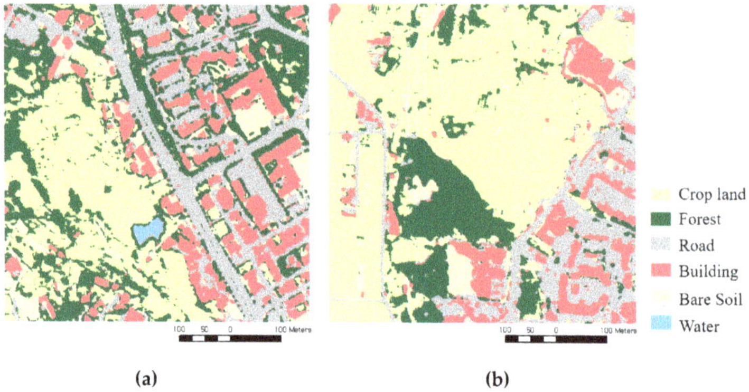

Figure 8. Classification results of hyperspectral UAV images at the two South Korea sites: (**a**) Site 1, (**b**) Site 2.

4.2. Detecting Inconsistent Parcels

According to the relevant regulations in South Korea [21], "cadastral inconsistent parcels" that need to be updated cover the following situations: (1) when the geometric information, such as parcel boundary and area, differs from the actual geometry; (2) when the parcel information is registered incorrectly in the cadastral system; (3) when the parcel information is registered differently from the land survey results; and (4) when the land owner requests an information change. Therefore, we detected the inconsistent parcels requiring update on discrepancies in the land category information.

For the inconsistency comparison, the cadastral maps were converted into a raster structure and compared with the land cover maps generated as the classification results in raster format. The rasterized cadastral map at each site was created by assigning the land category and parcel ID, as shown in Figure 9. The cadastral map comprised polygons containing the parcel boundary information; however, the geometric information representing the parcel boundary was lost during the rasterization process. As a parcel-wise comparison was required for detecting the inconsistent areas between two maps, the cadastral map was rasterized with the parcel ID, as shown in Figure 9b,d. Figure 9 shows the rasterized cadastral map with land category with the same color palette as that of the land cover map, whereas randomly selected colors were used to represent a rasterized cadastral map with parcel ID.

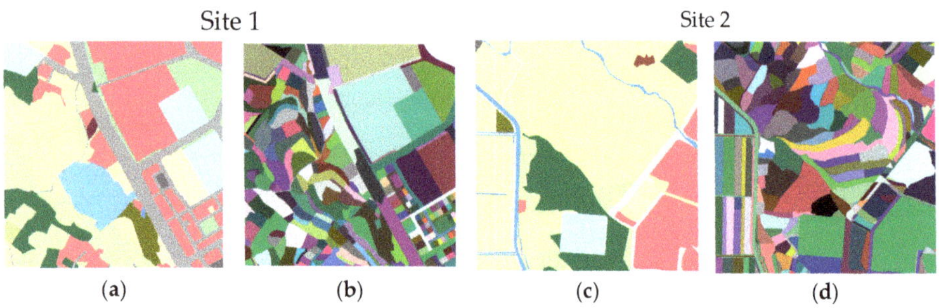

Figure 9. Cadastral maps rasterized with (**a**) land category at Site 1, (**b**) parcel ID at Site 1, (**c**) land category at Site 2, and (**d**) parcel ID at Site 2.

The query-based comparison could be automated because each polygon contained the land cover information from the HSI along with the land category information from the cadastral map. Classifying the actual land cover information from images in response to the cadastral map system, specifically to the land category framework, is a technically difficult problem [15]; the criteria for defining the land cover classes that can be extracted from HSIs are difficult to reconcile with the land category items in the cadastral map. Therefore, before identifying the inconsistent areas in the cadastral map and the actual land cover map, we must establish a mapping rule between the matched classes under each framework. Then, the consistency of the land cover and land category can be determined from the mapping rule. Because the land category items and land cover classes were classified by different criteria, they could not be matched one-to-one. Based on empirical investigations of the test sites, this study defines *M*:*N* matching pairs of land category items and land cover classes (Figure 10). Finally, a query for inconsistency comparison could be made using the mapping rule shown in Figure 10.

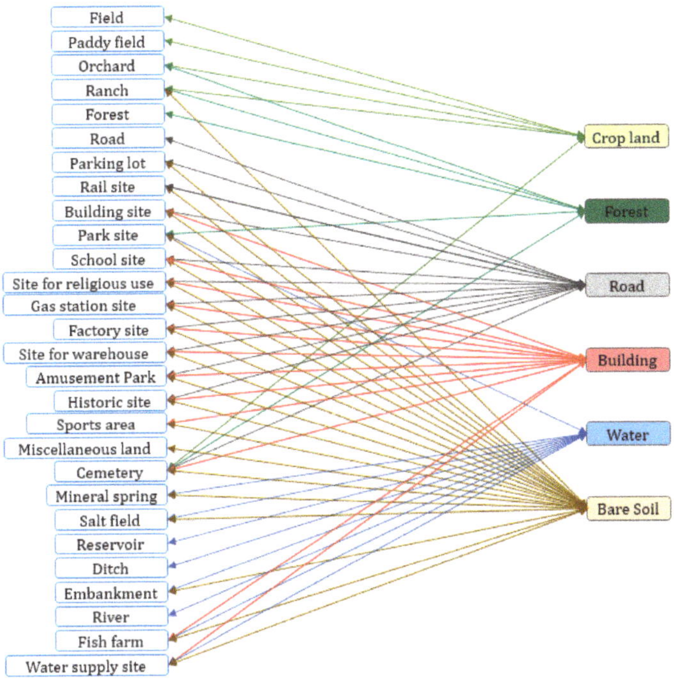

Figure 10. Mapping rule between land category and land cover.

Land cover was classified into crops, forests, buildings, roads, water, and bare soil, which can be distinguished in HSIs. Figure 10 is constructed from the 28 land category items used in the Korean cadastral system. These items and their rules can be adjusted to other cadastral systems, providing source information for other countries. Furthermore, because the query can be modified according to the mapping information, the proposed technique is applicable to the discrepancy analyses of other cadastral systems.

Panels (a) and (d) of Figure 11 present the combined raster maps of Sites 1 and 2, respectively. In these maps, the land cover, land category, and parcel boundary information were combined by encoding with the rasterized cadastral maps and the generated land cover map. The combined raster map was restructured in vector format, and the parcel boundary with the land cover and land category attributes was retrieved by decoding the assigned values (Figure 11b,e; note that each parcel contains many polygons). Finally, the discrepancy map was generated by dissolving the combined vector map based on the parcel ID, leaving only the parcel boundaries, as shown in Figure 11c,f.

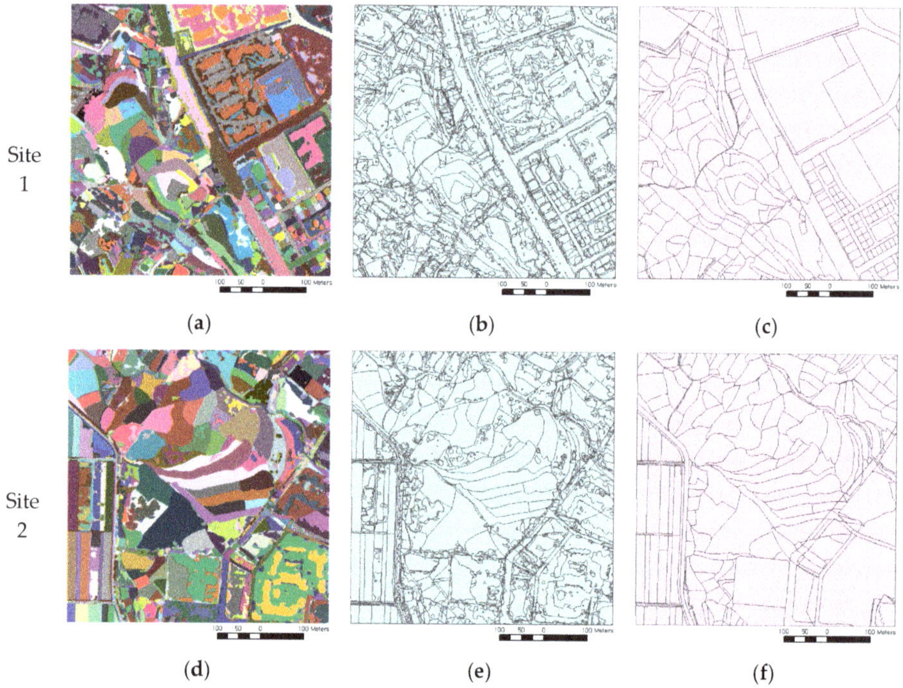

Figure 11. Result of the proposed process: (**a**) combined raster map of Site 1, (**b**) combined vector map with attributes of Site 1, (**c**) discrepancy map of Site 1, (**d**) combined raster map of Site 2, (**e**) combined vector map with attributes of Site 2, and (**f**) discrepancy map of Site 2.

As mentioned earlier, land categories in Korea's cadastral system should be registered based on the primary use of each parcel. The regulations [21] state that if the proportion of parcels used for purposes other than the primary use exceeds a certain level, the parcels must be divided. The polygons in each parcel in the vector layer (Figure 11b,e), which is the intermediate result of the current study, contained the detailed land cover information extracted from the HSI. Therefore, the potential areas to be divided could be concurrently investigated by calculating the polygon area per usage.

The discrepancy maps included the sum of the inconsistent areas as an attribute of each parcel. Therefore, the ratio of the inconsistent area to the total area of each parcel could be calculated. This discrepancy ratio represents the degree of discrepancy (in parcel units) between the land category information registered in the cadastral system and the actual land cover information. Parcels requiring an update of their land category were then identified as those with a large discrepancy ratio. Figure 12 shows the visualization result of the discrepancy ratios between the registered land category information and the actual land cover information extracted from the hyperspectral UAV imagery. If the discrepancy ratio in a parcel exceeds a certain ratio, and the land use of the parcel differs from the registered land category information, this parcel must be separately managed through a site survey. In Figure 12, the discrepancy degree is indicated by a red scale that ranges from white (no discrepancy) to deep red (high discrepancy). In this visualization, the target parcels to be managed can be clearly identified. However, because the threshold discrepancy ratio is not systematically defined, Figure 12 presents three maps of each site with different intervals of discrepancy ratio. More specifically, the inconsistent land parcels in Figure 12c,f are extracted under more rigid criteria than those shown in Figure 12a,d. The deep red regions are the areas that need updating.

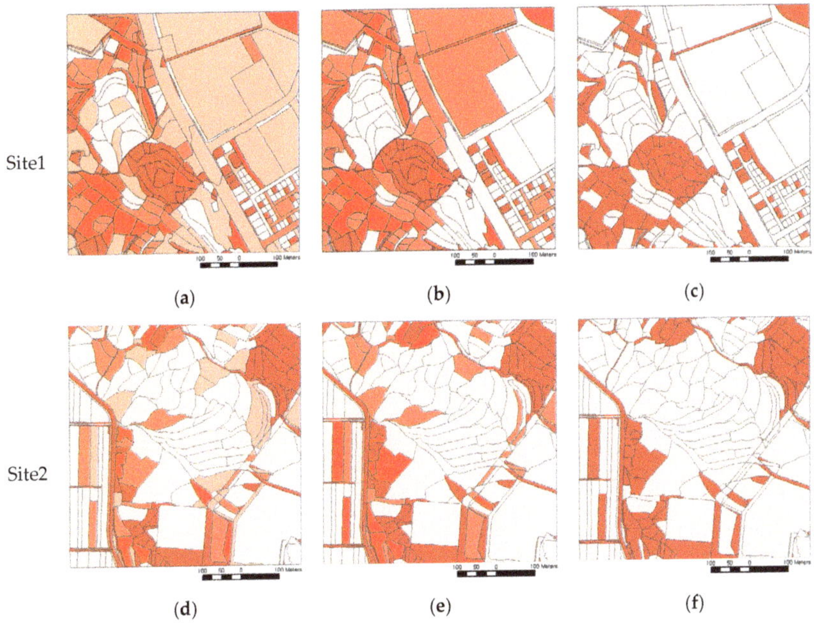

Figure 12. Visualization result of discrepancy ratios (the intensity of the red polygons directly relates to the discrepancy ratio): (**a**), (**d**) with five classes, (**b**), (**e**) with four classes, and (**c**), (**f**) with two classes.

Table 3 lists the numbers and land types of parcels with discrepancy ratios of 50% or higher. School sites, cemeteries, and factory sites have relatively low inconsistency probabilities because these land categories were mapped in a 1:*N* relationship over various land covers. However, paddy fields and bare fields encompass several inconsistent parcels, because many parcels classified as bare soil are actually crop lands that did not bear any crop at the image acquisition time. In particular, significantly fewer discrepancies of building sites and roads were found at Site 2 than at Site 1, because Site 2 covers many fields, paddy fields, and forests, and fewer urban areas such as building sites and roads.

Table 3. Numbers of parcels with discrepancy ratios of 50% or higher.

Land Category		Land Cover	Site 1	Site 2
	Building site	Road, Building, Bare soil	17	0
	Paddy Field	Crop land	24	44
	Park site	Forest, Water, Bare soil	0	0
	School site	Road, Building, Bare soil	0	0
	Road	Road	15	2
	Field	Crop land	22	25
	Forest	Forest	9	7
	Cemetery	Road, Building, Bare soil, Crop land, Forest	0	0
	Reservoir	Water	18	0
	Miscellaneous land	Bare soil	2	0
	Site for Religious use	Road, Building, Bare soil	0	0
	Gas station site	Road, Building, Bare soil	0	0
	Parking lot	Road, Bare soil	1	1
	Sport area	Building, Bare soil	1	0
	Ditch	Water	2	7
	Factory site	Road, Building, Bare soil	0	0
	Ranch	Crop land, Forest, Bare soil	1	0
	Total		112	86

5. Discussion

5.1. Analysis of Inconsistent Parcels

Figures 13 and 14 show the detailed results in two enlarged zones of Site 1 and Site 2, respectively. Some land parcels registered as "building site" on the cadastral system (i.e., designated for building construction) contained no buildings in the real dataset. These parcels often remained as a ground covered only by low vegetation. Thus, the discrepancy ratio was high in parcels of building sites in the cadastral map, but which were classified as crop land or bare soil in subset-1 of Site 1 (Figure 13a). Moreover, it was high in parcels of parking lots in the cadastral map, but identified as buildings in the land cover map. On "road," where the item of the cadastral map exactly matched the class of the land cover, the discrepancy ratios were increased to moderate because the boundaries were sharp on the cadastral map but fuzzy on the land cover map (the roadside trees were included in the road parcels). Although some parcels with more than 50% of discrepancy ratio were detected as candidates of updating, because of misclassified pixels, they showed a relatively low discrepancy ratio compared to other inconsistent parcels. In subset 2 of Site 1, the discrepancy ratio was high in the central part of the image because many parcels were registered as "reservoir" in the cadastral map but were actually crop land in the land cover map (Figure 13b).

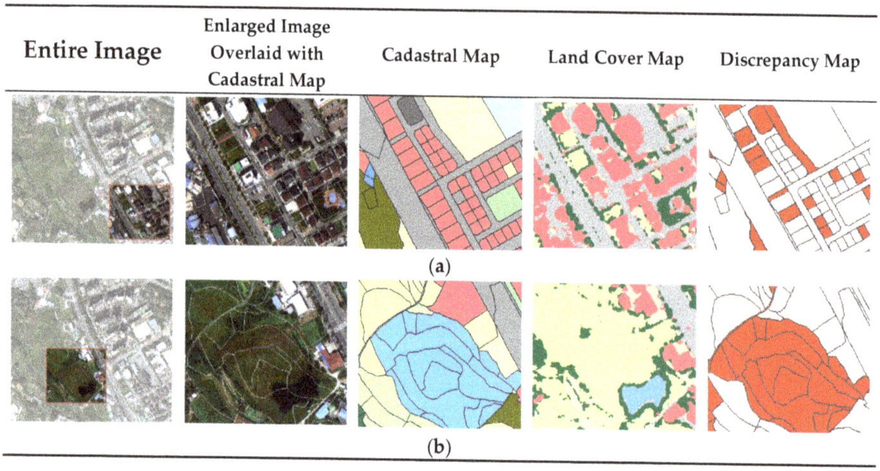

Figure 13. Enlarged hyperspectral UAV images, cadastral maps, land cover maps, and discrepancy maps in two zones of Site 1: (**a**) subset-1 and (**b**) subset-2.

In subset 2 of Site 2, some parcels were moderately inconsistent, identified as "paddy field" in the cadastral map, but as bare soil in the land cover map (Figure 14b). In a similar case of subset 1 of Site 2, the discrepancy ratio was high in areas identified as "paddy field" in the cadastral map, but classified as not only bare soil but also buildings in the land cover map. It seems that buildings were constructed on this site (Figure 14a). Moreover, if the vegetation index is low at the image acquisition time, paddy fields can be classified as bare soil.

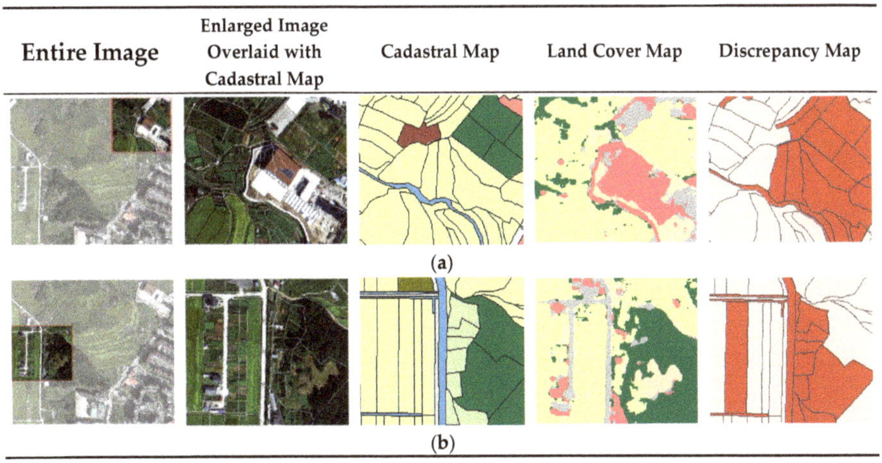

Figure 14. Enlarged hyperspectral UAV images, cadastral maps, land cover maps, and discrepancy maps in two zones of Site 2: (**a**) subset-1 and (**b**) subset-2.

5.2. Limitation and Future Work

Although the proposed method efficiently extracts inconsistent parcels by an automatic process, its performance depends on the acquisition time of the hyperspectral UAV image, which is input to the discrepancy analysis. For example, some crop parcels are erroneously classified as bare soil because the vegetation vitality is low when the image is captured. The results of the discrepancy analysis may depend on the matching criteria of the land category items and land cover classes, because no absolute rules for defining inconsistency can be established. In conclusion, changing queries for the comparison must be preceded according to the mapping criteria. On the plus side, the proposed method is generalizable to various cadastral systems through flexible modification of the matching criteria between the land categories and land coverage.

In a future, we will develop a classification network that distinguishes finer classes in hyperspectral UAV images with higher classification accuracy. For example, crop land can be divided into rice fields and other fields for finding complex matching relationships between land use and the land categories of cadastral maps.

6. Conclusions

Non-spatial data in cadastral maps, such as land use and land ownership is generally updated by field survey and updated manually after visual interpretation of source data, such as RS imagery. This study proposed an approach for analyzing the inconsistent areas between cadastral maps and hyperspectral UAV images. The proposed methods focus on the update of land category which is the attribute data that explain the characteristics of the parcel.

As a case study, the proposed discrepancy analysis was applied to the South Korea cadastral map, which includes 28 land categories. Land cover maps were generated from hyperspectral UAV images by using a hybrid CNN. The hybrid CNN outperformed previous 2D-CNN and 3D-CNN. The OAs of the land cover map using the hybrid CNN at Sites 1 and 2 were 99.93% and 99.75%, respectively. For comparing the two heterogeneous datasets, the existing cadastral map and the land cover map were encoded. After vectorization, the attributes of the combined vector map were decoded to recover the information of land categories and their coverage. The final discrepancy maps with different discrepancy ratios were generated through a query-based comparison. The discrepancy map reveals the inconsistent parcels, which are used illegally or which need to be subdivided. The discrepancy ratios of 39.4% and 34.4% of the parcels at Sites 1 and 2, respectively, were 50% or higher. The discrepancy was high in parcels containing building sites or newly constructed buildings on the cadastral map, but were being used as crop land. As our approach can automate the detection of inconsistent land parcels, it is expected to be applied to large areas and various scenarios. Therefore, they are time- and cost-effective alternatives to field surveys for cadastral map updates and the update cycle can be shortened because the required imagery is taken by UAVs.

Author Contributions: Conceptualization, Methodology, Software, Validation, Formal analysis, Data curation, Writing (original draft preparation), Writing (review and editing), A.S. and S.P.; Funding acquisition, Supervision, Project administration, A.S. All authors have read and agreed to the published version of the manuscript.

Funding: This research was supported by the National Research Foundation of Korea (NRF) funded by the Korean government (MSIT) (grant no. 2019R1A6A3A0109230211) and by a research and development program funded by the Spatial Information Research Institution of Korea Land and Geospatial Informatix Corporation.

Conflicts of Interest: The authors declare no conflict of interest.

References

1. Kumar, V.G.; Reddy, K.V.; Pratap, D. Updation of cadastral maps using high resolution remotely sensed data. *Int. J. Eng. Adv. Technol.* **2013**, *2*, 50–54.
2. Poornima, A.; Jagadeeswaran, R.; Kannan, B.; Sivasamy, R. Generating cadastral base for Kolathupalayam village in Tamil Nadu from high resolution LISS IV sensor data. *J. Appl. Nat. Sci.* **2016**, *8*, 2007–2010. [CrossRef]
3. Xia, X.; Persello, C.; Koeva, M. Deep fully convolutional networks for cadastral boundary detection from UAV images. *Remote Sens.* **2019**, *11*, 1725. [CrossRef]
4. Pržulj, D.; Radaković, N.; Sladić, D.; Radulović, A.; Govedarica, M. Domain model for cadastral systems with land use component. *Surv. Rev.* **2019**, *51*, 135–146. [CrossRef]
5. Ali, Z.; Tuladhar, A.; Zevenbergen, J. An integrated approach for updating cadastral maps in Pakistan using satellite remote sensing data. *Int. J. Appl. Earth Obs.* **2012**, *18*, 386–398. [CrossRef]
6. Rao, S.; Sharma, J.; Rajasekhar, S.; Rao, D.; Arepalli, A.; Arora, V.; Kuldeep, C.; Singh, R.; Kanaparthi, M. Assessing usefulness of high-resolution satellite imagery (HRSI) for re-survey of cadastral maps. In Proceedings of the ISPRS Annals of the Photogrammetry, Remote Sensing and Spatial Information Sciences, Hyderabad, India, 9–12 December 2014; Volume 2, pp. 133–143.
7. Jasińska, E.; Preweda, E. Determining the cadastral-tax areas for the real estate premises based on the model of qualitative and quantitative. In Proceedings of the Environmental Engineering 10th International Conference, Vilnius, Lithuania, 27–28 April 2017.
8. Heipke, C.; Woodsford, P.A.; Gerke, M. Updating geospatial databases from images. In *ISPRS Congress Book;* Taylor & Francis Grop: London, UK, 2008; pp. 355–362.
9. Koeva, M.; Muneza, M.; Gevaert, C.; Gerke, M.; Nex, F. Using UAVs for map creation and updating. A case study in Rwanda. *Surv. Rev.* **2018**, *50*, 312–325. [CrossRef]
10. Crommelinck, S.; Bennett, R.; Gerke, M.; Nex, F.; Yang, M.; Vosselman, G. Review of automatic feature extraction from high-resolution optical sensor data for UAV-based cadastral mapping. *Remote Sens.* **2016**, *8*, 689. [CrossRef]
11. Wassie, Y.A.; Koeva, M.N.; Bennett, R.M.; Lemmen, C.H.J. A procedure for semi-automated cadastral boundary feature extraction from high-resolution satellite imagery. *J. Spat. Sci.* **2018**, *63*, 75–92. [CrossRef]
12. Avramović, M.; Cvijetinović, Ž.; Mihajlović, D. Digital cadastral map updating status analysis in Serbia. In Proceedings of the International Scientific Conference and XXIV Meeting of Serbian Surveyors 'Professional Practice and Education in Geodesy and Related Fields', Danube, Serbia, 24–26 June 2011.
13. AL-Hameedawi, A.; Mohammed, S.; Thamer, I. Updating cadastral maps using GIS techniques. *Eng. Technol. J.* **2017**, *35*, 246–253.
14. Khadanga, G.; Jain, K.; Merugu, S. Use of OBIA for extraction of cadastral parcels. In Proceedings of the International Conference on Advances in Computing, Communications and Informatics (ICACCI), Jaipur, India, 21–24 September 2016.
15. Sung, C.J.; Lim, I.S. A objective study of non-coincidence between land category in cadastral map and land cover classification using satellite images. *J. Korean Cadastre Inf. Assoc.* **2008**, *10*, 177–190.

16. Crommelinck, S.; Bennett, R.; Gerke, M.; Yang, M.; Vosselman, G. Contour detection for UAV-based cadastral mapping. *Remote Sens.* **2017**, *9*, 171. [CrossRef]
17. Sim, S.; Song, D. Evaluation of cadastral discrepancy and continuous cadastral mapping in coastal zone using unmanned aerial vehicle. *J. Coast. Res.* **2018**, *85*, 1386–1390. [CrossRef]
18. Manyoky, M.; Theiler, P.; Steudler, D.; Eisenbeiss, H. Unmanned aerial vehicle in cadastral applications. In Proceedings of the International Conference on Unmanned Aerial Vehicle in Geomatics (UAV-g), Zurich, Switzerland, 14–16 September 2011.
19. Puniach, E.; Bieda, A.; Ćwiąkała, P.; Kwartnik-Pruc, A.; Parzych, P. Use of unmanned aerial vehicles (UAVs) for updating farmland cadastral data in areas subject to landslides. *ISPRS Int. J. Geo-Inf.* **2018**, *7*, 331. [CrossRef]
20. Gajendra, S.; Gopal Narayan, B. An assessment on cadastral map update technologies in Nepal. *Crisisonomy* **2017**, *13*, 133–142.
21. Ministry of Land, Infrastructure and Transport. *Act on the Establishment, Management, Etc. of Spatial Data*; MOLIT: Sejong-si, Korea, 2017.
22. Ministry of Land, Infrastructure and Transport. *Act on Land Survey, Waterway Survey and Cadastral Records*; MOLIT: Sejong-si, Korea, 2013.
23. Zhang, M.; Li, W.; Du, Q. Diverse region-based CNN for hyperspectral image classification. *IEEE Trans. Image Process.* **2018**, *27*, 2623–2634. [CrossRef]
24. Yang, X.; Ye, Y.; Li, X.; Lau, R.Y.; Zhang, X.; Huang, X. Hyperspectral image classification with deep learning models. *IEEE Trans. Geosci. Remote Sens.* **2018**, *56*, 5408–5423. [CrossRef]
25. Lee, H.; Kwon, H. Going deeper with contextual CNN for hyperspectral image classification. *IEEE Trans. Image Process.* **2017**, *26*, 4843–4855. [CrossRef]
26. Paoletti, M.; Haut, J.; Plaza, J.; Plaza, A. A new deep convolutional neural network for fast hyperspectral image classification. *ISPRS J. Photogramm. Remote Sens.* **2018**, *145*, 120–147. [CrossRef]
27. Li, Y.; Zhang, H.; Shen, Q. Spectral–spatial classification of hyperspectral imagery with 3D convolutional neural network. *Remote Sens.* **2017**, *9*, 67. [CrossRef]
28. Liu, X.; Sun, Q.; Meng, Y.; Fu, M.; Bourennane, S. Hyperspectral image classification based on parameter-optimized 3D-CNNs combined with transfer learning and virtual samples. *Remote Sens.* **2018**, *10*, 1425. [CrossRef]
29. Roy, S.K.; Krishna, G.; Dubey, S.R.; Chaudhuri, B.B. HybridSN: Exploring 3-D-2-D CNN feature hierarchy for hyperspectral image classification. *IEEE Geosci. Remote Sens. Lett.* **2019**, 1–5. [CrossRef]
30. Song, A.; Park, S.; Kim, Y. Updating cadastral maps using deep convolutional network and hyperspectral imaging. In Proceedings of the Asian Conference on Remote Sensing, Daejeon, Korea, 13–18 October 2019.
31. Arcgis Desktop. Available online: http://desktop.arcgis.com/en/arcmap/10.3/analyze/modelbuilder (accessed on 6 December 2019).
32. Arcgis Webmap. Available online: https://www.arcgis.com/home/webmap/viewer.html?useExisting=1 (accessed on 6 December 2019).
33. Guan, N.; Shan, L.; Yang, C.; Xu, W.; Zhang, M. Delay Compensated Asynchronous Adam Algorithm for Deep Neural Networks. In Proceedings of the IEEE International Symposium on Parallel and Distributed Processing with Applications and 2017 IEEE International Conference on Ubiquitous Computing and Communications (ISPA/IUCC), Guangzhou, China, 12–15 December 2017.

remote sensing

MDPI

Article

Innovative Remote Sensing Methodologies for Kenyan Land Tenure Mapping

Mila Koeva [1,*], Claudia Stöcker [1], Sophie Crommelinck [1], Serene Ho [2], Malumbo Chipofya [3], Jan Sahib [3], Rohan Bennett [4,5], Jaap Zevenbergen [1], George Vosselman [1], Christiaan Lemmen [1,5], Joep Crompvoets [2], Ine Buntinx [2], Gordon Wayumba [6], Robert Wayumba [6], Peter Ochieng Odwe [6], George Ted Osewe [6], Beatrice Chika [6] and Valerie Pattyn [7]

1 Faculty of Geo-Information Science and Earth Observation (ITC), University of Twente, 7514 AE Enschede, The Netherlands; e.c.stocker@utwente.nl (C.S.); s.crommelinck@utwente.nl (S.C.); j.a.zevenbergen@utwente.nl (J.Z.); george.vosselman@utwente.nl (G.V.); Chrit.Lemmen@kadaster.nl (C.L.)

2 Public Governance Institute, 3000 KU Leuven, Belgium; serene.ho@kuleuven.be (S.H.); joep.crompvoets@kuleuven.be (J.C.); ine.buntinx@kuleuven.be (I.B.)

3 Westfälische Wilhelms-Universität Münster, Institute for Geoinformatics is part of the Geosciences department of the faculty of Mathematics and Natural Sciences, 48149 Münster, Germany; mchipofya@uni-muenster.de (M.C.); sahib.jan@uni-muenster.de (J.S.)

4 Swinburne Business School, University of Technology Swinburne, Hawthorn campus BA1231, Australia; rohanbennett@swin.edu.au

5 Kadaster International, 7311 KZ Apeldoorn, The Netherlands

6 Technical University of Kenya, Nairobi P.O. BOX 52428–00200, Kenya; gwayumba@tukenya.ac.ke (G.W.); rwayumba@tukenya.ac.ke (R.W.); vc@tukenya.ac.ke (P.O.O.); George.Osewe@tukenya.ac.ke (G.T.O.); beatrice.chika@tukenya.ac.ke (B.C.)

7 Institute of Public Administration, Faculty Governance and Global Affairs, Leiden University, 2511 DP The Hague, The Netherlands; v.e.pattyn@fgga.leidenuniv.nl

* Correspondence: m.n.koeva@utwente.nl; Tel.: +31-53487-4410

Received: 16 December 2019; Accepted: 10 January 2020; Published: 14 January 2020

Abstract: There exists a demand for effective land administration systems that can support the protection of unrecorded land rights, thereby assisting to reduce poverty and support national development—in alignment with target 1.4 of UN Sustainable Development Goals (SDGs). It is estimated that only 30% of the world's population has documented land rights recorded within a formal land administration system. In response, we developed, adapted, applied, and tested innovative remote sensing methodologies to support land rights mapping, including (1) a unique ontological analysis approach using smart sketch maps (SmartSkeMa); (2) unmanned aerial vehicle application (UAV); and (3) automatic boundary extraction (ABE) techniques, based on the acquired UAV images. To assess the applicability of the remote sensing methodologies several aspects were studied: (1) user needs, (2) the proposed methodologies responses to those needs, and (3) examine broader governance implications related to scaling the suggested approaches. The case location of Kajiado, Kenya is selected. A combination of quantitative and qualitative results resulted from fieldwork and workshops, taking into account both social and technical aspects. The results show that SmartSkeMa was potentially a versatile and community-responsive land data acquisition tool requiring little expertise to be used, UAVs were identified as having a high potential for creating up-to-date base maps able to support the current land administration system, and automatic boundary extraction is an effective method to demarcate physical and visible boundaries compared to traditional methodologies and manual delineation for land tenure mapping activities.

Keywords: fit-for-purpose; land tenure; land administration; cadastre; UAV; feature extraction; needs assessment

1. Introduction

The first goal of the sustainable development goals (SDGs)—target 1.4—set by the United Nations (UN) aims to deliver tenure security for all [1]. Strategies to support this goal rely in part on the development of land administration systems (LAS) that formalize land rights that support secure land markets, facilitate poverty reduction and support national development [2]. Broadly speaking, LAS can provide the infrastructure for implementing land-related policies and management strategies and maintain information about people and land involving different organizations, processes, and technologies [3].

Contemporary land administration incorporates the concept of cadastre and land registration, often with a specific focus on the security of land rights [4]. It conceptually fits within the broader land management paradigm [5] with its four land administration functions (land tenure, land value, land use, and land development), ultimately seeking to deliver sustainable development. These functions utilize an underlying land information infrastructure including reliable remote sensing data. It should be noted that in this paper cadastre is considered synonymous with land registry and land administration system.

In sub-Saharan Africa, and in the other developing regions, numerous activities for land tenure recording have been, and continue to be, initiated. For example, in alignment with the SDGs, the Global Land Tool Network (GLTN), an international network of partners setting a global agenda for the improvement of land management and tenure security, develops the so called Social Tenure Domain Model (STDM), a tool for registering formal, informal, group, or individual rights [6].

However, it is estimated that only 30% of the world's population has documented land rights and has access to a formal cadastral system [7,8]. Cadastral mapping is proven as the most expensive part of the land administration system [5]; therefore, there is a clear need for innovation for fast, accurate, and cost-effective land rights mapping. Existing approaches using traditional methods including field surveys often prove to be time-consuming, costly, and labor-intensive.

In response, fit-for-purpose (FFP) land administration suggests technologies should be developed, adapted, selected, and applied to match the capacity and cost constraints of a specific context [4]. The main idea of the FFP approach is to ensure land tenure recording is delivered at scale on a regional and national level, rather than focusing on highly accurate solutions, with less coverage. Three main FFP characteristics are that the land administration systems should focus on the purpose, flexibility and upgradability. The concepts of FFP include principles that cover spatial, legal and institutional aspects on a country level. One of the key principles of the FFP approach is using "general" boundaries extracted by visual interpretation based on aerial images rather than "fixed" boundaries demarcated in the field and measured by a high precision Global Navigation Satellite System (GNSS) technology [9]. Some successful examples where the FFP approach was applied are Rwanda where a Land Tenure Regularisation Program (LTRP) was implemented, and Namibia and Ethiopia with their communal land registration and cadastral mapping [4]. To apply these principles of obtaining general boundaries cheaper and faster, there is a clear need for a new generation of tools and applications that are transparent and scalable [10].

In response, we are developing innovative, scalable methodologies, using remote sensing data and cadastral intelligence, based on fit-for purpose principles to respond to the continuum of land rights [11,12]. The aim of this paper is to assess user needs (in terms of land administration functions), and how the three remote sensing methodologies under development can meet these needs. It also considers how the adoption of these technologies may have governance implications. For the assessment we take into account the above mentioned land management paradigm [5] and FFP spatial and scalability requirements [4] for a case study located in Kenya. The combination of quantitative and qualitative results collected from fieldwork and workshops, taking into account both social (needs assessment and governance) and technical aspects (developed technologies), makes this paper a significant contribution.

The developed remote sensing methodologies for this study include (1) a unique ontological analysis approach using smart sketch maps, (2) unmanned aerial vehicles (UAV) for mapping procedures, and (3) Automatic Boundary Extraction (ABE), based on the acquired UAV images. The sketch maps mentioned above are hand-drawn either on a blank piece of paper or as annotations over existing spatial information, such as cartographic maps, aerial images, or other maps produced via community mapping. As people draw sketch maps based on observations and not based on measurements, the information is not georeferenced, but qualitative relations of the sketched information can usually be considered as correct (with respect to the background information) [13]. The usage of UAVs as cheap, affordable, and easy to use acquisition technology for obtaining high-resolution imagery is emerging for many applications [14,15]. Their applicability in the domain of land mapping was also explored extensively in a different contexts [16–21]. However, there is a lack of studies that evaluate the appropriateness of UAV technology considering the local context and the fit-for-purpose approach. The high-resolution images are usually used for manual delineation of visual boundaries with additional information attached including the ownership and value of the land [22]. However, manual delineation is time-consuming. To register unrecorded land rights more effective in terms of cost and time, innovative and scalable solutions were explored [23–25]. There are clear advantages in using ABE methodologies, therefore, new tools and techniques were developed for scaling up the mapping procedures in support of indirect cadastral surveying based on remotely sensed data [26,27].

The multidisciplinary nature of the current work, using different integrated approaches, and emerging remote sensing technologies, is novel and innovative to the land administration domain. The paper reports on the findings after fieldworks and workshops organized in Kenya, with the purpose of assessing the needs and end users' readiness, the applicability of the developed remote sensing methodologies, considering the needs and how they may affect the governance aspects. In the background section information related to the previous and current land administration system is explained and the study area is described. The overall methodology of the paper is explained in Section 3 and the concrete methods used for each of the assessed remote sensing methodologies are presented. The results are presented in section four, followed by critical discussion, conclusions, and suggested further steps.

2. Case Background and Study Area

Kenyan urbanization to date has been one of imbalanced growth due to ad hoc identification of urban areas, resulting in skewed distribution and inequality in development [28]. This challenge of land governance has been found to be a significant factor in constraining inclusive prosperity more generally across Africa's urbanization phenomenon [29]. It is particularly evident in contested peri-urban lands emerging as a result of metropolitan sprawl across sub-Saharan Africa [30,31]. In response, one of the key strategies consistently advocated by the international development community is the establishment (or improvement) of a formal land market. Such a techno-economic orientation and focus on market-driven urbanization is evident in many contemporary studies of land tenure that continues to pay limited attention to underlying political aspects of tenure regimes [32]. Land tenure is inherently social and political, and in Kenya, land is also overtly cultural. Attention to the cultural aspects of land is particularly relevant in urbanization in Kenya as, first, a majority of Kenya's land resources are held under customary tenure systems, and second, having remained unrecognized by formal systems since colonial rule, indigenous groups have long borne the burden of Kenya's structural adjustments, which have resulted in dispossession and longstanding tenure insecurity [33].

Situated in East Africa, Kenya covers almost 600,000 km^2 constituted of 47 counties with a population of almost 45 million [34]. Approximately 80% of Kenya's land is categorized as arid or semi-arid, with only 15% of this suitable—and fully used—for agricultural production [35]. Since 1963, the land administration in Kenya is under the Ministry of Lands, Housing and Urban Development. The organization structure is presented in [36]. Under colonial and post-independence governments,

Kenya has operated two land tenure systems simultaneously: statutory (based on English property law) and customary. The 2010 Constitution now recognizes customary tenure systems as Kenya's third type of legal tenure, but administrative implementation of this recognition remains in its infancy. The National Land Commission (NLC) is tasked with oversight for all planning processes in Kenya. In relation to FFP, Kenya is one of the countries that first introduced this approach in 1954. Under this major land reform program, land consolidation and systematic adjudication methods were used to determine the parcel boundaries in the rural parts of Kenya. These boundaries were identified, walked, and demarcated by the local inhabitants, based on aerial images, thus it was a participatory approach. As a result so called Preliminary Index Diagrams (PIDs) were produced, which were used to register the rural land parcels in Kenya for many years [37]. Generally, Kenyan cadaster consists of different types of maps, such as survey plans, field notes, registry index maps, aerial photographs, topo-cadastral maps, deed plans, and title deeds, sometimes with variety of names and accuracy [38]. However, most of them are in paper format and are kept in archives (Figure 1). There has been a research also on integrating the buildings into databased and adapting the existing land administration system according to the international ISO: 19152 Land Administration Domain (LADM) standard [36].

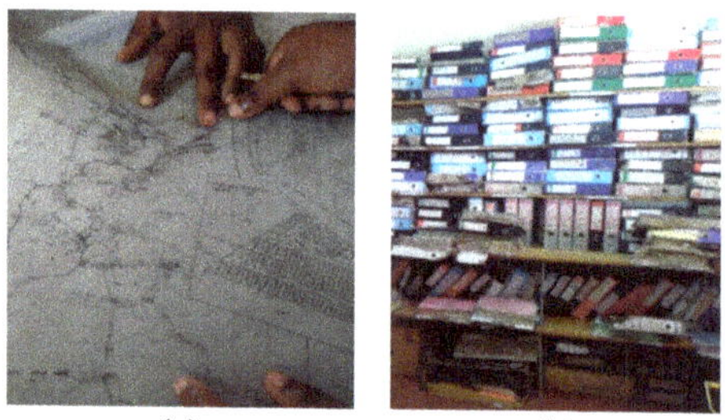

(a) (b)

Figure 1. (a) Cadastral map of part of Kajiado; (b) archive of land titles.

Kajiado is selected as the case study site for this study as it is part of the Nairobi metropolitan region and is the site of multiple contests for land (Figure 2). Its proximity to Nairobi and Amboseli national parks has also led to increasing human–wildlife conflict being experienced in Kajiado. Currently, approximately 25% of the county's population (of more than 800,000) is urban, almost 50% live below the poverty line, and the population growth rate of 5.5% is higher than the national average [39].

The current land registry map is riddled with information errors stemming from inappropriately scaled maps (resulting in scale errors and boundary overlaps)—the continued use of which introduces further errors in the land registry—which makes it now difficult or impossible to fit new development plans on the original map base [35]. These information issues have resulted in administrative challenges.

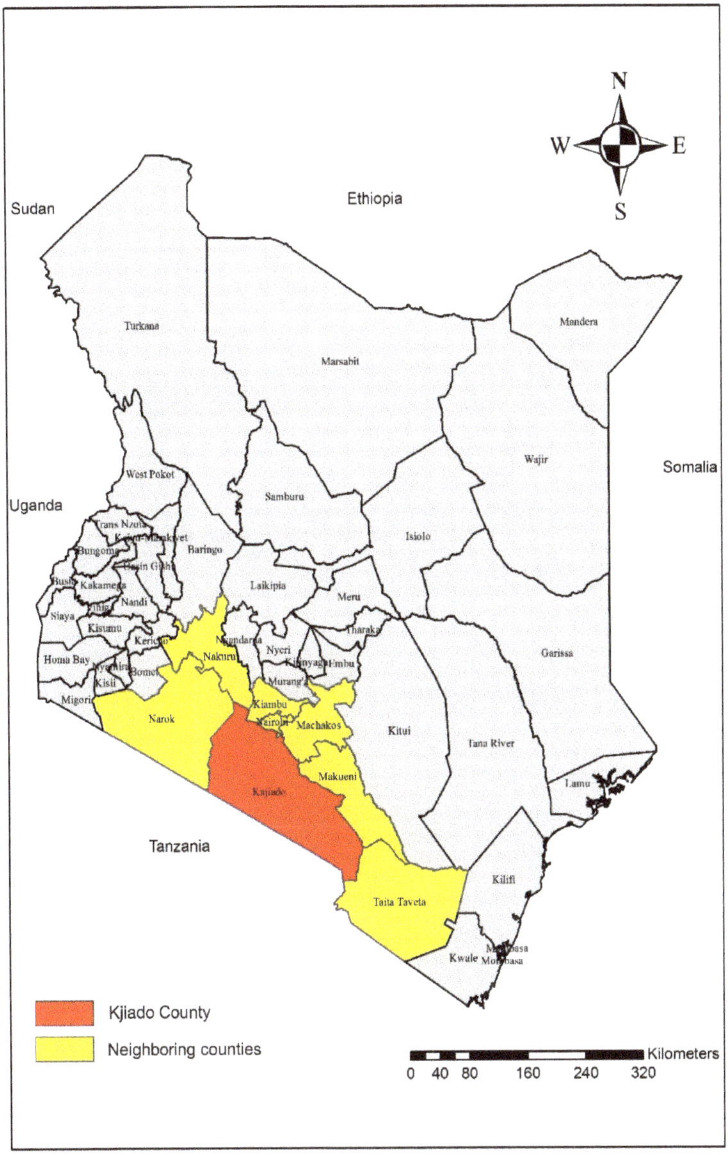

Figure 2. Location of Kajiado County and neighboring counties.

3. Materials and Methods

Land management activities rely upon three main components [5]: (1) land policy, (2) land administration functions, and (3) land information infrastructure. In the current research, the major focus is the land administration functions including (1) land tenure, (2) land value, (3) land use, and (4) land development. Specifically, this research is mainly seeking to contribute to improving land tenure including cadastral surveys of determining spatial information on parcel boundaries.

First, user needs were identified. Second, the identified needs were incorporated into a broader assessment of the three remote sensing methodologies, made up of 10 criteria, taken selectively from the land management paradigm [5] and FFP requirements [4]. Third, the governance aspects in relation to the new developments are also analyzed. The entire evaluation methodology is visualized as a conceptual framework in the following Figure 3.

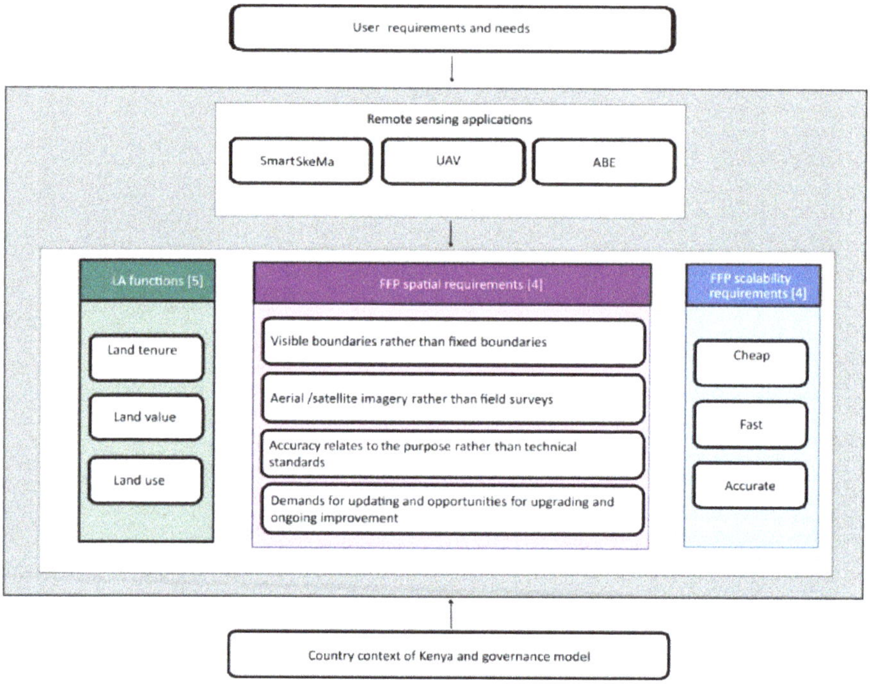

Figure 3. Conceptual evaluation framework.

3.1. Method for User Needs Assessment

First, for the user needs assessment, assessment was conducted using a behavioral-science-based method called the Nominal Group Technique (NGT). It was developed as a group process model to support the identification and prioritization of problems and/or solutions amongst groups of stakeholders by facilitating equal participation [40,41]. NGT was selected as it draws on individuals' knowledge and expertise while mitigating power dynamics in group-based data collection scenarios [42]. It also produces outcomes that have been found to be robust and meaningful while still being time- and resource-efficient process [43]. A more detailed explanation is provided in [44]. The assessment of the needs was completed prior to other fieldwork: Data was collected from representatives from a range of county government office functions (surveying, registration, planning), as well as county-level officials. Additional data was obtained from local communities with 35 community members from various Maasai families participating (25 men, 10 women).

3.2. Remote Sensing Methodologies

Second, fieldworks, workshops, semistructured interviews, and focus group discussions regarding the three remote sensing methodologies were then conducted. This primarily took place in Kajiado from 22th of September to 5th of October 2018. All the workshops were held at the Regional Centre for Mapping Resources in Kajiado. Overall, three one-day workshops with 58 land administration stakeholders from local government institutions, non-governmental organizations (NGOs), private companies, and national government institutions were organized. Each workshop followed the same structure: the project context was presented, participants were split into groups; activities, demonstrations, and discussions for the SmartSkema, UAVs, ABE, and governance aspects followed as shown on Figure 4 below. A follow-up discussion was held in plenary to produce a strengths, weaknesses, opportunities and threats SWOT analysis of each suggested methodology. Pictures from the workshops assessing the remote sensing methodologies are shown in Figure 4 below.

Figure 4. Workshops assessing the remote sensing methodologies.

3.2.1. SmartSkeMa—Sketch Map Data Collection Software

SmartSkeMa is a software application that we developed to support the documentation of land tenure information for communities with a focus on the actual land practices in the communities. SmartSkeMa supports land recording processes in two main ways [13]. First, it provides a means to document land related concepts as expressed within the local culture or context in a structured domain model [45]. Second, it supports sketch-based community mapping processes by providing a means to digitize, annotate, and geolocalize hand-drawn objects in a sketch map [46]. The method uses both qualitative and quantitative representations of a digitized sketch map and aligns features from the sketch map with corresponding features in the base map. For qualitative representations alignment of qualitative spatial configurations is done. In the case of quantitative (cartesian) representations, the alignment is performed by a coordinate transformation using predetermined control points. The latter approach allows SmartSkeMa to be used as a digitizer over aerial imagery (Figure 5).

Figure 5. Screencast of a SmartSkeMa live demonstration of processing the spatial information drawing on top of a satellite image, the vector representation of drawn features as a SmartSkeMa response, and georeferencing drawn features (web-link: www.smartskema.eu).

Sketch maps are uploaded into SmartSkeMa as raster images. SmartSkeMa then converts these images into vector form in two steps. First any symbols in an image are detected and recognized using a Convolutional Neural Networks (CNN) trained on a set of predefined hand-drawn symbols. The symbols form a visual language for representing land use concepts and land features. After symbol detection the system performs a stroke-based image segmentation wherein boundaries of sketched objects are traced and extracted. Finally, the concepts corresponding to the detected symbols are applied to the image segments based on distance and a fixed set of rules specifying spatial constraints on configurations of different types of features.

The data collection used for the current study needed for the SmartSkeMa system was completed during a series of fieldworks and workshops with male and female members of the Massai community starting from 2017 in Kajiado county and Nairobi, Kenya and running through to October 2018. The sessions included demonstrations of the three main functional parts of the SmartSkeMa (Figure 6), followed by discussions about the applicability of SmartSkeMa. Questions were posed through questionnaires to evaluate the applicability of SmartSkeMa in (1) standard (official) land information recording processes, (2) documenting local land tenure systems, and (3) other land administration tasks.

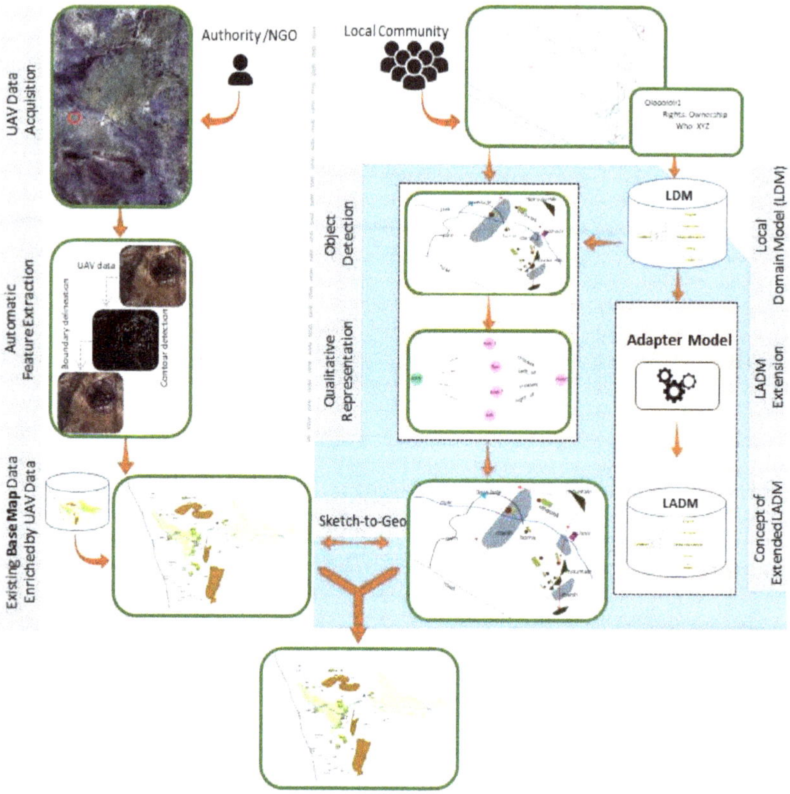

Figure 6. Workflow of SmartSkeMa: Right side: local communities provide spatial and nonspatial information via sketch maps. Nonspatial information is processed via local domain model (LDM) and connected via the adapter model to land administration domain model (LADM). Spatial information is recognized via the object detection technique, captured qualitatively via the qualitative representations, and aligned with existing dataset such as feature extracted from UAV data.

3.2.2. UAV Data Collection Methods

To prove the concept of UAV data capture as a remote sensing technology for land rights mapping in Kenya, an exploratory research investigation was undertaken. This included the entire UAV-based workflow, starting from the choice of UAV equipment, pilot and flight training, flight authorization, and the final data collection in the field which was carried out in two different sites in Kajiado County: a rural area in Mailua and the township of Kajiado. To accommodate the different characteristics of the flight locations, two different UAVs were chosen (see Figure 7) both with RGB sensors on board.

Figure 7. UAV data collection with the DT18 in Mailua and the Phantom 4 in Kajiado.

In Mailua, the DT18, a fixed-wing UAV with a long endurance and a large range was selected. In contrast, the vertical take-off and landing UAV DJI Phantom 4 was the preferred equipment to capture data of Kajiado, as the urban area did not provide large spaces for take-off and landing. Both study sites, were captured with indirect georeferencing (Figure 8), i.e., Ground Control Points (GCPs) were distributed within the field and measured with a Global Navigation Satellite System (GNSS) achieving a final accuracy of less than 2 cm. RGB orthomosaics and digital surface models (DSM) of approximately 6 cm Ground Sample Distance (GSD) were generated with Pix4DMapper. Three tiles of 300 × 300 m were selected to demonstrate the boundary mapping approach.

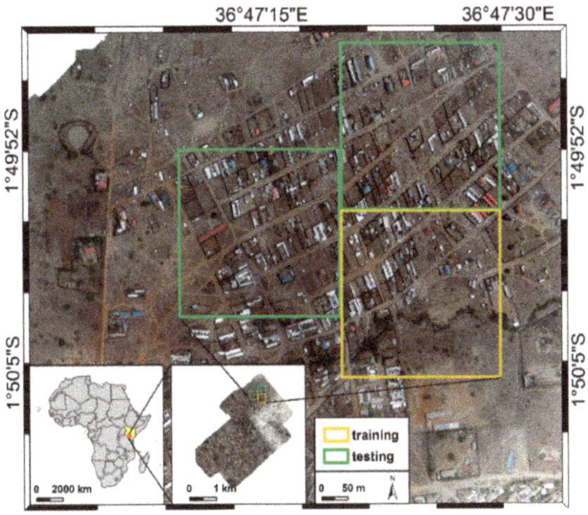

Figure 8. Areas of investigation of 300 × 300 m and a 6 cm GSD of Kajiado, Kenya.

The evaluation of the UAV workflow was based on the case study results from Kenya, as well as statistics of the UAV image processing, and resulted in a SWOT analysis. Further insights were gained from a stakeholder assessment of the potential of UAV-based technology to capture land rights in Kenya [47].

3.2.3. Automatic Boundary Extraction Methods

The method used for the current study is based on the work in [48] and shown in Figure 9. It supports the delineation of boundaries by automatically retrieving information from RGB data that is then used to guide an interactive delineation. It consists of three parts: (a) image segmentation, (b) boundary classification, and (c) interactive delineation. The source code is publically available [49].

(a) Image segmentation delivers closed contours capturing the outlines of visible objects in the image. Multiresolution combinatorial grouping (MCG) [50] has shown to be applicable on high-resolution UAV data and to deliver accurate closed contours of visible objects [48].

(b) Boundary classification requires labeling the contours from (i) into "boundary" and "not boundary" to generate training data. A set of features is calculated per line capturing its geometry (i.e., length, number of vertices, azimuth, and sinuosity) and its spatial context (i.e., gradients of RGB and DSM underlying the line). These features together with the labels are used to train a Random Forest (RF) classifier [51]. The trained classifier predicts boundary likelihoods for unseen testing data for which the same features have been calculated, as indicated with training and testing. An open-source RF implementation [52] is used.

(c) Interactive delineation allows a user to start the actual delineation process: the RGB orthomosaic is displayed to the user, who is asked to interactively delineate final parcels based on the automatically generated lines and their boundary likelihoods. A user can make use of four functionalities that simplify, vary and speed up the delineation process. We implemented (c) as publically available plugin [49] for the open-source geographic information system QGIS [53].

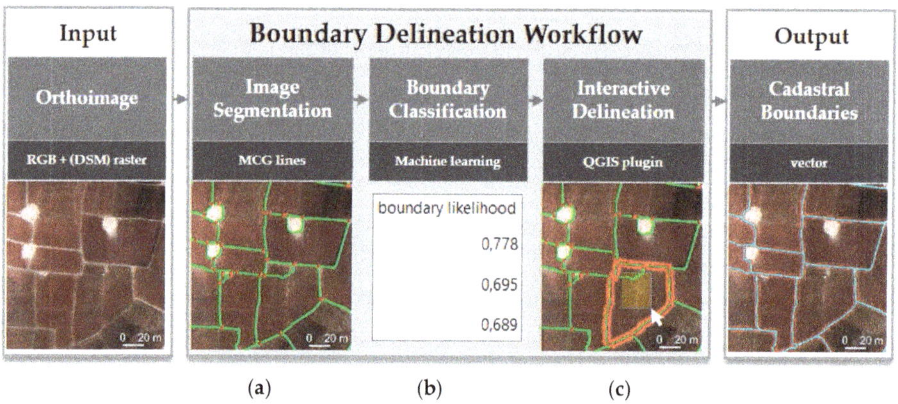

Figure 9. Boundary mapping method: (**a**) Multiresolution combinatorial grouping (MCG) image segmentation. (**b**) boundary classification that requires line labeling into "boundary" and "not boundary" for training. The labeled lines are used together with line-based features to train a Random Forest classifier that generates boundary likelihoods for testing. (**c**) interactive delineation guided by a QGIS plugin.

An analysis on our study area in Kajiado assessed the ABE application for extracting visible cadastral boundaries using the acquired UAV images [24]. During the workshops, we used the three tiles of 300 × 300 m shown in Figure 7 to demonstrate the boundary mapping method. In this way, we related the method to our conceptual framework (Figure 2). This allowed identifying and understanding bottlenecks: (i) the operational analysis questions when and why the method works better or worse compared to manual delineation and (ii) the feedback analysis investigates the method based on surveying stakeholders responses. The feedback analysis was based on discussing the strengths, weaknesses, opportunities, and threats (SWOT) of our proposed method compared to manual delineation as identified by the workshop participants.

3.3. Governance Aspects

To get an overview of the governance requirements to support the adoption and use of the above-mentioned remote sensing methodologies, in-depth semistructured interviews were conducted and focus group discussions were organized. The assessed governance aspects are strongly based on FFP demands [4] and dimensions of the governance assessment tool [54]. Three focus group discussions were organized. The first group focused on local government (with 18 participants), the second on private companies and NGOs (with 32 participants), and the third on national government (with three participants). During the focus groups, participants were able to map the different governance requirements (responsible actors, partners, levels, instruments, resources, organizational characteristics, capacity development characteristics, and cultural characteristics) needed to successfully adopt the technical applications. Besides these focus group discussions, we were also able to do 18 in-depth semistructured interviews in Kenya. For these interviews, we were making use of a guiding topic list to facilitate the extensive data collection to support the development of multi-sectoral profiles (e.g., socioeconomic characteristics, geospatial innovation trends, etc.) of the identified case areas pertaining to land tenure information. The topic list that guided the interviews was a compilation from an extensive literature reviews on governance and capacity development. The questions of the semistructured interviews were not only structured by the specific topics, but also open enough to allow for clarifications, new insights and deepening of the subjects by new, unexpected responses during the interviews.

4. Results

4.1. Results of Needs Assessment

A range of land administration needs were identified (Table 1) for Kajiado county stakeholders, categorized along the main land administration functions of tenure, value, use, and development (Figure 2); however, other needs (such as governance) also emerged.

Table 1. Assessing the needs.

Land Admin Needs	County Government Needs	Community Needs
Tenure	- Georeferenced land information connected to the registry index map - Land subdivision data - Updated land information - Resurvey of adjudicated areas of public utilities - Good practices related to surveying and mapping	- Updated land subdivision information - Right-of-way information about government wayleaves - Reduce fencing around properties owned by non-Maasai
Value	- No of properties (and its attributes) in the county	- Locate and protect culturally significant resources (e.g., important waterways) - Map areas of cultural value and culturally significant objects (e.g., trees).

Table 1. *Cont.*

Land Admin Needs	County Government Needs	Community Needs
Use	- Approval and placement of private boreholes - Environmental degradation - Documentation about the location of public utilities - Information about cottage industries (as these are damaging properties)	- Understand and respect Maasai land use practices - Improve poor animal husbandry practices - Manage overgrazing - Improve wildlife corridors to reduce loss of land - Information about wild animal infestations needed due to damage infrastructure and spread diseases
Development	- County spatial plan - Land use zoning and development controls need to be defined - Land fragmentation (subdivisions too small) - Map and demarcate roads to avoid informal development - Clearly mark ecologically fragile areas	- Define migratory corridors and restrict sale or give right-of-way to Maasai - Identify and document fertile grazing areas at waterways that need to be mapped and preserved - Improve drought mitigation - Mitigate deforestation
Governance	- Relationship between land laws - National vs. county land policy - Improving data management for multipurpose use - Community understanding of women's rights in land transactions - Improve understanding of community's land needs and improve engagement around land policy	- Need to integrate community knowledge with formal and/or statutory information. - Legal aspects of land conflict not well understood by community

Land tenure security is identified as a fundamental need in Kajiado, and is being challenged by urbanization. In interviews, the county estimated that almost 80% of its registry's resources were directed towards resolving land disputes (e.g., in ground truthing and reporting). Unsurprisingly, the needs assessment reflects this: land tenure needs identified included improving the quality of registry information, especially spatial information; associated with this is the need to improve subdivision data in general—both in terms of data and processes—and more generally, the need for updated information. Other tenure needs related to spatial and administrative information pertaining to public utilities. Similarly, communities identified the need for better information about subdivisions (especially within group ranches) and the spatial extent of government wayleaves and associated rights. This was important to support understanding how land is acquired to protect right-of-way for maintaining public infrastructure and compensation, provided around reduced use rights.

The information needs identified around land value, use and development were not as easy to differentiate: all three functions of land are interlinked. From the government's perspective, identified needs with direct implications for land value were around better-quality information about the number of properties in the county. For the community, the identified needs reflected the need for preservation of culturally significant areas (e.g., important waterways) and objects (e.g., trees).

There were greater needs identified around land use, and these related to lack of knowledge around where boreholes were being placed and used (which is draining the local water table), the increasing impact of drought, and mitigating general environmental degradation, especially resulting from unregulated cottage industries, such as charcoal production (where burning of trees also impacts the value of private properties). Similarly, for communities, needs identified were around mitigating unsustainable land use practices that were either impacting the Maasai way of life, or draining environmental resources that impacted on their ability to rear livestock. While not directly referencing land information, most of the needs certainly infer some type of spatially-enabled decision-making, e.g., knowing where overgrazing occurs, and knowing where existing wildlife corridors are, or where to situate new ones.

The broader issues implicated in the identified land use needs are also reflected in the land development needs. Kajiado is rapidly urbanizing; consequently, the county government would like to better understand how to plan and manage development. This included the need for better spatial planning (through production of a county spatial plan), better planning controls (through zoning) (especially as land fragmentation is becoming an issue), and defining ecologically fragile areas. For the Maasai, land information needs around migratory corridors (e.g., restrict sale or give right-of-way encumbrances in favor of Maasai) and fertile waterways reinforce their desire for land use practices that enable them to flourish culturally. Considering the rapid influx of "outsiders", i.e., non-Maasai, into Kajiado, the community emphasized the importance of understanding and respecting Maasai communal-based practices of resource sharing and the implications this will have on property boundaries. However, there are also needs around broader environmental issues caused by over and unmanaged development, such as drought and deforestation, and wild animal infestations which damage property (e.g., water pipes) and spread disease amongst herds.

Finally, the whole range of governance needs emerged, which reflected the disconnect in land information and land policies at national and county levels and the disconnect between government and communities (despite the Constitution enshrining participatory action in land development) around rights (e.g., women's land rights) and responsibilities (e.g., improving community engagement). For the Maasai, additional elements reflected the disconnect between formal and customary knowledge systems (and relevant data), but reinforced the fact that communities do not have a good understanding of legal and policy frameworks pertaining to land (e.g., land conflict), which leaves them vulnerable to poor decision-making.

4.2. Results from SmartSkeMa

Stakeholder impressions of the SmartSkeMa application were sought along three main dimensions: (1) ability to support conventional land tenure recording activities, (2) ability to facilitate community driven land tenure recording systems, and (3) applicability in other land administration functions. SmartSkeMa was generally judged to have the necessary functionality to support standard land tenure recording activities. Among 21 participants, 16 considered SmartSkeMa to be usable together with standard land administration systems, while two considered this to not be the case and three were ambivalent (they neither agreed nor disagreed to the statement). In addition, of the 21 participants, 18 agreed that the functionality of SmartSkeMa is useful for recording land tenure information while three mentioned that it was only partly useful for that purpose. The participants also indicated the reasons for their judgements or choice. Table 2 shows a summary of these data coded into themes as presented by the participants.

Table 2. Summary of participants perceptions in the usefulness of SmartSkeMa for land tenure recording.

Usefulness for Land Tenure Recording	Reasons	Comments
Partly useful	Poor geometric accuracy or poor precision	Poor accuracy or precision will lead to legal impediments. May not work in densely populated areas.
Very useful	Can be used to delimit communal land rights; physical planning; updating official maps; delimiting communal land rights; consultation and public participation; reach consensus when recording land rights; record information from community perspective.	Requires interoperability with government systems, government buy-in, and may face legal impediments.

In terms of facilitating community driven land tenure recording systems SmartSkeMa was considered more favorably. Of the 21 participants, 18 believed that SmartSkeMa could support communities to register and govern their lands according to local customs. There was no clear agreement on which other land administration tasks the SmartSkeMa application could be applied to. Several tasks stood out with land use documentation and land use planning mentioned by six participants; recording of historical and inaccessible information was mentioned by four participants; and aiding surveying and other traditional land information collection was mentioned by three participants. Finally, we asked the participants to perform a SWOT analysis of the tool based only on the functionalities that have been presented to them during the demonstration. The results of this analysis are shown in Table 3 below.

Table 3. SWOT results on SmartSkeMa.

Strengths	Weaknesses
- Has multiple applications (incl. collecting historical information; creating land use plans; documenting land rights) - Has low barrier to entry - Is participatory - May create trust among community participants - Can help reduce conflicts after parceling - Can produce preliminary data for land surveys	- Time-consuming in the field and during preparation - Cannot yet associate attributes to boundaries - Difficult to get community engaged - Requires background knowledge - Collected data may not be accepted as meeting legal standards for land adjudication due to poor accuracy
Opportunities	**Threats**
- Use in implementation of the Community Land Act's community land registers [55]. - Incorporation of Satellite imagery to prepare the base map data and sketches	- Difficult to get community engaged - Misalignment with official records or existing systems

The feedback obtained from the workshops laid the foundation for the development of the second method: use an aerial image as the background for a sketching exercise. This is expected to increase the precision and provide measurable accuracy. The alignment of a sketch traced on top of an aerial image is done by a 6-parameter affine transformation. The parameters for the transformation are estimated by ordinary least squares linear regression quadratic features.

The new method was tested on a small sample of parcels and three metrics were taken as shown in Table 4. The time for delineation cannot be compared with traditional method since the time to produce the parcels is mostly consumed by the field work. As field work is required to collect parcel information in other approaches as well, we conclude that the automatic delineation of sketch maps results in a faster process.

Table 4. Performance metrics of parcel delineation using SmartSkeMa's sketch-on-map.

Manual Delineation		
Number of parcels in the sample	Sketching time per parcel	Mean Deviation from cadastral boundaries in meters (sampled)
9	6 min	1.29

4.3. Results from UAVs

The case study revealed many opportunities but also a number of challenges for UAV data capture as a technical solution to provide a spatial database for capturing land rights and cadastral boundaries in Kenya. In most countries, before commencing a UAV flight mission, regulatory clearance has to be in place to ensure the safety of airspace users, people, and property on the ground [56]. In that regard, Kenyan UAV legislation underwent changes during the case study. Before the official regulations were gazetted [57], the use of UAVs was heavily restricted, with a mandate to seek flight permission from Ministry of Defense and Kenya Civil Aviation Authority. At the time the regulations were passed, processes for flight authorizations seemed to be straight forward. However, a reality of a too costly and restrictive procedure largely impeded the rise of UAV technology in Kenya. Soon after release in June 2018, the regulations were nullified by the Government, leaving a regulatory vacuum in the country. Both data acquisition flights were carried out with a temporal flight authorization and awareness of the local government.

After an extensive sensitization of the local government and community, the UAV data, as well as GNSS measurements, were completed in March 2018 (Mailua) and September 2018 (Kajiado). The RGB pictures were processed with Pix4D to create an orthophoto (Figure 10). Flight specifications and information on geometric accuracy are summarized in Table 5.

Figure 10. Overview of UAV datasets.

Table 5. Flight characteristics and geometric accuracy of Kajiado and Mailua dataset.

	Mailua	Kajiado
UAV equipment/sensor	DT18 PPK/ DT18 3bands	DJI Phantom 4/inbuilt sensor
UAV Type	Fixed-wing	Fixed-wing
UAV Sensor [mm]	8.45 × 7.07	8.45 × 7.07
Resolution	5 MP	5 MP
Flight time	4 flights á 45 min	15 flights á 10–15 min
Area captured	3.32 km^2	8.28 km^2
Flight height	200 m	200 m
Overlap forward/side [%]	80/70	80/70
Ground Sampling Distance	5.72 cm	5.8 cm
Ground Control Points (GCPs)/	8	16
RMS Error (X/Y)	0.022 cm/0.024 cm	0.040 cm/0.042 cm

However, our case study showed that UAV workflows are easy to transfer to different contexts: data acquisition always follows a standard procedure following the steps of flight planning, data collection and postprocessing. Prices of UAV equipment vary largely, offering technical platforms for almost every budget without compromising too much on data quality. Nevertheless, the purchasing costs might give an indication of the longevity and the reliability of the UAV components, which is beyond the results that the case study currently provides. Similar to the price for UAVs, the accuracy of the final orthomosaic can differ from several centimeters to meters, as it depends on the GNSS sensor of the UAV, the availability of a geodetic network, the visibility of satellites during data acquisition, and the strategy of ground control measurement. The insights from the workshop can be concluded in a SWOT analysis (Table 6).

Table 6. SWOT results on UAV data acquisition.

Strengths	Weaknesses
- Provides reliable data products (orthophoto, 3D point cloud, and digital surface model) for multiple purposes in land administration - Various UAV platforms and sensors can be utilized depending on the context and geographical conditions - Immediate data collection to gather up-to-date base data, if flight permission is granted by the authority - Automated flight planning and image processing reduces training effort - High spatial resolution and geometric accuracy	- Dependent on weather conditions - Limited to small to medium scales - Real-Time Kinematic (RTK) or Post-Processed Kinematic (PPK) workflows require professional GNSS equipment for static observations - Time-consuming measurement of Ground Control Points to reach a high level of geometric accuracy if RTK or PPK workflow is not supported by the UAV equipment
Opportunities	**Threats**
- Community engagement as data is being collected directly in the field - Ease of use allows capacity development at the local level (e.g., for bottom-up initiatives) - High-resolution orthorectified images for cadastral mapping in urban contexts - Low investment costs for decent UAV equipment	- The unclear legal situation that potentially prohibits or restricts UAV flights - Maintenance services of UAV equipment not available in the country - UAV technology not included in current surveying act—a high barrier to adopt the technology

4.4. Results from ABE

Delineating boundaries with indirect surveying from the remote sensing imagery requires knowledge about the boundaries. To recognize boundaries in an image, it helps to be familiar with their appearance on the ground. We, therefore, went to the area for which UAV data was captured and took images of example boundaries. A team of village elders and a local researcher joined us to communicate with land owners when passing and capturing their boundaries. The team explained which objects were typically used to demarcate boundaries and provided insights on local boundary demarcation challenges.

During fieldwork in Kajiado, we obtained an understanding of local boundary characteristics and demarcation challenges. The letters used in the following refer to Figure 11.

Figure 11. (**a–d**) Examples of visible boundaries in Kajiado. (**e–h**) Boundary demarcations challenging to identify correctly from remote sensing imagery collected during the field survey.

A majority of boundaries are demarcated by visible objects such as (a) stone walls, (b) corrugated metal fences, (c) vegetation, or (d) ditches. The following examples are extractable from remote sensing imagery though require local knowledge or context for a correct identification: (h) ditches can be confused with soil erosion when extracted from imagery alone. (d) Some fences demarcating boundaries are challenging to differentiate from its surrounding. High-resolution digital surface models (DSMs) can support the identification of such fences. (f) Beacons demarcate boundary corner points and (g) can be used in parallel with linear boundary demarcations, or as control points for hosting measurements.

The cadastral boundary has often remained on the connection of the beacons, instead of on the visible boundary. Based on the local knowledge obtained during fieldwork, and the large portion of cadastral boundaries in Kajiado being visible following the FFP principles, the boundary mapping approach could be applied to the captured UAV data (Figure 12).

Figure 12. (**a**,**b**) Cadastral boundaries delineated from UAV data.

Some challenges that we observed during delineation are shown on Figure 13 below.

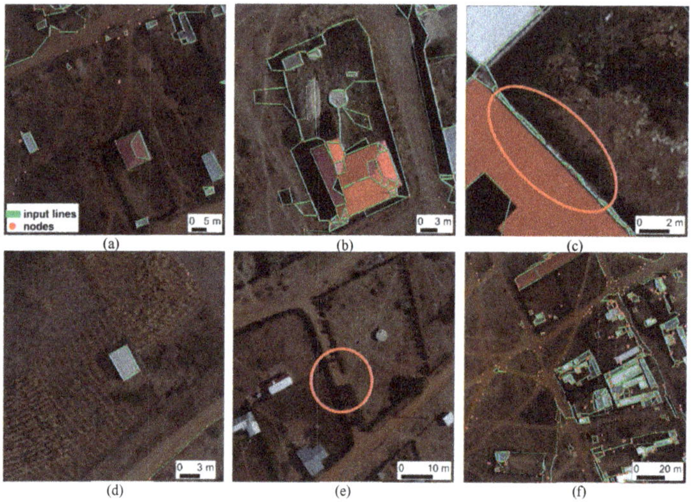

Figure 13. Challenges observed during delineation: (**a**) undersegmentation, (**b**) oversegmentation, (**c**) fragmented segmentation, (**d**) redundancy of least-cost-path calculation, (**e**) visible boundary not demarcated by objects, but by context, and (**f**) identification of delineation areas through boundary mapping approach.

Existing reference maps for our area would mostly consist of Registry Index Maps (RIMs). RIMs show the outline of land parcels within a given jurisdiction using general boundaries along visible features. The boundaries' position is only indicative and not legally binding. RIMs and survey plans for urban areas have the highest accuracy specifications of 30 cm nominal positional accuracy [38]. Different types of RIMs exist that partly allow positional errors of up to 200 cm [58,59]. As the digital cadaster coverage in country is low the local experts shared that even meter level accuracy can be acceptable for certain areas. However, we observed how time-consuming and tiring this procedure is. The expert should zoom in and out continuously in searching for visible boundaries. Then, the accuracy for delineation of each parcel will be different depending on the skills and precision of the operator. The automatic approach that was proposed speeded this process. The automatically detected and suggested boundaries just have to be checked by the operator and with several clicks to be adjusted and approved. It was observed that for long curved objects manual delineations is much slower and requires continuous clicking and the automatic one requires to click only on the starting and ending point. For a small rectangular object it is required to click only inside of the object and the boundary will be automatically delineated. Using the new proposed method, we reduced the number of clicks with 80%, saved 38% of the time and achieved 71% accuracy compared to manual delineated boundaries [48].

The operational analysis showed that the approach is most suited for the delineation of visible cadastral boundaries demarcated through physical objects. In our study side area, walls and fences were partly covered by vegetation and not built consistently. From 211 parcels, 21 could be delineated without further editing, 24 required minor editing on <20% of the outline length, and the remaining parcels were digitized through snapping to the automatically generated lines and generating new ones. In general, the approach obtains the highest time savings for areas in which boundaries are visible, long and curved, whereas boundaries in our study side are often covered, short, and straight.

The feedback analysis investigated the strengths, weaknesses, opportunities, and threats (SWOT) of the approach. Feedback is derived from three one-day workshops for 57 land administration stakeholders from local government institutions, NGOs, private companies, and national government institutions. The SWOT feedback from the three workshops is shown on Table 7.

Table 7. SWOT results on automated boundary extraction approach.

Strengths	Weaknesses
- Image-based delineation facilitates participatory mapping - Allows accurate delineation of georeferenced boundaries without fieldwork - Adaptable per area and its characteristics - Fast data processing and clear visualization - Intuitive usability - Less zooming and clicking - Open-source implementation - Less monotonous delineation work - Easy to implement for existing workforce with surveying background - Strong match with Kenyan land challenges (subdivision, digitization, transformation from general to fixed boundaries, correction of overlapping boundaries, mapping of unrecorded boundaries)	- Dependence on visible boundaries - Time-consuming for large areas - Dependence on knowledge, interpretation and skills of delineator - Superiority over manual delineation depends on image segmentation and its match with cadastral boundaries - Varying data quality due to lack of standardized image capture/processing - Interactive design limits reproducibility - High initial costs (set up of digital infrastructure, capacity development) - Image acquisition requires equipment, training and permissions - Open-source solution requires acceptance - Update of Registry Index Maps (RIM) not included
Opportunities	**Threats**
- Potential to increase superiority over manual delineation by adding functionalities (geometric checks for output lines, creation of polygons/attributes and change protocols) - Modular workflow can be updated in case of future innovations (e.g., on image segmentation or classification) - Applicable in further object delineation applications (e.g., land use mapping) - Implementable in existing systems due to modular design - Potential to reduce land-related disputes through clear visualization and identification of boundaries	- Possible inability to cope with rapid technology changes despite modular design - Superiority over manual delineation too small (reduced efficiency when object outlines do not match with cadastral boundaries, high percentage of invisible boundaries, or beacon demarcation) - Possible non-acceptance of open-source solution (no guarantee for long-term source-code maintenance) and threat from commercial solutions - Resistance to innovative approaches (fear of job loss due to automation) - Uncertain legal allowance to capture/use aerial imagery - Unstable Kenyan digital infrastructure

The methodology was tested and works also for remote sensing data with different resolutions (0.02–0.25 m) acquired from other platforms such as satellite and aerial cameras on board of an airplane. Advantages are strongest when delineating in rural areas due to the continuous visibility of monotonic boundaries. Manual delineation remains superior in cases where the boundary is not fully visible, i.e., covered by shadow or vegetation. Although our methodology has been developed for cadastral mapping, it can also be used to delineate objects in other application fields, such as land use mapping, agricultural monitoring, topographical mapping, road tracking, or building extraction.

4.5. Results from Analysis of Governance Aspects

In the focus group discussions and individual semi-structured interviews held around each remote sensing application, several governance aspects were raised. As most of these apply to two or all three remote sensing methodologies, and are discussed here jointly along the lines of six aspects derived from the discussions: (1) legal versus informal rights, (2) government versus non-governmental actors, (3) the national versus regional/local government, (4) digital versus paper way of working, (5) use of open source software, and (6) lack of clear legislation for specific new tools and applications esp. UAVs.

Many different definitions of the term "governance" exist, but most important is that it stands for a broader concept than government, and also includes the influence of other actors on processes that affect all. Within the context of the research, a definition was developed where governance is "the process of interactively steering the land tenure society to sustain the use of the its4land tools" [60].

In Kenya the 2010 Constitution brought a number of changes that affect the governance aspects of our remote sensing methodologies. As mentioned earlier, customary tenures are now explicitly recognized in the Constitution, although the attention to them in specific laws and regulations is still lagging, and in peri-urban (and informal) areas, other forms of non-statutory tenure rights exist that are not specifically mentioned. The formal systems for land administration, that tend to only serve statutory rights, are embedded in laws and regulations, but also in the way the different formal land sector actors operate in practice; which tends not to focus on innovation or broadening of the beneficiary group. There is, currently, a lack of participatory mechanisms that can support the collaboration between the different governmental levels and the non-governmental actors. Political interests or corrupt practices were mentioned during the workshops and interviews. These practices happen due to both the lack of transparency in the decision making process and lack of an enforcing institutional environment. Further, there is no specific legislative framework that supports innovative approaches as the ones offered via our developed applications.

Allowing non-governmental organizations (such as private companies, NGOs and professional network associations) to take the lead in implementing the more participatory and innovative technical applications is also difficult. There is not really a tradition to do so, which is partly due to lack of resources: financial, human and technological. Further, the fear of losing jobs due to introduction of new ways of work make the street level bureaucrats wary, whereas the higher level workers fear of loss of the control of the currently used methods which involve political interests and corruption practices. As most of the national government and counties lack basic infrastructure, one way the national government could support the implementation of technical applications is by providing financial or legal incentives to non-governmental actors, as in many cases there are consultancies who have the expertise and could support the adoption of technical applications within a short time frame. However, neither governmental actors nor private companies are used to this type of participatory approaches. Until now, according to the different actors who participated in the workshops, there have not been real participatory approaches that could support their implementation. The capacity of the local levels to implement technical applications like ABE and SmartSkeMa face the challenge of variety in capacity among the counties, and some cases were reported where governmental employees need to use their personal computers to carry out their daily job activities.

The 2010 Constitution brought the devolution of powers to the 47 counties. There is still lack of clarity relating to the division of responsibilities between the county and national government level, and the different governmental levels currently often lack resources to implement, maintain or upgrade the use of innovative technical applications, especially when those require the specific IT knowledge that comes with geospatial techniques. The current governance structure favors a top-down implementation process where the national government is the main actor. While some counties have the capacity to support the implementation of the technical applications, others clearly lack infrastructure, financial resources or knowledge.

In addition to the limited capacity, it also became clear that only some governmental actors see the transition from paper-based data to digital based data as a priority. The transition from paper-based data to digital data is already set in some counties, but is not always perceived as a priority by all governmental actors. Due to the lack of political will, the implementation of our technical applications cannot be expected to occur in the short-term. Political interests or corruption practices around the possible implementation of the technical applications were also mentioned by the different interviewed actors. This situation is due to the lack of legislation for digital data and the current prioritization of paper-based data.

5. Discussion

This paper was designed to assess user needs (in terms of land administration functions), how the three remote sensing methodologies under development meet these needs, and finally what governance aspects would be critical in widespread update. As a result of the workshops in Kenya, a SWOT analysis was created for each developed application. The results of those SWOT analyses as well as from the fieldwork are summarized and visualized in Table 8: the adherence to 10 aspects of the assessment criteria, derived from the user needs assessment, land management paradigm [5], and FFP requirements [4] is shown.

Table 8. Assessment of remote sensing methodologies with regard to fit-for-purpose land rights mapping in Kenya. Green indicates compliance with an aspect, yellow indicates that the application partially complies with an aspect.

	Sketch Maps (SmartSkeMa)	UAV-Based Data Collection	Automated Boundary Delineation
1. Land Tenure	Green	Green	Green
2. Land Value	Yellow	Green	Green
3. Land Use	Green	Green	Green
4. Visible boundaries rather than fixed boundaries	Green	Green	Green
5. Aerial imagery rather than field surveys	Yellow	Green	Green
6. Accuracy relates to the purpose	Green	Green	Green
7. Updating and ongoing improvement	Green	Green	Green
8. Cheap	Green	Yellow	Yellow
9. Fast	Green	Yellow	Yellow
10. Accurate	Yellow	Green	Green

With regards to SmartSkeMa, is seems clear that this is not a methodology aiming to replace data collection via aerial images or other surveying techniques, but sketch maps can be used to complement and support collecting data about the relationship of people with respect to land. When SmartSkeMa is considered as methodology for documenting community land tenure in Kenya, its ease of use makes it a cheap option as, once set up, it allows communities to document their land with little additional cost. Its level of accuracy can also be tailored to the task at hand since communities can sketch on top of an aerial image allowing higher precision than is obtained using a plain sketch map. Finally, because a community can use SmartSkeMa with relative independence it may produce data faster than would be possible using traditional land survey methods where the skilled personnel in Kenya are scarce. From the results obtained, SmartSkeMa's functionalities contribute to meeting most of the 10 aspects. The wide range of spatial precision covered by SmartSkeMa presents a great opportunity for incremental and progressive land data acquisition. However, data produced by SmartSkeMa is not very well suited for land valuation in the sense of calculating objective monetary equivalents. The data however may include information about relative values as perceived by land users within a cultural context. More work is needed to determine the extent to which these land values can be captured in the data and how they can be interpreted.

For UAVs, during the workshop most interest was conveyed in the provision of an up-to-date map. Various local government entities such as the department of urban planning and spatial development identified the potential of UAV data to derive information on the current land use and for monitoring urban developments. Furthermore, the immediateness of the data provision was seen to be very beneficial to investigate and solve land disputes within group ranches. However, since the registry index maps are paper-based, the entry barrier to adopt UAV technology is very high. Good visibility of rooftops and information on the height of buildings was found to support land valuation processes. The exploratory case study in Kenya showed that most of the 10 aspects can be met. As an indirect surveying technique, the concept of using UAV technology in cadastral mapping is based on the existence of visible boundaries which can either be extracted by automated image analysis or manual delineation. However, it was also found that a precise and accurate generated orthophoto allows extraction of boundaries that are not necessarily visible, such as combining features that demarcate the corner point of the parcel even though the line in between is not visible. The ease of use and the flexible setup in terms of the technical standards of the sensor and platform allows covering a large range of different purposes. In terms of scalability, UAV technology only serves a limited range of different scales as costly and lengthy flight authorization procedures hinder an efficient application. Furthermore, in many countries current regulations require to fly missions which are in visual line of sight, allowing only some hundred meters of a possible flight trajectory. According to Kenyan stakeholders, limited battery capacity was found to be the second bottleneck currently impeding large scale implementation.

For ABE, the results from the workshop proved that our proposed automatic boundary extraction approach facilitating the delineation of visible objects and cadastral boundaries can be used to collect information on land tenure, land value, and land use. It further aligns well with the FFP spatial and scalability requirements: it allows a cheap, fast, and accurate delineation of visible boundaries from aerial imagery. However, costs, speed, and accuracy can vary depending on the capture and processing of the aerial imagery and the implementation of the automated boundary extraction: the approach is currently open source, which seems low-cost, but might require more time in acceptance as the SWOT analysis revealed. Given the complexity of cadastral boundaries, automating their delineation remains challenging: the variability of objects and extraction methods reflect the problem's complexity, consisting of extracting different objects with varying characteristics. These circumstances impede the compilation of a generic model for a cadastral boundary and thus the development of a generic method. These remarks come back to the limitations of general boundaries: no standardized specifications exist for boundary features, boundaries are often not marked continuously and maintained poorly [59]. To further develop automated boundary extraction in indirect surveying, we suggest considering the extractable boundary rather than the visible boundary alone (Figure 14): instead of focusing on the visible boundary comprising of outlines of physical objects, automated boundary extraction should focus on the extractable boundary that incorporates local knowledge and context. This information is not inherent in the concept of the visible boundary, but it is extractable from remote sensing imagery.

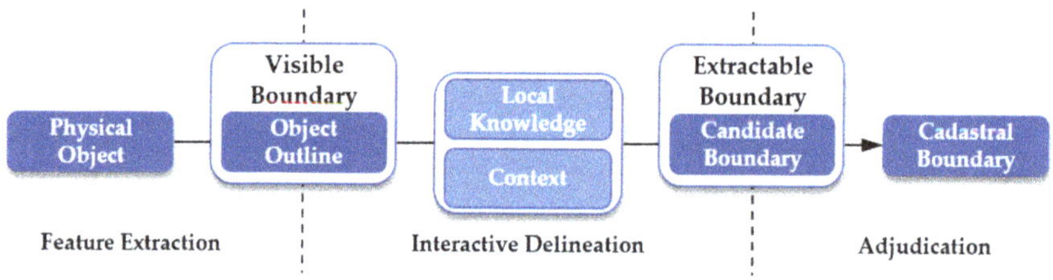

Figure 14. From physical object to cadastral boundary: reformulated boundary concepts for indirect surveying.

Overall, our approach that couples a machine-based automatic feature extraction with a delineator-based interactive delineation can be used to map extractable boundaries. The delineation cannot be fully automated at the current state since the extracted outlines require (legal) adjudication and incorporation of local knowledge from human operators to create final cadastral boundaries. Image-based approaches bear potential to automatically extract use rights, which do not necessarily represent legal rights. These circumstances limit the scope of automated approaches. We observed that automating boundary extraction dealing with sensitive land rights can only be successful, when the interactive part that bridges the gap between automatically generated results and the final cadastral boundary is designed and implemented in correspondence to user needs. Our work revealed limitations of the current approach and ideas for improvements to be addressed in future work, in order to advance the current approach regarding efficiency and acceptance. This would promote the paradigm shift towards cadastral intelligence that integrates human-based expert knowledge with automatically generated machine-based knowledge. Additionally, future studies should provide approaches to capture requirements from existing technical, legal, financial, and institutional frameworks to be considered when aiming to implement innovative cadastral mapping procedures successfully.

Finally, for governance aspects, further notes on legislative and financial aspects are worth expanding upon. Implementing any of the remote sensing methodologies at any scale, without an appropriate legislative framework, appears fraught. This partly relates to modernizing existing laws and regulations to open up to innovative approaches, and partly to new rules for new challenges. Especially when this new legislation would give clarity on the responsibilities of the different actors, prioritize cheap and open source technologies and stimulate and facilitate partnerships between the governmental and non-governmental actors that would make the uptake and upscaling of the remote sensing methodologies much more likely. Without this an occasional "pilot" might continue and show what can and cannot be achieved within a certain setting, but for true upscaling, a supportive environment will be needed; appropriate laws and regulations and a collaborative attitude among national and local government, as well as with non-government actors. As Kenya has a long land administration history, there is the human capacity in the field, however it seems this country is lacking the political will to introduce that supportive environment to a large extent. Focusing specifically on UAV legislation, getting to balanced legislation that allows a responsible use of UAVs without truly compromising the other issues is not easy. This can be seen worldwide, but even more in countries like Kenya, which struggle to get political will to make clear instructions for UAVs. The implementation of the UAVs could be improved by increasing collaboration between the national and local governments with the non-governmental actors. This collaboration could help to solve the lack of important financial resources. Resources are needed to hire new staff, training, certification, among others. In this sense, the national level can also play an important role as a facilitator to allow private companies to participate. Some non-governmental actors such as private companies could have the resources to use the UAVs; however, they require certain governmental support such as the issuing of the permits or incentives to invest. Meanwhile, on the financial aspect, both proprietary and open source options present challenges: actors payments for software, licenses, and the required updates prohibitive; however, even with open source software, the lack of IT infrastructure and internet access still impacts negatively on scaled uptake. In Kenya, the current resources are not enough to establish a sustainable implementation.

6. Conclusions

The paper has described challenges around land tenure mapping in Kenya and presented potential remote sensing methodologies that respond to current end user needs and that are furthermore investigated from a governance perspective. Although the 2010 Constitution resulted in a land policy reform setting out a framework to better respond to the needs of large customary groups in Kenya, actual implementation is slow and both county governments and communities themselves continue to grapple with a multitude of issues relating to rapid urbanization, unmanaged development, and unregulated land activities. Communities are not engaged with land policies, and spatial planning and needs are not being met. For counties like Kajiado, these challenges are further exacerbated by issues of scale, high levels of corruption and poor-quality of existing land data. In future, since the needs are changing with time the new technologies in support of land administration should definitely be adapted accordingly.

With regards to the demonstrated remote sensing methodologies, SmartSkeMa was revealed as a versatile land data acquisition tool that requires little expertise to be used and is based on community participation; UAVs were identified as having a high potential for creating up-to-date base maps to support the current land administration system; the automatic boundary extraction approaches designed for areas demarcated by physical objects and are thus visible were found to be useful for collecting information on land tenure, land value, as well as land use (aligned with the 10 aspects).

Finally, with regards to ensuring responsible governance related to the scaled implementation of the remote sensing methodologies, as there is no appropriate legal framework for applying them, the viability and medium timeframe for increased usage in the sector remains unclear. A more robust legal framework could strengthen authority; operationalize in administrative orders, rules and planning; and serve as the basic control system (for possible sanctions). After establishing the framework, if not during, serious attention needs to be given to the cooperation between all relevant actors, where interorganizational relations are ruled by the acknowledgement of mutual interdependencies, trust and the responsibilities of each actor.

Author Contributions: Conceptualization: M.K., C.S., S.C., S.H., M.C., J.S., R.B., J.Z., C.L., and J.C.; Funding acquisition, R.B.; Visualization: M.K., C.S., S.C., S.H., M.C., and J.S.; Writing—original draft: M.K., C.S., S.C., S.H., M.C., J.S., and I.B., Review & editing: M.K., C.S., S.C., S.H., M.C., J.S., R.B., J.Z., G.V., C.L., J.C., I.B., G.W., and V.P., Fieldwork support: G.W., R.W., P.O.O., G.T.O., and B.C. All authors have read and agreed to the published version of the manuscript.

Funding: The research described in this paper was funded by the research project "its4land", which is part of the Horizon 2020 program of the European Union, project number 687828.

Acknowledgments: Its4land team would like to express also their acknowledgements to Gordon Wayumba, Robert Wayumba, Peter Ochieng Odwe, George Ted Osewe, Beatrice Chika, and their colleagues for their support during the fieldworks and workshops.

Conflicts of Interest: The authors declare no conflicts of interest.

References

1. Jāhāna, S. *Human Development Report 2015: Work for Human Development*; United Nations Development Programme: Nova Iorque, NY, USA, 2015.
2. Williamson, I. The justification of cadastral systems in developing countries. *Geomatica* **1997**, *51*, 21–36.
3. Dale, P.; McLaughlin, J. *Land Administration*; Oxford University Press: Oxônia, Inglaterra, UK, 1999.
4. Enemark, S.; Bell, K.C.; Lemmen, C.; McLaren, R. *Fit-For-Purpose Land Administration*; International Federation of Surveyors (FIG): Copenhagen, Denmark, 2014.
5. Williamson, I.; Enemark, S.; Wallace, J.; Rajabifard, A. *Land Administration for Sustainable Development*; ESRI Press Academic: Redlands, CA, USA, 2010.
6. Lemmen, C. *Social Tenure Domain Model—A Pro-Poor Land Tool*; International Federation of Surveyors (FIG): Copenhagen, Denmark, 2010.
7. Zevenbergen, J.; Augustinus, C.; Antonio, D.; Bennett, R. Pro-poor land administration: Principles for recording the land rights of the underrepresented. *Land Use Policy* **2013**, *31*, 595–604. [CrossRef]
8. Enemark, S.; McLaren, R.; Lemmen, C.; Antonio, D.; Gitau, J. *Guiding Principles for Building Fit-for-Purpose Land Administration Systems in Developing Countries*; UN-Habitat: Nairobi, Kenya, 2016.
9. Rahmatizadeh, S.; Rajabifard, A.; Kalantari, M.; Ho, S. A framework for selecting a fit-for-purpose data collection method in land administration. *Land Use Policy* **2018**, *70*, 162–171. [CrossRef]
10. De Vries, W.; Bennett, R.; Zevenbergen, J. Toward Responsible Land Administration. In *Advances in Responsible Land Administration*; CRC Press: Boca Raton, FL, USA, 2015; pp. 3–14.
11. Koeva, M.N.; Crommelinck, S.; Stöcker, C.; Crompvoets, J. Its4land—Challenges and Opportunities in Developing Innovative Geospatial Tools for Fit-For-Purpose Land Rights Mapping. In *FIG Congress 2018*; International Federation of Surveyors (FIG): Copenhagen, Denmark, 2018; pp. 1–17.
12. Koeva, M.N.; Bennett, R.M.; Gerke, M.; Crommelinck, S.C.; Stöcker, E.C.; Crompvoets, J.; Ho, S.; Schwering, A.; Chipofya, M.; Schultz, C.; et al. Towards innovative geospatial tools for fit-for-purpose land rights mapping. In Proceedings of the ISPRS Geospatial Week 2017, Wuhan, China, 18–22 September 2017.
13. Chipofya, M.; Jan, S.; Schultz, C.; Schwering, A. Towards Smart Sketch Maps for Community-driven Land Tenure Recording Activities. In Proceedings of the Societal Geo-Innovation: Short Papers, Posters and Poster Abstracts of the 20th AGILE Conference on Geographic Information Science, Wageningen, The Netherlands, 9–12 May 2017. Available online: https://agile-online.org/index.php/conference/proceedings/proceedings-2017 (accessed on 8 June 2019).
14. Colomina, I.; Molina, P. Unmanned aerial systems for photogrammetry and remote sensing: A review. *ISPRS J. Photogramm. Remote Sens.* **2014**, *92*, 79–97. [CrossRef]
15. Nex, F.; Remondino, F. UAV for 3D mapping applications: A review. *Appl. Geomat.* **2014**, *6*, 1–15. [CrossRef]

16. Mumbone, M.; Bennett, R.; Gerke, M.; Volkmann, W. *Innovations in Boundary Mapping: Namibia, Customary Land and UAV's*; University of Twente Faculty of Geo-Information and Earth Observation (ITC): Enschede, The Netherlands, 2015.
17. Ramadhani, S.A.; Bennett, R.M.; Nex, F.C. Exploring UAV in Indonesian cadastral boundary data acquisition. *Earth Sci. Inform.* **2018**, *11*, 129–146. [CrossRef]
18. Jazayeri, I.; Rajabifard, A.; Kalantari, M. A geometric and semantic evaluation of 3D data sourcing methods for land and property information. *Land Use Policy* **2014**, *36*, 219–230. [CrossRef]
19. Koeva, M.; Muneza, M.; Gevaert, C.; Gerke, M.; Nex, F. Using UAVs for map creation and updating. A case study in Rwanda. *Surv. Rev.* **2016**, *50*, 1–14. [CrossRef]
20. Manyoky, M.; Theiler, P.; Steudler, D.; Eisenbeiss, H. Unmanned Aerial Vehicle in Cadastral Applications. In Proceedings of the ISPRS Zurich 2011 Workshop, Zurich, Switzerland, 14–16 September 2011.
21. Maurice, M.J.; Koeva, M.N.; Gerke, M.; Nex, F.; Gevaert, C. A photogrammetric approach for map updating using UAV in Rwanda. In Proceedings of the Conference: GeoTechRwanda 2015, At Kigali, Rwanda, 18–20 November 2015.
22. IAAO. *Standard on Digital Cadastral Maps and Parcel Identifiers*; Int. Association of Assessing Officers (IAAO): Kansas City, MO, USA, 2015.
23. Crommelinck, S.C.; Bennett, R.M.; Gerke, M.; Koeva, M.N.; Yang, M.Y.; Vosselman, G. SLIC superpixels for object delineation UAV data. In *Peer Reviewed Annals, Proceedings of the International Conference on Unmanned Aerial Vehicles in Geomatics, Bonn, Germany, 4–7 September 2017*; Stachniss, C., Förstner, W., Schneider, J., Eds.; The International Archives of the Photogrammetry, Remote Sensing and Spatial Information Sciences, International Society for Photogrammetry and Remote Sensing (ISPRS): Bonn, Germany, 2017; Volume IV-2/W3, pp. 9–16. [CrossRef]
24. Crommelinck, S.; Höfle, B.; Koeva, M.N.; Yang, M.Y.; Vosselman, G. Interactive cadastral boundary delineation from UAV images. *ISPRS Ann. Photogramm. Remote Sens. Spat. Inf. Sci.* **2018**, *4*, 81–88. [CrossRef]
25. Crommelinck, S.; Bennett, R.; Gerke, M.; Yang, M.; Vosselman, G. Contour detection for UAV-based cadastral mapping. *Remote Sens.* **2017**, *9*, 171. [CrossRef]
26. Xia, X.; Koeva, M.; Persello, C. Extracting Cadastral Boundaries from UAV Images Using Fully Convolutional Networks. In Proceedings of the IGARSS 2019–2019 IEEE International Geoscience and Remote Sensing Symposium, Yokohama, Japan, 28 July–2 August 2019; pp. 2455–2458. [CrossRef]
27. Xia, X.; Persello, C.; Koeva, M. Deep Fully Convolutional Networks for Cadastral Boundary Detection from UAV Images. *Remote Sens.* **2019**, *11*, 1725. [CrossRef]
28. Wunzala, J. As Pastoralist Land Shrinks, Maasai Women Take Livestock Lead—Reuters. Available online: https://www.reuters.com/article/us-kenya-herders-women-idUSKCN1060Q4 (accessed on 8 June 2019).
29. Deininger, K.; Hilhorst, T.; Songwe, V. Identifying and addressing land governance constraints to support intensification and land market operation: Evidence from 10 African countries. *Food Policy* **2014**, *48*, 76–87. [CrossRef]
30. Mbiba, B.; Huchzermeyer, M. Contentious development: Peri-urban studies in sub-Saharan Africa. *Prog. Dev. Stud.* **2002**, *2*, 113–131. [CrossRef]
31. Obeng-Odoom, F. The State of African Cities 2010: Governance, inequality and urban land markets. *Cities* **2013**, *31*, 425–429. [CrossRef]
32. Davies, S.J. The Political Economy of Land Tenure in Ethiopia. Ph.D. Thesis, University of St Andrews, Setor Andrews, Escócia, UK, 2008.
33. Moyo, S.; Yero, P. (Eds.) *Reclaiming the Land: The Resurgence of Rural Movements in Africa, Asia, and Latin America*; Zed Books: London, UK, 2005. [CrossRef]
34. KNBS. Kenya National Bureau of Statistics, Nairobi, Kenya. Available online: https://www.knbs.or.ke/ (accessed on 8 June 2019).
35. Mwathane, I. Counties Need Up-To-Date Land Registry Maps to Resolve Disputes—Daily Nation. Available online: https://www.nation.co.ke/oped/Opinion/Counties-need-up-to-date-land-registry-maps/440808-3379856-format-xhtml-rku6y9z/index.html (accessed on 8 June 2019).
36. Siriba, D.N.; Mwenda, J.N. Towards Kenya's Profile of the Land Administration Domain Model (LADM). In Proceedings of the 5th Land Administration Domain Model Workshop, Kuala Lumpur, Malaysia, 24–25 September 2013.

37. Wayumba, G.; Wayumba, R.; Odwe, P. Early Implementation Of Fit-For-Purpose Land Administration In Kenya: The Case Of Mass Adjudication And Registration Of Customary Rights In Kenya. *Int. J. Sci. Res. Eng. Stud.* **2018**, *5*, 1–11.
38. Siriba, D.; Voß, W.; Mulaku, G. The Kenyan Cadastre and Modern Land Administration. *Geod. Info* **2011**, *3*, 177–186.
39. COUNTY INTEGRATED DEVELOPMENT PLAN 2013–2017. Available online: https://www.google.com/search?q=GOK+%282013%29%2C+Government+of+Kenya.+County+Government+of+Kajiado%2C+County+Integrated+Development+Plan+2013-2017&ie=utf-8&oe=utf-8&client=firefox-b (accessed on 8 June 2019).
40. Delbecq, A.L.; Van de Ven, A.H. A group process model for problem identification and program planning. *J. Appl. Behav. Sci.* **1971**, *7*, 466–492. [CrossRef]
41. Gallagher, M.; Hares, T.; Spencer, J.; Bradshaw, C.; Webb, I. The nominal group technique: A research tool for general practice? *Fam. Pract.* **1993**, *10*, 76–81. [CrossRef]
42. Langford, B.E.; Schoenfeld, G.; Izzo, G. Nominal grouping sessions vs focus groups. *Qual. Mark. Res. Int. J.* **2002**, *5*, 58–70. [CrossRef]
43. Lloyd, S. Applying the nominal group technique to specify the domain of a construct. *Qual. Mark. Res. Int. J.* **2011**, *14*, 105–121. [CrossRef]
44. Ho, S.; Pattyn, V.; Broucker, B.; Crompvoets, J. Needs Assessment in Land Administration: The Potential of the Nominal Group Technique. *Land* **2018**, *7*, 87. [CrossRef]
45. Karamesouti, M.; Panagos, P.; Kosmas, C. Model-based spatio-temporal analysis of land desertification risk in Greece. *Catena* **2018**, *167*, 266–275. [CrossRef]
46. Schwering, A.; Wang, J.; Chipofya, M.; Jan, S.; Li, R.; Broelemann, K. SketchMapia: Qualitative Representations for the Alignment of Sketch and Metric Maps. *Spat. Cogn. Comput.* **2014**, *14*, 220–254. [CrossRef]
47. Stöcker, C.; Koeva, M.; Bennett, R. Evaluation of uav-based technology to capture land rights in kenya: Displaying stakeholder perspectives through interactive gaming. In Proceedings of the 20th Annual World Bank Conference on Land and Poverty, Washington, DC, USA, 25–29 March 2019.
48. Crommelinck, S.; Koeva, M.; Yang, M.; Vosselman, G. Application of Deep Learning for Delineation of Visible Cadastral Boundaries from Remote Sensing Imagery. *Remote Sens.* **2019**, *11*, 2505. [CrossRef]
49. Crommelinck, S. Delineation-tool GitHub. Available online: https://github.com/its4land/delineation-tool (accessed on 13 May 2019).
50. Pont-Tuset, J.; Arbelaez, P.; Barron, J.; Marques, F.; Malik, J. Multiscale combinatorial grouping for image segmentation and object proposal generation. *IEEE Trans. Pattern Anal. Mach. Intell.* **2016**, *39*, 128–140. [CrossRef] [PubMed]
51. Breiman, L. Random Forests. *Mach. Learn.* **2001**, *45*, 5–32. [CrossRef]
52. Pedregosa, F.; Varoquaux, G.; Gramfort, A.; Michel, V.; Thirion, B.; Grisel, O.; Blondel, M.; Prettenhofer, P.; Weiss, R.; Dubourg, V.; et al. Scikit-learn: Machine Learning in Python. *J. Mach. Learn. Res.* **2011**, *12*, 2825–2830.
53. QGIS Development Team QGIS Geographic Information System, Open Source Geospatial Foundation. Available online: https://www.qgis.org/en/site/ (accessed on 16 December 2019).
54. Bressers, H.; Bressers, N.; Kuks, S.; Larrue, C. *The Governance Assessment Tool and Its Use*; Springer: Berlin, Germany, 2016. [CrossRef]
55. Government of Kenya. The Community Land Act, No27. Available online: http://kenyalaw.org/kl/fileadmin/pdfdownloads/Acts/CommunityLandAct_27of2016.pdf (accessed on 6 August 2019).
56. Stöcker, C.; Bennett, R.; Nex, F.; Gerke, M.; Zevenbergen, J. Review of the Current State of UAV Regulations. *Remote Sens.* **2017**, *9*, 459. [CrossRef]
57. Civil Aviation (Unmanned Aircraft Systems) Regulations Legal Notice No. 259. Available online: https://www.kcaa.or.ke/index.php?option=com_content&view=article&id=207:drones-regulations&catid=92:newsandevents&Itemid=742 (accessed on 8 June 2019).
58. Wayumba, G. The Structure of Cadastral System in Kenya. *J. Land Admin. Eastern Africa* **2013**, *1*, 6–20.

59. Mwenda, J. Spatial information in land tenure reform with special reference to Kenya. In Proceedings of the International Conference on Spatial Information for Sustainable Development, Nairobi, Kenya, 2–5 October 2001.
60. Buntinx, I.; Crompvoets, J.; Ho, S.; Timm, C.; Wayumba, G. Deliverable 7.1: Governance and Capacity Development Definition. Available online: https://its4land.com/wp-content/uploads/2016/06/D7-1-Governance-and-capacity-development-definition.pdf (accessed on 9 July 2019).

remote sensing

MDPI

Article

Bridging the Semantic Gap between Land Tenure and EO Data: Conceptual and Methodological Underpinnings for a Geospatially Informed Analysis

Cheonjae Lee * and Walter Timo de Vries

Chair of Land Management, Department of Aerospace and Geodesy, Technical University of Munich (TUM), Arcisstrasse 21, 80333 Munich, Germany; wt.de-vries@tum.de

* Correspondence: Cheonjae.lee@tum.de; Tel.: +49-(0)89-289-25790

Received: 4 November 2019; Accepted: 6 January 2020; Published: 10 January 2020

Abstract: When spatial land tenure relations are not available, the only effective alternative data method is to rely on the agricultural census at the regional or national scale, based on household surveys and a participatory mapping at the local scale. However, what if even these are not available, which is typical for conflict-affected countries, administrations suffering from a lack of data and resources, or agencies that produce a sub-standard quality. Would it, under such circumstances, be possible to rely on remotely sensed Earth Observation (EO) data? We hypothesize that it is possible to qualify and quantify certain types of unknown land tenure relations based on EO data. Therefore, this study aims to standardize the identification and categorization of certain objects, environments, and semantics visible in EO data that can (re-)interpret land tenure relations. The context of this study is the opportunity to mine data on North Korean land tenure, which would be needed in case of a Korean (re-)unification. Synthesizing land tenure data in conjunction with EO data would align land administration practices in the respective parts and could also derive reliable land tenure and governance variables. There are still many unanswered questions about workable EO data proxies, which can derive information about land tenure relations. However, this first exploration provides a relevant contribution to bridging the semantic gap between land tenure and EO data.

Keywords: land administration; geospatially informed analysis; land tenure; land tenure relations; remote sensing; earth observation (EO) data

1. Introduction

Land tenure data contain geospatial, anthropological and socioeconomic attributes since it builds on both the physical delineation of land and the identification of social relations governing land use, land access and land ownership [1,2]. Collecting land tenure data is, however, neither administratively straightforward nor always technically feasible or financially affordable. There are even many challenges which make collecting land tenure data complex, such as data availability and data accessibility [3,4]. However, new data collection technologies, including, amongst others, voluntary geographic information in connection to social media technologies, Unmanned Aerial Vehicles (UAVs) and big data mining may overcome some of these barriers. Yet, there is a dearth of the methodological reflections in how such geospatial technologies can identify and formalize land tenure relations. What these technologies are currently able to do includes: (1) underpinning land tenure-enabling environments; (2) mining land tenure data; (3) transforming land tenure relations [5]. However, the quality of all these heavily depends on the completeness and full access of the terrain and the data sources. In many cases these basic criteria cannot be guaranteed, leaving the land tenure information scarce [1].

A promising and yet unexplored technology to derive socio-legal land tenure information is Earth Observation technology. The utilization of Earth Observation (EO) data has increased significantly in many disciplines. Literature shows applications ranging from environmental and regional studies to economics, and peace and conflict research, for example [6–9]. More specific to the interest of this paper, there is growing body of literature on methods to extract and map cadastral boundaries using EO data [10–19]. However, this literature rarely effectively bridges the knowledge gap between social land tenure and spatial descriptions of boundaries. In other words, the (even automated/machine-learning based) spatial descriptions do not identify the underlying social or legal relations to land, such as effective land ownership, private or communal land use or land access rights or presumed land claims.

The methods in detecting, extracting and identifying land tenure relations always require both geometric or topographic characteristics and ground-truth information of land tenure. However, spatially explicit land tenure relations through EO data remains one of the foremost challenges. As a societal institution, land tenure has a great influence on how people decide on land use. Such decisions are observable in land cover changes and spatio-temporal patterns of land use (inducted from similarities, differences, repetitions or sudden changes in space and time). The dynamics of landscape changes are intrinsically linked to land tenure relations and decisions [20–22]. Detecting and extracting physical features is possible by connecting spectral reflectance values, shapes, and texture features of ground components to be pre-defined. By sampling and generalizing these connections, one can construct algorithms, which detect and predict spatio-temporal patterns with EO data, such as the (rate of) land fragmentation, land ceiling and urban encroachment [16,23,24]. Such spatio-temporal processes could be connected to land tenure information if these are aligned with automated identification and reconstruction of cadastral boundaries. For example, the morphology of a cadastral boundary is associated with the spatial nature of land tenure on the aspects of, physical realm of land interests, temporal practices of land use rights and the legal nature of boundaries [25].

Then, how do we derive the features or characteristics of land tenure if we only have access to the physical objects or spectral changes in objects in time and in space? According to [26], land tenure aspects may cross multiple spatially observable boundaries in a given landscape. Additionally, tenure and land right boundaries are also not always visible through specific elements in the landscape or through specific spectral reflectance values. One still needs to combine the location of specific landscape elements to alternative data source, such as agricultural census data at the regional or national scale, and/or household surveys and a participatory mapping at the local scale [1,27]. Nevertheless, what if these locally collated datasets are not available? Is it in such cases still possible to rely on EO data only, combined with a set of basic assumptions about the spatial nature of land tenure? We hypothesize that this is possible; however, this requires a set of fundamental proxies connected to specific documented knowledge on land tenure. This article will describe how this is possible and under which conditions this is possible.

The first challenge to overcome this problem is to address the degree of semantic information connected to spatial information. When it comes to extracting socio-spatial aspects of land tenure using EO data, the formalized and proven semantic rules do not yet exist. Or more precisely, the rules and assumptions, which induce a land tenure relation type, do not yet exist. EO data only distinguishes "low-level semantic features" of land cover information such as physical features, spatial objects and configuration of ground components. In contrast, land tenure information requires "high-level semantic features" connected to knowledge-based information, and reflecting institutionalized human-land relationships and based upon the varying human socio-economic activities on land such as land use and ownership trajectories. In other words, the low-level semantic features provided by direct EO data acquisition methods are insufficient for the derivation of land tenure relations. One needs some sort of socialization of the pixels, i.e., a high-level semantic data collection and interpretation procedure which represents knowledge epitomized by indirect access to EO data. In practice, there is a discrepancy between the levels of detected low and high-level semantic features and it is labelled as the "semantic gap" [24,28,29]. Therefore, it is important see how the process of socialization of pixels can take place and how EO data can be (re-)interpreted into semantic land tenure relations with a rational and rigorous methodology. Only then, it is possible to identify, bridge and close the semantic gap.

Hence, this paper makes a review of the challenges posed by the identification of land tenure relations from Earth Observation data. In order to overcome some of these challenges, we propose to use a mix of methods and information fusion to identify proxies that may help derive unknown land tenure relations. This illustrates our approach by constructing proxies for land tenure relations over North Korea. The research questions are:

- Which kind of land tenure-related data can one derive and acquire when information access is limited?
- Which proxies can help to derive currently unknown land tenure relations in conjunction with EO data?

We first present the conceptual foundations of EO data applications for identifying land tenure relations. The next section addresses substantive and methodological considerations. Then, we explore a set of proxies in relation to five land tenure related questions. Finally, the conclusion gives brief summary and provides recommendation on how to proceed with this research.

2. Fundamentals of EO data Applications for Identifying Land Tenure Relations

2.1. The Conceptual Models of Semantic Land Tenure Relations

The lists in Table 1 are a number of key models and concepts capturing land tenure relations. Henssen [30] depicts land tenure as institutionalized people-to-land relationships with his "Subject-Right-Object model" [31]. This basic model of land administration has been further modified by for example highlighting the dynamics of land tenure [32]. The "Land Administration Domain Model (LADM)" is to a large degree an extended and more sophisticated model of the basic model, and has become both a conceptual and descriptive standard (ISO 19152). The LADM covers all land tenure-related data components including parties, legal/administrative units, spatial objects, and data on surveying and spatial representation. The LADM can bridge the gap between land policies and information management opportunities and is adaptable to local situations [33,34].

Table 1. The conceptual models of semantic land tenure relations.

Semantic Land Tenure Relations	Land Tenure Data Specification	EO Data Application
Subject-Right-Object Model [30]	The model only distinct three categories: "subject-rights-objects". Subjects are persons, groups, firms or States. Rights are ownership, use, control, access and transfer rights. Objects are physical features. The model puts in principle the accent on the relation "subject-right (who and how)", and on the relation on "right-object (where and how much)".	Scalability: currently EO data only looks at physical objects. This includes identifying cadastral (parcel and building) boundary-mapping approaches and land use attributes. However, other attributes can be derived using technical advances of Earth Observation (EO).
Land Administration Domain Model (LADM) [33]	The Land Administration Domain Model (LADM) facilitates the management of different tenures in "one environment"; it covers all land tenure-related data components including parties (person or organization), legal/administrative units (right, responsibility and restrictions), spatial objects (parcel, buildings and utility networks), and data on surveying and spatial representation (geometric/topological data).	Inter-operability: to capture semantics of the land administration and data-related components, a range of data acquisition methods is emphasized (e.g., satellite images, Unmanned Aerial Vehicles (UAVs) and automatic feature extraction).
Continuum of Land Rights [35]	It refers to recognizing, recording, administering a variety of appropriate and legitimate land tenure data. It, thus, focuses on the "tenurial pluralism" (diversity of tenure arrangements) and duality in subjects.	Flexibility: underlining importance of data robustness and accuracies using more sophisticated technologies to systemically accumulate land tenure data

Table 1. *Cont.*

Semantic Land Tenure Relations	Land Tenure Data Specification	EO Data Application
Fit-For-Purpose Land Administration [36]	Capturing spatial land tenure data should be "flexible and participatory" that covers all tenure data in scope. Moreover, acquired land tenure data is used affordable technologies and needs to provide adequate reliability within a limited time and resources. All land tenure data should be kept up-to-date.	Accuracy: application of general boundary mapping (rural); the use of high resolution satellite imagery (urban); high accuracy of information; on-going updating, sporadic upgrading and incremental improvement
Responsible Land Administration [37]	It addresses changes in people to land relations based on "socio-technical and institutional advances". New geoICT-driven and thought-restructuring land data capture, visualization, and sharing techniques with a clear understanding of a legal, organizational, and governance context can acquire specific characteristics of land tenure.	Legitimacy: emerging geospatial technologies including high-resolution satellite imagery for data collection and management offers new insights on legitimizing land rights and documentation as well as acknowledging different forms of land tenure.

Furthermore, the "Continuum of Land Rights" approach emphasizes that land tenure arrangements vary along a continuum of land rights. Not only documented formal land rights are legitimate, but also undocumented informal land rights may exist, and society may accept or condone these alongside formal rights. The continuum of land rights approach is useful in describing de facto land tenure, which is much more fluid and flexible than the static and unchanging (spatio-temporal) description of land rights. It allows more flexibility to define and recognize land tenure based on evidence from the field [35].

The "Fit-For-Purpose land administration" approach mainly focuses on building geospatial data framework of large-scale mapping that can address emerging land tenure issues where no reliable land information exists. This framework highlights following constituent principles in order to not only improve recognition of "value-of information" and maximize "cost-effectiveness", but also decrease "capacity-demanding": (1) general boundaries rather than fixed boundaries; (2) aerial imageries rather than field surveys; (3) accuracy relates to the purpose rather than technical standards and; (4) opportunities for upgrading and improvement [36].

"Responsible land administration" expands the conventional notions of land administration with a normative framework. What is distinctive about this concept is that it takes the following aspects into account: the requirement for any land administration system to ensure the representation of multi-stakeholders in order to foster institutional innovation and inclusion; the incorporation of a broad array of scholarly disciplines into the methodological repertoire, in particular, connecting from technical and information sciences to the social sciences and humanities; the need for a proactive stance in laying the foundation of cutting-edge land administration systems design; connecting the global context to the local and vice versa; the continued need to transfer knowledge into the practice, and vice versa [37]. Technical and operational designs of land information systems can only be innovative if particular societal needs embedded in the design process and in the manner in which land administration is based on shared responsibilities.

2.2. *Advancement of EO and AI Applications in Identifying Land Tenure Relations*

One of the significant discussion in EO applications for land tenure relations is to provide the institutional and spatial aspects of cadastral boundaries by identifying relationships between physical objects and visual boundaries based on the notion of cadastral morphology [15,16] and cadastral intelligence [17]. The early experiment demonstrates that over 80% of cadastral boundaries coincide with visible physical objects [15]. In line with a previous endeavor, more tailored object-based workflows using extraction algorithms delineate about 50% of parcel boundaries successfully [16]. Investigating technically transferable workflows is a continuing concern within UAV-based cadastral mapping. For instance, both the *gPb* contour detection method and the ENVI feature extraction

(FX) module has proven accurate results of visible object delineation that coincide with cadastral boundaries at completeness and correctness of up to 80% [11,13]. To extract visible cadastral boundaries within Object-Based Image Analysis (OBIA) environment from High Resolution Satellite Imagery (HRSI), the (semi-)automatic feature extraction methods have been employed and tested in rural areas: mean-shift segmentation with the buffer overlay method [18], and both multi-resolution segmentation (MRS) and estimation of scale parameter (ESP) (only able to automatically extract 47.4%) [17].

In light of state-of-the-art methods in land administration, a deep-learning is becoming highly prominent for the detection of cadastral boundaries [12,19]. Recent evidence suggests that deep fully convolutional networks (FCNs) ensures the high accuracy rather than *gPb* and MRS, with results of 0.79 in precision, 0.37 in recall and 0.50 in F-score [19]. For optimizing image segmentation, one study by [12] not only introduced the interactive boundary delineation workflow, but also examined the better suitability of the deep learning in cadastral mapping with convolutional neural networks (CNNs) by comparing random forest (RF) in machine learning: RF-derived boundary likelihoods (accuracy: 41%, precision: 49%), CNN-derived boundary likelihoods (accuracy: 52%, precision: 76%).

Several attempts make to extract, classify and quantify cadastral boundaries using EO data in association with AI technologies (see Table 2). Along with these varying workflows and its image segmentation techniques that employed, however, there is increasing concern over further investigating deep-learning driven image analysis in land administration including image fusion, image registration, scene classification and retrieval and object detection. For remote-sensing image interpretation, the most applicable deep-learning models in remote sensing are: supervised CNN, recurrent neural network (RNN), unsupervised authencoders (AE), deep belief networks (DBN), and generative adversarial networks (GAN) [38]. Although research on effective use of spatial contextual information in remote sensing for land administration is still in infancy, it can substitute the interpreter to a certain extent (not completely) by delving deeply into AI technologies with computer-vision and deep-learning algorithms.

Table 2. Earth Observation (EO) data and Artificial Intelligence (AI) delves deeper into the future of land administration and the advanced techniques substitute to a certain extent the feature and boundary extraction for cadastral mapping. However, a number of critical questions remain about the interpretation of semantic land tenure relations using both EO and AI.

Technologies			Techniques	Sources
EO	AI			
	CV	DL		
Aerial imagery (Orthophoto)	No	No	Cadastral morphology investigation: visual interpretation from the overlay of the cadastral map over orthophotos	[15]
Airborne Laser Scanning (ALS)	√		Semi-automatic boundary extraction: Alpha shape (α-shapes), Canny, and Skeleton algorithm	[16]
Unmanned Aerial Vehicles (UAVs)	√		Automatic feature extraction: Globalized Probability of Boundary (*gPb*) contour detections	[11]
High Resolution Satellite Imagery (HRSI)	√		Semi-automatic boundary feature extraction: mean-shift segmentation plug-in QGIS, the buffer overlay methods	[18]
Unmanned Aerial Vehicles (UAVs)	√		Automatic boundary extraction: ENVI feature extraction (FX) module	[13]
High Resolution Satellite Imagery (HRSI)	√		Automatic boundary extraction: Multi-Resolution Segmentation (MRS), estimation of scale parameter (ESP)	[17]
Unmanned Aerial Vehicles (UAVs)		√	Automatic cadastral boundary detection: deep Fully Convolutional Networks (FCNs)	[19]
Aerial imagery and UAVs		√	Automatic boundary classification: Random Forest (RF), Convolutional Neural Networks (CNN)	[12]

3. Methodological considerations

3.1. A Difficult-to-Access Region: North Korea in the Contexts of Fragile and Conflict-Affected Areas

There is increasing concern that remotely obtained data using disruptive technologies in fragile and conflict-affected areas (See Figure 1), where has been named hard-to-reach areas, is more worthwhile in optimal decision-making rather than a limited groundtruthing provided by direct observation (for example, [39–41]). According to [42], some countries such as Somalia, North Korea, and some Caribbean and Pacific island economies do not consistently render an account of internal data owing to conflict, lack of data capacity, or other reasons including quality of sources. An additional encounter with data scarcity and reliability is associated with national security issues in any contexts of fragile and conflict-affected regions worldwide.

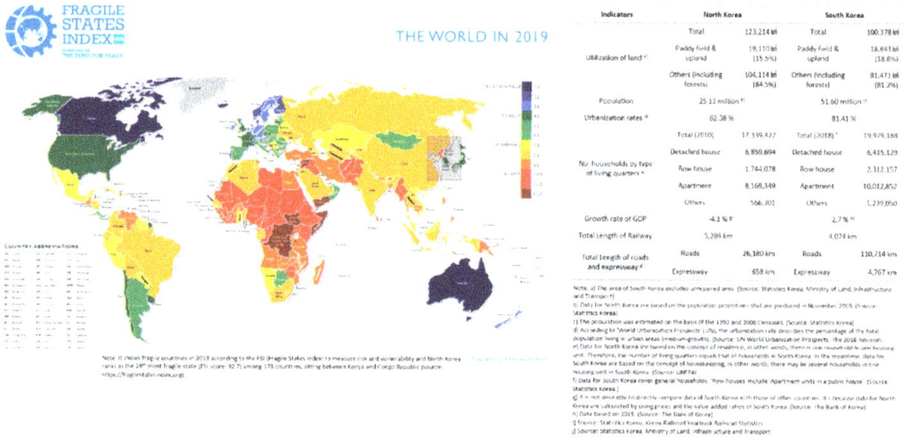

Indicators	North Korea		South Korea	
Utilization of land a)	Total	123,214 ㎢	Total	100,378 ㎢
	Paddy field & upland	19,110 ㎢ (15.5%)	Paddy field & upland	18,893 ㎢ (18.8%)
	Others (including forests)	104,114 ㎢ (84.5%)	Others (including forests)	81,471 ㎢ (81.2%)
Population	25.13 million b)		51.60 million c)	
Urbanization rates d)	62.38 %		81.41 %	
No. households by type of living quarters e)	Total (2010)	17,339,422	Total (2018) f)	19,979,188
	Detached house	6,859,694	Detached house	6,415,129
	Row house	1,744,078	Row house	2,312,157
	Apartment	8,168,349	Apartment	10,012,852
	Others	566,301	Others	1,239,050
Growth rate of GDP	-4.1 % g)		2.7 % h)	
Total Length of Railway	5,289 km		4,074 km	
Total length of roads and expressway j)	Roads	26,180 km	Roads	110,714 km
	Expressway	658 km	Expressway	4,767 km

Figure 1. Contextualization of Area of Interest (left: Fragile States Index 2019 (https://fragilestatesindex.org), right: background information of South and North Korea based on Major Statistics Indicators of North Korea 2019 (https://kosis.kr/bukhan/index/index.do) (devised by authors).

Gathering and establishing reliable information for policy-making in pursuing Korean (re-)unification is more significant than ever during a peace-building process. In this process, re-shaping land governance are a fundamental question focusing on land tenure security, transferability, legitimacy and identity in (re-)unification setting [43,44]. Despite the passive attitude of the North Korean government to disclose information, it is possible to obtain data in a direct or indirect manner, such as [45]: official government reports (e.g., Korean Central News Agency: KCNA and Rodong Sinmun); materials from international organizations dealing with humanitarian aid (e.g., FAO, UNDP, UNFPA, WFP, WHO, UNICEF (For the resources, [46] etc.); information from external agencies or observers in cooperation with local authorities or residents (e.g., Hanns Seidel Foundation for EU-funded project on sustainable forestry in North Korea etc.); data acquired through the joint projects (e.g., the North Korean Ministry of Land and Environment Protection: MoLEP, Swiss Agency of Development and Cooperation: SDC, World Agroforestry Centre's East and Central Asia Office for the Sloping Land Management Program (See further details for the project, [47] etc.).

However, there are still great difficulties in collecting land tenure-related information in North Korea since the government rarely discloses or distributes any policy-related documents, data and statistics. The scope of current research on mapping land tenure relations has been very limited such as restoring historical cadastral maps [48–50] that include both geographical and textual land information. However, most research is still highly dependent on secondary data sources. Many studies using EO data for North Korea have been proposed for monitoring land use and land cover (LULC) over the past several decades (for example, [51–55]). Furthermore, a number of government institutes and think-tanks have already established different types of thematic maps in North Korea using EO data (e.g., agricultural maps; deforestation maps; land cover maps, etc.). More internationally, a platform called 38 NORTH (https://www.38north.org), provides informed analysis of events in and around North Korea using HRSI, as well as develops the digital atlas that was built in the Google Earth platform. Nevertheless, researchers and policy-makers still have faced with difficulties in incorporating land tenure-related data with EO data due to: (1) levels of accessibility: the limited access to North Korean data; (2) methodological levels: complexity of integrating land tenure attributes with EO data; (3) analytical levels: its lower reliability and validity of acquired information.

3.2. Existing Rules for Defining Land Tenure Relations and LULC classifications

This section provides an overview of the existing rules for land tenure relations in South and North Korea that can identify the data gap between them. The classification methods of land tenure relations are based on diverse land trajectories: by land ownership, land (use) categories, 3Rs, land characteristics and urban planning facilities (See Table 3).

In South Korea (SK), land is divided into private land, government-owned land including State, province and county land, land owned by corporation or judicial person and land owned by non-judicial person, according to the land ownership trajectories. Contrastingly, the Constitution and the Land Law in North Korea (NK) does not tolerate private transactions with land. The State, collective farms, institutions, enterprises and organizations only govern land and local residents have land use rights (LURs). According to the Constitution (Article 21; 22) and the Civil Law (Article 45; 53) in NK, there is no restrictions on the subject of State ownership and only the State can own land. The cooperative entity refers to the form of collective ownership in which cooperatives assume the ultimate authorities for the land that are restricted by the State.

According to the Act on the Establishment and Management of Spatial data in SK (Article 2), a land category means a type of land that is classified under its primary use, and registered in the cadastral record. Land is currently classified into 28 categories to represent the nature, purpose and status of the land. Meanwhile, in NK the Land Law (Article 7) distinguishes six categories of land use classes: agricultural-purpose land; settlement land; forestry land; industry land; waterstock land; and special-purpose land. However, it is not yet clear whether these land use categories correspond to the zoning system or land category system in SK [56]. Based upon the Civil Law and LADM (focusing on 3Rs), the right type in SK includes co-ownership, servitude, lease, ownership, partitioned ownership, superficies, sectional superficies, tenancy, usufruct, and fishing. The responsibility types include keeping a snow free pavement and cleaning a ditch, and the restriction type includes servitude and servitude partly [57]. For NK, we assume that there is no land use regulation through the restriction of private rights (3Rs), since NK does not recognize private land ownership.

A land characteristics survey investigates land-related data from physical, spatial and socio-economic conditions in SK [58]. The 45 types of land use indicators are basic data for the land classification aforementioned. Moreover, land infrastructure (urban planning facilities) refers to facilities determined by urban management plan among infrastructures. The legal grounds is the National Land Planning and Utilization Act (Article 2) [59]. In NK, land use classifications are different. It follows the land characteristics in the same way as the six types of land prescribed in the Land Law. The diversity of land use appears to be very simple when compared to that in SK, although the Urban Management Law in NK does not explicitly stipulate land infrastructure, it identifies buildings and facilities, which need to be managed. These include residential and factory buildings, water and sewage and heating operation, urban roads and river arrangements, landscaping and urban beautification. Roughly, one can distinct 29 types of land infrastructure elements in NK.

Earth observation (EO) is one of the most essential methods for monitoring the earth's surface and its dynamics at regional to global scales [29]. The term land use defines how a certain portion of the surface is being utilized. In other words, a particular land use label identifies the purpose for which humans exploit the land cover [60,61]. The land cover denotes to the biophysical appearance on the land and determined by the elements of the Earth's (sub) surface. For example, a State park may be used for recreation but have a deciduous forest cover [60,61]. In some countries, a formal/ government LULC classification system exists which can easily describe the actual condition and changes of spatial structures of the land and its attached attributes: the U.S. National Land Cover Database (NLCD, USA [62]), the national Dynamic Land Cover Dataset (DLCD, Australia [63]), the European CORINE land cover (CLC, EU [64]), the Land Cover Map (LCM, Korea), and the National Land Use/Cover Database of China (NLUD-C, China [65]). Although these datasets have been developed with different mapping methodologies and criteria (e.g., variations in the classes and thresholds applied, time of data collection, sensor types, classification techniques, use of in situ data etc.) [65], one can utilize it as basic spatial data to support the design of scientific and efficient policies.

Table 3. Existing rules for defining land tenure relations in the context of Korean (re-)unification (functional classifications).

Categories	Existing Rules for Identifying Land Tenure Relations in SK	Existing Rules for Identifying Land Tenure Relations in NK	Legal Grounds
3Rs (rights, responsibility and restrictions)	Private land; State land; province land; county land; land owned by corporation (judicial person); land owned by a clan; land owned by a religious group; land owned by other groups; others (9 types)Common ownership; lease; ownership; partitioned ownership; tenancy; superficies; partitioned superficies; usufruct; easement; fishing; keeping a snow-free pavement; cleaning a ditch; servitude; servitude partly (14 types)	State land; collective farmland (2 types)(cf. Since North Korea does not recognize private ownership; there is no land use regulation through the restriction of private rights. Although all land belongs to the State, both the State and the individual or collective can restrict the use by restricting the access. Nature reserves, military sites, public heritage are typically locations where the State wants to restrict access, use and control through such restrictions.)	The Constitution (NK) The Civil Law (NK) The Civil Act (SK) LADM (SK);
Land (use) categories	Building site; dry paddy-field; paddy-field; orchard; forestry; pasture site; mineral spring site; saltern; factory site; school site; parking lot; gas station site; warehouse site; road; railway site; water supply site; river; ditch; fish-farm; park; historic site; gymnasium site; recreation area; religious site; graveyard; miscellaneous land (28 types)	Agricultural-purpose land (arable land); settlement land (construction land and its attached land in local labor areas as well as public land); forestry land (land used in the hills and fields); industry land (sites of industrial facilities such as mine, factories, and the land pertaining to it); waterstock land (land for coast, territorial waters, river and streams, lake, reservoir and irrigation ditch); special-purpose land (cultural heritage sites, historical landmarks, sanctuary and military) (6 types)	The Act on the Establishment, Management, etc. of Spatial data (SK); The Land Law (NK)
Land (use) characteristics	detached-house lot, row-house lot, multiplex-house lot, apartment lot, residential vacant lot, other residential lots; commercial lot, office lot, commercial/office lot, other commercial/office lots; mixed-use lot, mixed-use vacant lot, other mixed-used lots; industrial lot, industrial vacant lot, other industrial lots; dry paddy-field, orchard, other dry paddy-fields; paddy-field, other paddy-fields; afforestation, natural forest, forest land, pasture, other forestry; mineral spring site, mining site, saltern site, recreation area, cemetery park, golf course, racecourse, passenger transport terminal, condominium, other special-purpose lands; roads etc., rivers etc., parks etc., playgrounds etc., parking lot etc., high-risk establishments, obnoxious facilities and Others (45 types)	Agricultural-purpose land (arable land); settlement land (construction land and its attached land in local labor areas as well as public land); forestry land (land used in the hills and fields); industry land (sites of industrial facilities such as mine, factories, and the land pertaining to it); waterstock land (land for coast, territorial waters, river and streams, lake, reservoir and irrigation ditch); special-purpose land (cultural heritage sites, historical landmarks, sanctuary and military) (6 types)	The Act on the Public Announcement of Values and Appraisal of Real Estate (SK); The Land Law (NK)
Land (use) infrastructure	Road; park; railway; (public) open space; waste treatment facilities and water-pollution preventive facilities; heat/gas/oil supplying and storing installations; electric supplying installations; slaughterhouse; graveyards; markets and distribution facilities; recreation area; parking lot; car stations; square; playground and sport facilities; water supplying instalments; public buildings (e.g., school and library); communication facilities; cultural, research, social welfare, public vocational training, youth training facilities, others (21 types)	Dwelling house; public buildings; production buildings; water supplying instalments; heat/gas/oil supplying facilities; road; street green; footpath; streetlight; bridge; tunnel; underground passage; road safety facilities; road markings; bus/tram station; car washing facilities; river (stream); park; recreation area; open space; urban forest; protection forest; zoo/botanical gardens; greenhouse; tree nursery; flower garden; cultural facilities; sanitation facilities; cremation facilities (assumed 29 types)	The National Land Planning And Utilization Act (SK); The Urban Management Law (NK)

3.3. Adopting a New Methodology: Mixed Methods Design and Information Fusion Approach

A number of image segmentation techniques and workflows have been developed to detect visible land tenure relations with EO data. Each has its advantages (e.g., automation, coverage, up-to-date, cost-effective etc.) and drawbacks (e.g., technological bias, methodological rationale, lack of social sensing etc.). To especially overcome these constraints, a further advancement with more focus on the idea of 'triangulation' is therefore considered which is particularly associated with methods of investigation and sources of data [66]. A triangulation logic is chosen because land tenure relations are unknown in a given context and monoscopic EO data interpretation without valid inferences would misguide to identify transferrable and applicable proxies. Data integration is at the heart of discerning epistemological assumptions from multiple sources to attain narrative illustration, convergent validation and analytic density [67]. In this regard, this study makes an original contribution to when land tenure and EO data integration occurs, what types of EO data are integrated and how we integrate them.

In aiming to derive informative land tenure relations from EO data, one has to rely on both the characteristics of tenure itself and a number of proxies derived from EO data and spatio-temporal combinations of EO that may capture a particular land tenure characteristic. Our research started by adopting the subjects (e.g., who is the right holder?), rights (e.g., what is included in a certain right?), and objects (what physical extension of a right has?) model of land administration [68]. In parallel with ownership, rights may embrace complex set of rules related to the access, use, develop or transfer [69]. In other words, a household can be associated with a particular land parcel where people can live, own, rent or have the right to use [70]. Thus, the analytical premise has been questioned on the basis of underlying assumptions: (1) is it possible to distinguish collective farmland from State land?; (2) can one see land use rights (LURs)?; (3) is there a use right that can be linked to an individual or group?; (4) are there land transfer rights (LTRs)?; (5) are there land access rights (LARs) and restrictions?

However, methodological difficulties, using EO data in accurately conjoining a household and physical extension of a right over land parcels and measuring the quality of linking information, have existed. There have been only few empirical investigations into decision-making what proxies shall be operationalized based on both theoretical and practical grounds. In terms of using terms 'proxy' rather than similar terminologies such as 'interpretation key', 'index', 'indicator' or 'variable', we follow a definition labelled by [71] that refers to "use of observable physical features or directly measurable variables to understand and extract what actually exists on the ground, but what is not directly observable or measurable from remote sensing data."

Our approach comes from multiple sources, namely EO data, especially focusing on LULC information, *prior* (expert) and contextual knowledge on land tenure relations acquired through previous experiences and perceptions. In view of all that has been mentioned so far, one may suppose that 'information fusion' approach must be considered in order to extract and conciliate significant elements for the semantic (re-)interpretation and, subsequently, for decision-making [72]. Generally accepted disciplines for the notion of information fusion include: psychology, human factors, knowledge representation, artificial intelligence, mathematical logic, and signal processing [73]. It has been noted that transforming data into knowledge is most striking feature of information fusion and must be converted into a certain language or presented by other means such as visualization techniques [73]. Moreover, this method emphasizes that a wide range of structured/unstructured or primary/secondary data sources address semantic relationships and co-occurrence between them [74].

According to [75], an information element is regarded as "an entity composed of a definition set and a content set linked by a functional relationship called informative relation, associated with internal and external contexts". This highlights that one given single data set do not qualify or quantify to make it informative. When answering the research questions or testing relationship between proposed proxies and ground-truth, it is important to quantify recurring spatial attributes and uniformity or distinctiveness in qualitative data allowing rigorous analysis and to determine rational and optimal proxies. It is therefore considered that triangulation logic and information fusion approach would usefully supplement and extend the methodological and epistemological assumptions of semantic land tenure relations through EO data interpretation. Having defined what EO data proxies and information element meant, different types of information element, thus, should be included that enable the EO data proxies to identify land tenure relations logically. Figure 2 depicts the workflow and the main components of the information acquisition and interpretation process of the semantic land tenure relations.

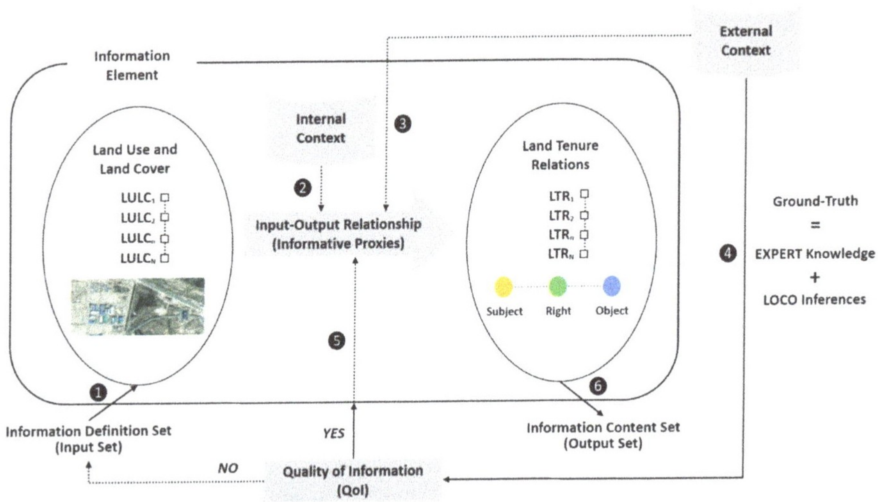

Steps	Elements	Descriptions	Instructions
1	Definition set	Providing the well-defined or categorized land-related information input elements	Define what the land use and land cover information refers to
2	Internal context	Clustering basic principles, characteristics, constraints, or controls about the EO data extraction and land tenure essence itself	Investigate spatial resolution demands; possible sensors; mono or multi-temporal data usages; and documented insights into the nature of spatial arrangements etc.
3	External context (Inner-loop)	Incorporating both documented and undocumented knowledge on EO data and land tenure	Discover when land tenure and EO data integration occurs, what types of EO data are integrated and how we integrate them
4	External context (Outer-loop)	Reflecting data, information, or knowledge whether elaborated meaning or the interpreted information element is valid or not	Measure accessibility, appropriate amount, believability, completeness, concise representation, consistent representation, ease of operation, free of error, interpretability, objectivity, relevancy, reputation, security, timeliness, understandability based on the AIMQ methodology
5	Input-Output relationship	Identifying proxies that bridge the gap between the input (LULC) and output (LTR) information elements	Decide informative proxies based on the elements of image interpretation: color, shape, size, texture, pattern, shadow, height, site, association, density
6	A content set	Converting to explicit LTR knowledge provided by the LULC information	Answer the unknown land tenure related questions (e.g., based on land rights classification, land subject-right-object entities, and 3Rs-related information etc.)

Figure 2. The general structure of an information element and its processes for the interpretation of the semantic land tenure relations (devised by authors).

4. Deriving Workable EO Data Proxies for Interpreting Land Tenure Relations

This section explores a number of workable proxies based on the land related categories defined in Table 2, whereby the proxies are derived from the existing EO data. We, then, discuss hereunder how one can evaluate the five key land tenure-related questions defined in Section 3.3. The exact spatial information and point of interest (POI) in this section were pre-identified from the openly accessible platform.

4.1. Is It Possible to Distinguish Collective Farmland from State Land?

A characteristic and distinctive feature of collective land (as compared to State land) is both the type and number of buildings/dwellings adjacent to the land and the spatial distribution of buildings/dwellings. Another characteristic is the state and density of infrastructure. Clusters of buildings suggest the presence of a collective only if: buildings look similar and simultaneous changes in structures occur. If spatial patterns of buildings and farm sizes and shapes re-occur in different places, it is probably part of a joint collective spatial planning strategy, so it is more likely collective land than State land. (See Figure 3b,c).

Figure 3. Spatial characteristics of cooperative farmlands extracted from high-resolution EO data. (**a**) stands for the Unha collective farmlands surrounded by (dry) paddy-fields (georeferenced: Onchon county, South Pyeongan); (**b**) represents planned spatial arrangements and a centralized cluster of buildings and dwellings; (**c**) highlights buildings and houses utilize homogeneous materials and retains its similar physical shapes and simple roof structures; (**d**) includes textures of irrigation channels and features of rice (dry) paddy-fields. (Image sources: Google Earth, date of access: 9 October 2019).

It has commonly been stated that North Korea (NK) collective farmland plays a pivotal role in major food production (approximately 85%~90% of total production), such as rice, corn, beans, and potatoes [76]. In this regard, most collective farmlands are utilized as a (dry) paddy-field, so it can therefore be assumed that the collective farmland can be confirmed through the presence of (dry) paddy-fields. An area linked to or surrounded by a substantial portion of (dry) paddy-fields can be considered as a collective farmland, which is following the association element of the image interpretation (See Figure 3a). According to [77], rice fields include periodically flooded flat surfaces with the rice plant, open water surfaces on fields, stubble or rice, irrigation channels between land parcels and embankments between rice fields. These can be interpreted with the rough (or coarse) image texture caused by variation in tonal values of an image that helps to identify single objects (See Figure 3d).

The collective farmlands include ranging between 80 and 300 households and operates on a large-scale from approximately 1,300,000 to 5,000,000 m^2 [76]. Thus, the relative size and high density or compactness of the settlement helps to distinguish when compared with State (farm) lands. Collective farmlands accompany a farming equipment, materials, and production facilities from the State and benefit from all the new building construction including rural dwellings (See Figure 3c).

The characteristics of rural dwellings in collective farmlands are homogenous building colors in grey scales, a signature line of the tiled roof, and densely built-up block structure with single-story detached houses. This indicates a need to understand physically detectable proxies that the farming-related objects will be more captured on the ground (in spring/summer) or stored in warehouses (autumn/winter) rather than State farmlands. In addition, since agricultural production is mainly concentrated in the springtime, changes of agricultural activities and its densification, which implies collective farmlands, may be compared using time-series analysis. However, these proxies tend to be unreliable unless used with other complementary sets of proxies. It therefore requires a rigorous image interpretation of EO data in combination with other interpretation elements as well as secondary data.

On the other hand, State-owned (farm)land in NK refers to nationalized (farm)land in the process of land reforms in the past, consisting of agricultural testbed, farms for the seed-production and livestock [76]. As far as this assumption is concerned, the combination of the geometric properties of an object such as shape and (building and roof) size, orientation, density, height as well as (building and roof) colors/tones that identifies agriculture-based patches or infrastructures can be considered as workable proxies. These include small dot-shaped (for orchards) and smooth (for pastures) textures, out-buildings (sheds), dispersion value (low building density), irregularly shaped object boundaries, complex, elongated or irregular building shapes, and distinctive roof colors (e.g., blue, green, yellow and red as well as brightness etc.) and the association with agriculture-based infrastructures, monumental buildings, and welfare facilities (See Figure 4). However, the association elements should be synthesized with documented or local knowledge as the exact points of information has not yet clarified.

Figure 4. Spatial characteristics of State farmlands extracted from high-resolution EO data; (**a**) describes whole region of the No. 5 State farmland in Taehongdan county in Ryanggang; (**b**) shows key spatial arrangements of State farmlands embracing: agriculture-based infrastructures (e.g., fertilizer and processing factories and colleges and research institutes etc.); monumental buildings (e.g., revolutionary museums etc.); welfare facilities (e.g., house of culture, markets and shop, and kindergarten etc.); (**c**) is a site of microbial compound fertilizer factory (upper) and agricultural testbed or greenhouses (middle); (**d**) a site for potato processing factory (image sources: Google Earth, date of access: 9 October 2019).

4.2. Can One See Land Use Rights (LURs)?

A typical feature of LURs is that it usually relates to consistent patterns in space and there exists regularity in time such as seasons. The right itself must be inducted or assumed if such patterns and consistencies exist. Reversely, finding such consistencies is an indication of the right. This implies that from EO data over a number of years one can see similarities each similar season. If all indicate the same type of land use, at the same points in time over a number of years, then one may assume consistent land use, and LURs. If conversely there is a large variation in this, one has to assume that the LURs are not consistent, or do not belong to a single person or group. The variations in land use itself suggests an allocation of what the land may be used for over a longer period of time, yet in a specific time of the year. This suggest the presence of a consistent LUR of single land user. Any land use which is not observed suggests a restriction in land use or a specific allocation of land use.

The LULC in most areas between 1990 and 2000 remains unchanged, assuming existing consistent LURs due to strict land use restrictions for nature reserves, military sites, and public heritages, or unplanned and poor land management (see Figure 5). However, Figure 5a reveals more intensely developed lands (red color) are shown in POI and we note that the development is mostly associated with constructing new dwellings (with LURs) in fallow land. Moreover, Figure 5b highlights that land for agricultural use (yellow color) has significantly increased in POI. This is due to the increase of farmlands to cultivate, especially with collective farmlands, and hence the State provide new houses for the farmer households that grants LURs. In addition, Figure 5c shows more intense LULC changes in an urban area with the development of water bodies (blue color) and these provide additional evidence with respect to LURs. It demands sufficient quantity and quality of water resources for increasing urban households, and it can be inferred that developing water bodies are correlated with LURs since the increased number of urban households represents an increase in granting LURs. Lastly, Figure 5d underlines that LULC changes have occurred more in a border region than inland areas. This, we assume, is because the border in NK started allowing LURs to be sold, transferred, and leased to foreign corporations in Special Economic Zones (SEZs) by modifying socialist land tenure system for economic recovery.

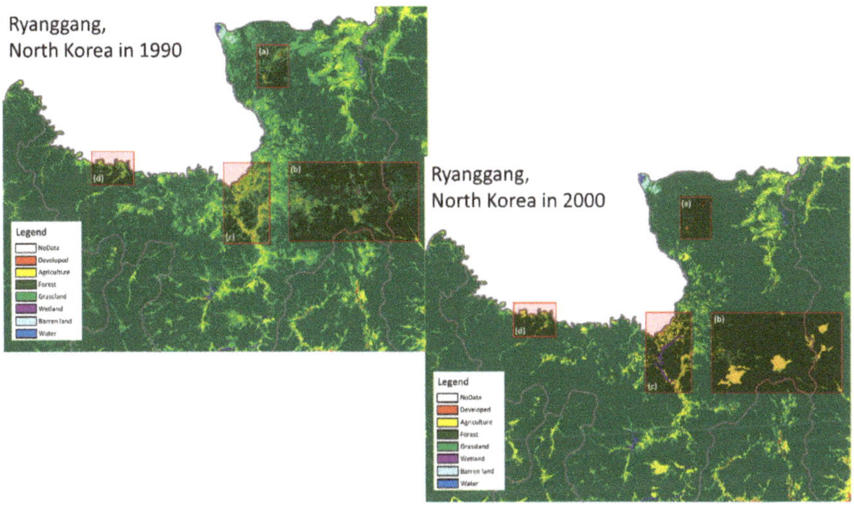

Figure 5. The example of land use and cover changes in Ryanggang between 1990 (upper) and 2000 (lower) with the currently available data set produced by the Ministry of Environment (MoE) using Landsat TM imagery (1991–1999) and Landsat 7 ETM+ imagery (2008–2010) (image sources: [78] and revised by authors).

All the means of production and socio-cultural facilities, including land, are jointly used in NK. Meanwhile, housing and the allotment with an average size from 60 to 130 m^2 are owned by the State [79], but LURs are granted to individuals, and the product is allowed to belong to them. We therefore assume that the presence of all types of houses and their accompanying allotments can be chosen to confirm the existence of LURs. To identify the (semi-) detached houses, it includes the following proxies: low building density, 1 or 2 storied houses, uniformly shaped settlement, proximity to roads, and low to intermediate imperviousness. In terms of condominium-related proxies, we consider large rectangular simple form buildings, regular alignment, more than three stories, and low to intermediate imperviousness and shadow silhouettes (see Figure 6). When it comes to the workable proxies for allotments, the following "if" statements are considered: if the land (or site) has detached small-sized buildings, if it is low built-up land, if it is low imperviousness, if it has plants or vegetation, and if it is used as buffer between houses (see Figure 6b, Figure 7, and Figure 8b).

Figure 6. Different types of housing and their morphological features. (**a**) Condominium or residential block buildings; (**b**) detached houses; (**c**) (semi-)detached houses; (**d**) showing different forms of housing shapes (e.g., linear, curved, rectangular patterns and different colors of roofs etc.) (Image sources: Google Earth, date of access: 16 October 2019).

Before 1998, a new housing reverted to the State, and only the right to use was given to individual households by permission. However, after the amendment of the Constitution, the building was excluded from collective ownership, enabling the possession of new housing. For the sake of legal certainty, the form of housing is divided into State-owned, cooperative-owned and individual-owned, but its ownership is very limited regarding use and transaction. According to [80], three types of housing have been investigated, with a semidetached house (or row house) being the highest proportion at 43.9%, a detached house (or single-family house) has been estimated as 33.8%, and apartments (or condominium) account for nearly one-fifth (21.4%) of housing [81].

More specifically, in the rural areas of North Korea, the 'harmonica houses' have often been observed where two or three households, and even five to six or more households, live together in a detached house. A variety of identifiable proxies such as a small roof with slate materials, chimneys on rooftops (small dot-shaped objects or a light shadow Silhouette) and a fence (with line-shaped objects) installed to distinguish garden plots have been detected in the images. The evidence reviewed here seems to suggest that the physical attributes of varying forms of dwellings through EO data acquisition have significant correlations with granting LURs. Moreover, the growing new construction/extension of residential buildings and expansion of construction activities in certain regions over time may confirm the significant increases in authorities' awareness on LURs. As shown in Figure 6, (a) shows varying geometrical attributes of apartments along Ryomyong Street in Pyeongyang in 2019, while (b) provides a typical example of detached houses that displays the roofing, chimneys and a fence for defining spatial boundaries between neighborhoods. Furthermore, (c) presents a normative sample of semidetached houses and the building shadows that determine building heights. In addition, (d) demonstrates different forms of residential building shapes such as linear, curved, rectangular fit and different colors of roofing.

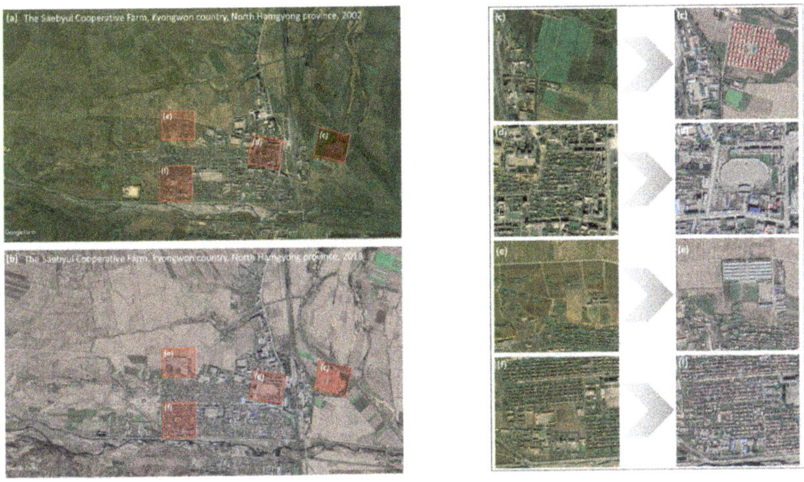

Figure 7. Morphology and urban structures of the Socialist lifestyle are discovered in the cooperative farmland in 2002 (**a**) and 2018 (**b**). (**c**) Changes in land use from farmland to residential areas; (**d**) changes in land use with more community-oriented development; (**e**) showing newly-built agricultural facilities; (**f**) changes of residential development at higher densities (image sources: Google Earth, date of access: 17 October 2019).

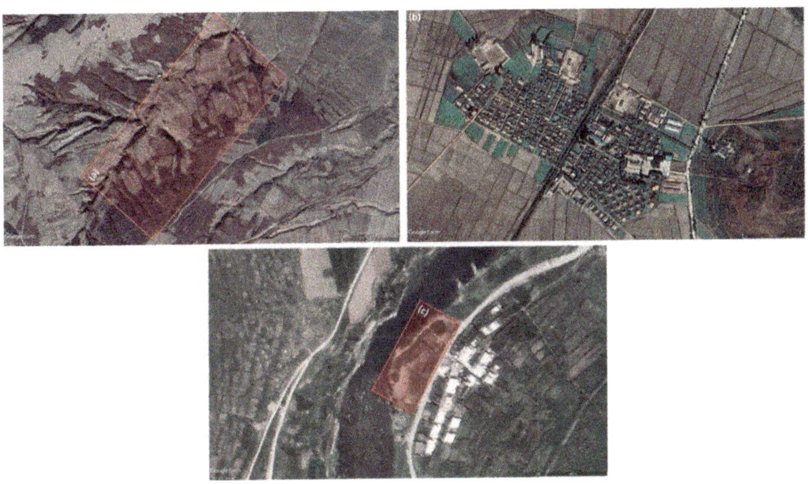

Figure 8. Spatial characteristics of the 'sotoji'. (**a**): between Onsong and Sambong, North Hamgyong; (**b**) the Sambong cooperative farm, Phyeongwon; (**c**) Hyesan, Ryanggang (Image sources: Google Earth, date of access: 17 October 2019).

4.3. Is There a Use Right that Can be Linked to an Individual or Group?

Another possible explanation for the linkage of individual or group-based LURs might be that land use is homogenous between adjacent land parcels and buildings. A characteristic feature of an individual right as compared to a group right is that there is a large variety in land use between neighboring parcels. In a group land tenure, people tend to converge to similar crops or building and housing types. The observed land use can be connected to an observed set of combined proxies on the land rights: similarity or dissimilarity of neighboring land use in space and over time; changes in the adjacent buildings or houses; changes in the road infrastructure; number of buildings in a certain vicinity.

Cooperative farms in NK are based on collective ownership. Along with economic activities, cooperative farms function as rural communities that manage collective socio-cultural activities. However, property rights are exercised by the State, and households only have exclusive LURs. Therefore, all the means of production and socio-cultural facilities, including land, are entitled to group-based LURs. The multiple objects of socio-cultural facilities incorporate 'cultural houses' (rural houses with welfare facilities), various community amenities, and nurseries and kindergartens. These possible proxies could define multiple LURs of groups over the same piece of land with specific characteristics such as building geometry, arrangement pattern, roofing color, and site characteristics with EO data.

Figure 7 presents an amalgamation of diverse community amenities: government offices (e.g., a cooperative farm management committee and a party committee), educational institutions (e.g., a middle school and a kindergarten), medical facilities (e.g., hospital), and socio-cultural facilities (e.g., station, a park, a skate park, a restaurant and hotel, revolutionary museum, monuments—see Figure 7a,b). These regular arrangement patterns of building objects are a common feature that appears in collective farms, thus representing group-based LURs. As (c) indicated, we found that significant land use changes from the cultivated farmlands to residential areas (confirmed by the presence of multiple building objects with similar looks, a high density of settlement, simple rectangular forms and same roof colors in red scales) occurred. The objects with a similar appearance are perceived as a group figure or shape, and thus the similarity/dissimilarity of neighboring land use in space and over time with other contextual knowledge (such as relationship, adjacency, inclusion, composition, and neighborhood) can be regarded as workable proxies for defining group LURs.

Indeed, the site, situation and structure of objects in the urban/rural landscape on the image helps identify their significance and (d) depicts changes of former settlements as a newly built community asset (for a skate park) in line with improved access to road (types: paved road and wider widths) surrounded areas. Therefore, the (re-)construction/extension of community buildings or infrastructure by the existing building removal could become proxies for LURs linked to group tenure. In the same vein, (e) indicates that agriculture infrastructure is not newly located where it was fallow or barren land, but also adjacent to the residential dwellings. With the acquired EO data, (vinyl) greenhouses as a particular form of non-irrigated arable land have been identified with some elements of image interpretation: building materials (plastic or glass), roof colors (white or grey), brightness (light), and texture (rough). Hence, the changes of association elements with the close proximity or adjacency to the agriculture-related objects or neighborhood and specific characteristics of the objects might quantify group-based LURs. Then, (f) interprets the increase of the number of houses (high building density) in a certain vicinity over time. While this phenomenon can be seen as an increase in an individuals' LUR, it can also be regarded as an increase in the group-based LUR, as NK's housing supply is mainly carried out on cooperative farms to improve agricultural productivity. Therefore, there have to be more proxies to make this argument complete.

The existence of undivided shared areas of the common property or public infrastructures between the roads or buildings can be regarded as a proxy that can be related to the collective LURs. Under the socialist urban planning system in NK, the arrangement standards for housing and neighboring residential structures are based on the sub-district plan that housing and production facilities should be located adjacent and in the vicinity of the planned area. Within the sub-district, diverse socio-cultural facilities are located and observed with relatively low building density in scope. This is a proxy that emphasizes the straightforward approach and characterizes the socialist lifestyle based on a community unit rather than the individual [81].

4.4. Are There Land Transfer Rights (LTRs)?

When selling, mortgaging, or conveying the land to others, it exhibits a considerable variation in transfer of land. However, what is actually transferred is not the land or building, but rather a bundle of rights pertaining to it [82], and this could be interpreted as land transfer rights (LTRs). A characteristic feature of when land or houses are being bought or sold is that the LTRs are accompanied by changes in the objects, which are not occurring in the neighboring objects. One can think of constructing a new roof, painting the house, construction of fences, construction of new objects or infrastructural works on the land. Moreover, the changes in structure, type and shape of the object are occurring in a relatively short time span. We would then assume that there might have been an LTR related to these objects.

According to [83], there is a tendency for households or farmers, those who suffered from tenure insecurity, to utilize tools of land conversion or reclamation as a way of building informal LTRs. In the same vein, in the mid-1990s, as a result of the massive food insecurity in the NK, unauthorized households reclaimed and cultivated vacant land as well as cleared the forests and occupied so-called 'small-land (sotoji)'. 'Sotoji' is located in the mountain slope and its transactions are being made publicly among the households. Three types of 'sotoji' existed: garden plot (GP); side-job plot (SJP); and tiny patch of land (TPL) [84]. In principle, the law prohibits the sale of land in NK. In reality, GP, having a large share to produce their foods, is explicitly recognized by the Constitution, the Civil Law and the Land Law (Article 13). GP was originally allowed to use in individual households within the collective farmlands and it was common to situate at the front yard of houses or on an empty space between them. The size of GP is approximately 66 to 100 m^2, but, in fact, it covers about 100 to 165 m^2. In addition, if the house is sparse and the vacant land is immense in size, it is reported that an even larger-sized GP is allowed. Likewise, since the mid-1990s, the GP was built in a vacant land attached to a detached house or a balcony of an apartment in urban areas.

SJP was developed from the early 1980s to cultivate the barren land, which has not cultivated by the farmers of collective farmlands. If the GP is individual farming units, the SJP is a group units (cf. the scale is approximately 3000 to 6600 m^2). Unlike the SJP, TPL is illegally cultivated private land. This originally refers to a small-scale farmland rather than linking the subject or illegality of cultivation. TPL is deemed an object of their own since households put considerable effort into cultivation. The authorities, however, investigate the TPL and impose land use fees to place it under the State control. In this process, land transaction has actually occurred and there are also certain cases where it is handed over to the neighboring landowner, or it is exchanged with other corresponding goods or cash. Although it is different from SK's land transaction that transfers ownership through such transaction processes, it can be assumed that land transaction of TPL would inevitably occur.

Figure 8 reveals that (a) describes the major feature of TPL located in the mountain slope with the evidence of forest farming. TPL may be located in a relatively lower elevation and land parcels where slopes have gentle or stable slopes. As the population grows in areas where arable land is scarce, people tend to take advantage of reclaiming land in fallow on the terraced hillsides that are easily accessible. Therefore, the proxies for TPL include: lower elevation using a digital elevation model (DEM) and slopes have a gradient less than 15%. With HRSI, the small patches of vegetation cover between neighboring lands on the mountain can be considered as a proxy to detect and identify TPL. Meanwhile, (b) illustrates GP where individual households officially are allowed to cultivate and produce SJP in their front yard or at the rear of a house. Other indicators of SJP in the cooperative farmlands where a group of farmers can cultivate the barren land are the length or width of image features (small parcel size for GP and large parcel size for SJP), location-specific features (front/back yard or attached to each other), and natural colors of features (GP and SJP are often depicted in green on the imagery). In addition, (c) is an exemplary attribute of TPL where individual households cultivate vacant lands along the streams, so adjacency to the streams or ditches and the small and regular/irregular patches of vegetation cover along the streams or ditches can be interpreted as transferrable and applicable proxies.

To sum, the authorities in NK drag 'sotoji' within the public land management sphere to restrict (illegal) land use. On the other hand, households generate more income out of the management area with the sense of personal land tenure. Therefore, 'sotoji' can be important proxies that prove the existence of LTRs with some of the aforementioned elements of image interpretation.

4.5. Are There a Land Access Rights (LARs) and Restrictions?

What is known about LARs is largely derived from a private land tenure system that gives priority to the rights of individuals. The main segment of LARs frequently addressed are: an easement (servitude) and rights of way. A characteristic of LARs is that multiple objects are connected to single or multiple types of objects. An easement generally places an emphasis on allowing for separate usage of land which could refer to the right to use another household's land for different purposes. There may be varying activities over the single parcel of land or an entire property over the land that represents LARs. For instance, one of the best known is installing public utilities into a certain land parcels [85]. Another major illustration is reaching inaccessible properties or linking two separated objects through road-related infrastructure.

Turning now to the LARs in the given context of this paper, it is important to bear in mind that we may accompany a possible bias in describing unknown land tenure relations since private land tenure is not recognized in NK and there is no land use regulation through the restriction of private rights. Although all land belongs to the State, we assume that both the State and collectives can restrict the use by restricting access for public purpose. We thus, for the purpose of analysis, assume that the State, collectives and the households in NK may acknowledge LARs.

If our assumptions are to be accepted, they enable us to provide a number of available proxies (see Figure 9). We also need to derive whether the object under consideration is connected to any form of infrastructure or not, as well as if other objects (such as silos) are connected to infrastructure. In this regard, identifying several utility networks offer an effective way of confirming the presence of LARs in NK. Moreover, when newly creating a parcel (called "division") and two or more parcels of land from the existing one (called "subdivision") for the commercial or residential purpose, these properties commonly generate different types of easements under certain physical characteristics of the objects. Therefore, the subdivision of land parcels or (in)consistent land use may implicate whether LARs exist or not.

As shown in Figure 9, we were able to detect and label what objects are especially linked to the public rights of way and servitude (focusing on restrictions on the use of land) rather than individual rights of way. These include: solar panels, railroads, drinking water production facilities, a transmission tower, pipelines, footpaths, military site, and reservoir. We noted that there are some site characteristics near public utility networks, nature reserves, and a public heritage site. However, defining which proxies are workable for LARs is highly context-dependent due to the lack of formalized and proven rules as well as its application in different contexts. It also requires forming an ensemble with other types of proxies to describe the socio-legal status of the objects. Among the elements of image interpretation, only site or situation elements are valid and reliable in detecting the public rights of way. We then produced only few proxies by deriving similar site and situational features from nine images as follows: proximity to hazardous or isolated locations, poor accessibility (lack of access roads), elongated object shapes, and less green and open spaces.

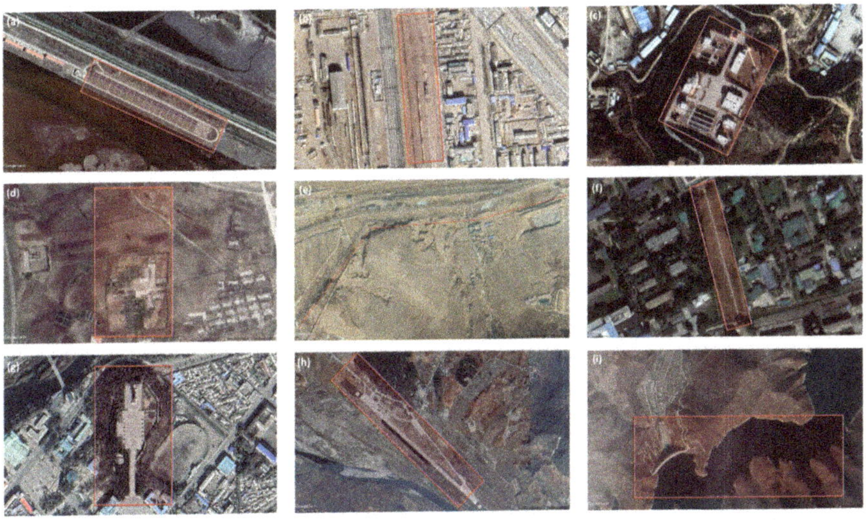

Figure 9. Multiple examples of urban structures on certain land parcels: (**a**) solar panels (in Kumsanpho solar power station), (**b**) railroads (in Rajin), (**c**) drinking water production facilities (in Kaesong), (**d**) a transmission tower (in Electric power transmission office in Sepho), (**e**) pipelines (in Seungri chemical complex refinery in Rason), (**f**) footpaths (in Pyeongyang), (**g**) public heritage (in Hyesan), (**h**) military site (in Pukchang), (**i**) reservoir (in Ryongrim) (image sources: Google Earth, date of access: 17 October 2019).

4.6. Summary of Discussion

This section summarizes the findings to identify proxies to derive unknown land tenure relations over North Korea (see Table 4). The first set of questions aimed to address that whether the observed land is State or collective farmlands. There was no significant difference between two groups in both the general spatial arrangements in scope. However, we found that the location-specific features in line with physical and temporal characteristics helps to identify single objects on (dry) paddy fields. This is one of the most significant characteristics for detecting visually analogues arrangements of collective farmlands. On the other hand, what stands out in the State (farm) land is the combination of the geometric properties of objects characterizes a common feature of agriculture-based patches and infrastructures.

In order to assess the feasible proxies of LURs without having access to the ground, the different EO data sources have been utilized to not only detect LULC changes, but also ascertain a variety of dwellings and its morphological features. These EO datasets present a plausible interpretation with the association element that addresses the probable occurrence among different sets of entities as well as socio-legally documented local knowledge that leads to confirmation of LURs.

The cooperative farms under the collective ownership regime typically accompany a socialist morphology, with the unification of forms and construction to distinguish whether individual or group-based LURs, proxies such as building geometry, arrangement pattern, and site characteristics could define multiple LURs of a group over the same piece of land. The regular arrangement patterns of sites and building objects with other contextual knowledge is a common feature, thus representing group-based LURs that can be jointly used by the multiple groups of people.

Another question sought to determine whether there are LTRs or not. With respect to this subject, it was hypothesized that households reclaimed and cultivated vacant land as well as cleared the forests, and made transactions between households after illegal occupation. Based upon the normative concept of 'small-land (Sotoji)', the discernable proxies that prove the existence of LTRs with following elements of image interpretation are: the low elevation, slopes have gentle slopes less than 15%, small and regular/irregular patches of vegetation cover, the length or width, location, colors and adjacency to the specific objects.

Table 4. Identifying proxies to derive unknown land tenure relations over North Korea in conjunction with EO data (devised by authors).

No.	Land Tenure Relations	Proposed Proxies	Elements of EO Data Interpretation
4.1	Collective (farm) land	**Presence of (dry) paddy fields**	
		Area linked to or surrounded by (dry) paddy fields	
		(Dry) paddy fields: rough (or coarse) image texture	
		Settlements: high density or compactness	
		Rural dwellings: object colors in grey scales, a signature line of the tiled roof, densely built-up block structure with single-story detached houses	
		Presence of portable farming-related objects on the ground	
		Seasonal changes of agricultural activities	
	State (farm) land	Orchards: small dot-shaped patch	
		Pastures: smooth textures	
		Warehouses (or sheds): out-buildings	
		Building sites: low building density	
		Buildings: complex, elongated or irregular object boundaries	
		Roofs: blue, green, yellow and red and light (brightness)	
		Agriculture-based infrastructures, monumental buildings, and welfare facilities	
4.2	LURs	Land uses: intense land development	
		Land uses: an increase of agricultural land	
		LULC changes: urban areas with the development of water bodies	
		LULC changes: in border regions than inland areas	
		Presence of different types of houses (and allotments)	
		(Semi-)detached houses: low building density, 1 or 2 storied houses, uniformly shaped settlement, in close proximity to roads, low to intermediate imperviousness	
		Apartments: large rectangular simple form, regular alignment, more than three stories, and low to intermediate imperviousness and shadow silhouettes	
		Allotments: detached small-sized buildings, low built-up land, low imperviousness, buffer between houses	
		Harmonica houses (in rural areas): small roof with slate materials, chimneys on rooftops (small dot-shaped objects or a light shadow Silhouette) and fences (line-shaped objects)	
		New construction or extension of residential building and expansion of construction activities	
4.3	Group LURs	Amalgamation of diverse community amenities	
		Conversion: presence of multiple building objects with similar patterns, high density of settlement, simple rectangular forms, and same roof colors	
		Adjacent land uses: similarity or dissimilarity	
		Construction/extension of community buildings or infrastructure by the existing building removal	
		Accessibility: improved access to roads (paved road and wider widths)	
		Greenhouses: new construction in a barren land and adjacency to dwellings (materials: plastic or glass, roof colors: white or grey, brightness: light, and texture: rough)	
		Increase of the number of houses in a certain vicinity (high density)	
		Existence of undivided shared areas of the common property or public infrastructure	
4.4	LTRs	Presence of small land (sotoji): garden plot (GP), side-job plot (SJP), and tiny patch of land (TPL)	
		Garden plot (GP): small parcel size, in front/back yards or attached to each other, green color	
		Side-job plot (SJP): large parcel size, in front/back yards or attached to each other, green color	
		Tiny patch of land (TPL): lower elevation, gentle slope less than 15%, the small patches of vegetation cover between neighboring lands on the mountain (hillsides) or along the streams or ditches	
4.5	LARs	Public utility networks, nature reserves, and heritage sites: in close proximity to hazardous or isolated locations, poor accessibility (lack of access roads; low to intermediate imperviousness), elongated object shapes, and less green and open spaces (fewer green colors and rough textures)	
		Subdivision of land parcels	

Note. Color, Shape, Size, Texture, Pattern, Shadow, Height, Site, Association, Density (color-coded bars).

The combination of multiple man-made structures over a single parcel of land or entire property over the land provides some support for the conceptual premise. Although all land belongs to the

State, both the State and the collectives can restrict the use by restricting access for public purpose. Among the elements of image interpretation, only site or situational elements are valid and reliable in detecting the public rights of way. We then produced only few proxies by deriving similar site and situational features from nine images as shown in Table 4.

While some progress has been made for cadastral mapping, very little was found on the question of how we bridge the semantic gap between land tenure and EO data. Thus, this account, in methodological terms, seeks to propose a new notion of remote-sensing based proxies for interpreting land tenure relations that could be transferable and applicable in land administration domain at a semantic level. With regard to the research findings, some limitations need to be acknowledged. A first limitation is that since this study was only conducted from steps 1 to 3, validating was beyond the scope of this work (see Figure 2). However, the preliminary investigations indicated that the subsequent steps for validation (steps 4 to 6) will further move us closer to developing a full picture of the identification of transferrable and applicable proxies for geospatially informed analysis. In other words, it is possible that these results are only valid when a holistic methodological approach takes place. This experiment also has not suggested any technological advancements yet and the proposed proxies require a rigorous AI-based (semi-)automated image interpretation of EO data with other complementary sets of proxies. One possible implication of this is that algorithmic approaches and methodologies concerning deep-learning networks will be able to mine land tenure relations from EO data and these are divided into: supervised learning approaches trained from scratch, pre-training and fine-tuning approaches, advanced learning techniques, and novel technologies developed by the remote-sensing community [86].

5. Concluding Remarks

Land tenure relations, which are relevant in the Korean (unification) context, include the difference between private, State and collective land, the type and location of land use rights, the spatial allocation of rights and restrictions, the ability and spatio-temporal changes of transferring land rights, and the spatial restrictions in access and use. So far, such land tenure relations could only be derived when combining topographic data with agricultural census data at the regional or national scale, and household surveys and a participatory mapping at the local scale. However, given documented insights into the nature of spatial arrangements and the similarities and patterns when observing in features of typical land use structures in North Korea, it was possible to derive proxies for particular types of land tenure from openly accessible EO data.

The proxies consist of specific combinations and patterns of shapes, colors, textures related to physical structures such as roofs, types of buildings, infrastructures, types of land use and vicinity of comparable features. The assumptions connected to these proxies relate to fundamental notions of tenure claims and interests such as collective ownership, land lease and use, occupation (reclamation), transaction (sell and convey) and land access (servitude and rights of way). Overall, this study strengthens the idea that data mining for North Korea related land (tenure) information in the context of Korean (re-)unification is possible and feasible.

The application of EO data involves image processing and data mining technologies which can help to generate a better insight in current land and property interests (such as land tenure, land rights, land responsibilities and duties related to land and properties), and to better prepare, execute, enforce, assess and monitor land interventions. In the context of (re-)unification, the sample tests are particularly relevant for re-uniting countries where different land tenure systems exist and where the data are not coherent. For example, prior to the (re-)unification in Germany, there were two different systems of land tenure, which co-existed next to each other. Unifying the system in Germany was difficult at first partly because each of the previous countries had recorded and administrated in a significantly different manner. A similar situation exists in anticipation of a unified Korea, especially considering that little information is available about the varieties of land tenure and the missing links to individual people in North Korea.

One way to overcome this challenge is to detect land tenure with the use of remote sensing and open access aerial and satellite images. Normally, this technique is possible when having access to ground control points, civil registers and semantic interpretation of both the tenure and the people's components. When this information is however missing or this data source is unreliable—as is the case in North Korea—one has to rely on a number of assumptions and a set of test trials, which if proven right, can be generalized with artificial intelligence connected to image processing. In other words, one has to understand the socio-legal relations to land with pixel. This socio-legalizing the pixel is still largely an idea rather than an available set of techniques. In order to develop such techniques, which will ultimately facilitate the land tenure unification process in Korea, and possibly also improve existing land tenure records (including both public/private land rights, restrictions and responsibilities), one needs a collaborative research development.

The next step is to enrich and test the information quality of the above assumptions and proxies with empirical data tests, inclusion and reflections of local knowledge on the ground (focusing on North Korean defectors' perspectives) and expert knowledge in EO and land administration sciences in North and South Korea. The findings in Section 4 could also help for creating more machine-learning and deep-learning algorithms that provide reference to other papers. The construction of these algorithms was, however, beyond the scope of this paper.

Author Contributions: Conceptualization, C.L. and W.T.d.V.; Data curation, C.L.; Formal analysis, C.L.; Investigation, C.L.; Methodology, C.L.; Project Administration, C.L.; Resources, C.L.; Supervision, W.T.d.V.; Validation, C.L. and W.T.d.V.; Visualization, C.L.; Writing—original draft, C.L.; Writing—review and editing, C.L. and W.T.d.V. All authors have read and agreed to the published version of the manuscript.

Funding: This work was supported by the German Research Foundation (DFG) and the Technical University of Munich (TUM) in the framework of the Open Access Publishing Program.

Acknowledgments: We thank three anonymous reviewers for their insightful comments in narrowing the gap between our claims and the actual content of the manuscript. We are also immensely grateful to colleagues who provided insight and expertise that greatly assisted the research, although any errors are our own and should not tarnish the reputations of these esteemed persons.

Conflicts of Interest: The authors declare no conflict of interest.

References

1. Robinson, B.E.; Holland, M.B.; Naughton-Treves, L. Does secure land tenure save forests? A meta-analysis of the relationship between land tenure and tropical deforestation. *Glob. Environ. Chang.* **2014**, *29*, 281–293. [CrossRef]
2. Payne, G.; Durand-Lasserve, A. Holding on: Security of Tenure-Types, Policies, Practices and Challenges. Research Paper Prepared for an Expert Group Meeting of Tenure Convened by the United Nations Special Rapporteur on Adequate Housing, October 2012. Available online: https://www.ohchr.org/Documents/Issues/Housing/SecurityTenure/Payne-Durand-Lasserve-BackgroundPaper-JAN2013.pdf (accessed on 18 January 2019).
3. Petersen, R.; Stevens, C. *Putting People on the Map: The Land Tenure Data Challenge and Global Forest Watch.* Global Forest Watch. Available online: https://blog.globalforestwatch.org/data-and-research/land-tenure-gfw (accessed on 2 October 2018).
4. Chaturvedi, R.; Shelar, K.; Singh, K.K. *MAPTenure: Enabling Tenurial Clarity for Orange Areas of Central India;* WRI: Mumbai, India, 2018.
5. Biscaye, P.; Callaway, K.; Chen, K.; McDonald, M.; Morton, E.; Reynolds, T.; Anderson, L. *Land Tenure Technologies Summary of Services and Implementation;* Evans School Policy Analysis and Research (EPAR): Washington, DC, USA, 2017.
6. Schweik, C.M.; Thomas, C.W. Using Remote Sensing to Evaluate Environmental Institutional Designs: A Habitat Conservation Planning Example. *Soc. Sci. Q.* **2002**, *83*, 244–262. [CrossRef]
7. Hall, O. Remote sensing in social science research. *Open Remote Sens. J.* **2010**, *3*, 1–16. [CrossRef]
8. Donaldson, D.; Storeygard, A. The View from Above: Applications of Satellite Data in Economics. *J. Econ. Perspect.* **2016**, *30*, 171–198. [CrossRef]

9. Patino, J.E.; Duque, J.C. A review of regional science applications of satellite remote sensing in urban settings. *Comput. Environ. Urban Syst.* **2013**, *37*, 1–17. [CrossRef]
10. Crommelinck, S.; Bennett, R.; Gerke, M.; Nex, F.; Yang, M.; Vosselman, G. Review of Automatic Feature Extraction from High-Resolution Optical Sensor Data for UAV-Based Cadastral Mapping. *Remote Sens.* **2016**, *8*, 689. [CrossRef]
11. Crommelinck, S.; Bennett, R.; Gerke, M.; Yang, M.; Vosselman, G. Contour Detection for UAV-Based Cadastral Mapping. *Remote Sens.* **2017**, *9*, 171. [CrossRef]
12. Crommelinck, S.; Koeva, M.; Yang, M.Y.; Vosselman, G. Application of Deep Learning for Delineation of Visible Cadastral Boundaries from Remote Sensing Imagery. *Remote Sens.* **2019**, *11*, 2505. [CrossRef]
13. Fetai, B.; Oštir, K.; Kosmatin Fras, M.; Lisec, A. Extraction of Visible Boundaries for Cadastral Mapping Based on UAV Imagery. *Remote Sens.* **2019**, *11*, 1510. [CrossRef]
14. Konecny, G. Cadastral Mapping with Earth Observation Technology. In *Geospatial Technology for Earth Observation*; Li, D., Shan, J., Gong, J., Eds.; Springer: Boston, MA, USA, 2009; pp. 397–409. [CrossRef]
15. Luo, X.; Bennett, R.; Koeva, M.; Lemmen, C.; Quadros, N. Quantifying the Overlap between Cadastral and Visual Boundaries: A Case Study from Vanuatu. *Urban Sci.* **2017**, *1*, 32. [CrossRef]
16. Luo, X.; Bennett, R.M.; Koeva, M.; Lemmen, C. Investigating Semi-Automated Cadastral Boundaries Extraction from Airborne Laser Scanned Data. *Land* **2017**, *6*, 60. [CrossRef]
17. Nyandwi, E.; Koeva, M.; Kohli, D.; Bennett, R. Comparing Human Versus Machine-Driven Cadastral Boundary Feature Extraction. *Remote Sens.* **2019**, *11*, 1662. [CrossRef]
18. Wassie, Y.A.; Koeva, M.N.; Bennett, R.M.; Lemmen, C.H.J. A procedure for semi-automated cadastral boundary feature extraction from high-resolution satellite imagery. *J. Spat. Sci.* **2018**, *63*, 75–92. [CrossRef]
19. Xia, X.; Persello, C.; Koeva, M. Deep Fully Convolutional Networks for Cadastral Boundary Detection from UAV Images. *Remote Sens.* **2019**, *11*, 1725. [CrossRef]
20. Lambin, E.F.; Turner, B.L.; Geist, H.J.; Agbola, S.B.; Angelsen, A.; Bruce, J.W.; Coomes, O.T.; Dirzo, R.; Fischer, G.; Folke, C.; et al. The causes of land-use and land-cover change: Moving beyond the myths. *Glob. Environ. Chang.* **2001**, *11*, 261–269. [CrossRef]
21. Cumming, G.S.; Barnes, G. Characterizing land tenure dynamics by comparing spatial and temporal variation at multiple scales. *Landsc. Urban Plan.* **2007**, *83*, 219–227. [CrossRef]
22. Donnelly, S.; Evans, T.P. Characterizing spatial patterns of land ownership at the parcel level in south-central Indiana, 1928–1997. *Landsc. Urban Plan.* **2008**, *84*, 230–240. [CrossRef]
23. National Research Council. *People and Pixels: Linking Remote Sensing and Social Science*; The National Academies Press: Washington, DC, USA, 1998. [CrossRef]
24. Liu, X.; He, J.; Yao, Y.; Zhang, J.; Liang, H.; Wang, H.; Hong, Y. Classifying urban land use by integrating remote sensing and social media data. *Int. J. Geogr. Inf. Sci.* **2017**, *31*, 1675–1696. [CrossRef]
25. Bennett, R.; Kitchingman, A.; Leach, J. On the nature and utility of natural boundaries for land and marine administration. *Land Use Policy* **2010**, *27*, 772–779. [CrossRef]
26. Whittal, J. A new conceptual model for the continuum of land rights. *S. Afr. J. Geomat.* **2014**, *3*, 13–32.
27. Herrera, M. Land Tenure Data and Policy Making in Latin America. In Proceedings of the Conclusive Expert Meeting on FAO Normative Work on Land Tenure Data, Rome, Italy, 22–23 September 2005.
28. Bratasanu, D.; Nedelcu, I.; Datcu, M. Bridging the Semantic Gap for Satellite Image Annotation and Automatic Mapping Applications. *IEEE J. Sel. Top. Appl. Earth Obs. Remote Sens.* **2011**, *4*, 193–204. [CrossRef]
29. Yang, D.; Fu, C.-S.; Smith, A.C.; Yu, Q. Open land-use map: A regional land-use mapping strategy for incorporating OpenStreetMap with earth observations. *Geospat. Inf. Sci.* **2017**, *20*, 269–281. [CrossRef]
30. Henssen, J. Basic Principles of the Main Cadastral Systems in the World, Modern Cadastres and Cadastral Innovations. In Proceedings of the One Day Seminar Held During the Annual Meeting of FIG Commission 7, Cadastre and Rural Land Management, International Federation of Surveyors (FIG), Delft, The Netherlands, 16 May 1995; pp. 5–12.
31. Zevenbergen, J.A.; Frank, A.; Stubkjaer, E. *Real Property Transactions: Procedures, Transaction Costs and Models*; IOS Press: Amsterdam, The Netherlands, 2007.
32. Van der Molen, P. The Future Cadastres—Cadastres After 2014. In Proceedings of the FIG Working Week 2003, Paris, France, 13–17 April 2003.
33. Lemmen, C.; van Oosterom, P.; Bennett, R. The Land Administration Domain Model. *Land Use Policy* **2015**, *49*, 535–545. [CrossRef]

34. Lemmen, C.H.J. A Domain Model for Land Administration. Ph.D. Thesis, Delft University of Technology, Delft, The Netherlands, 2012. Available online: http://www.itc.nl/library/papers_2012/phd/lemmen.pdf (accessed on 29 March 2019).
35. UN-Habitat. *Secure Land Rights for All*; UN Habitat: Nairobi, Kenya, 2008.
36. Enemark, S.; Bell, K.C.; Lemmen, C.; McLaren, R. *Fit-For-Purpose Land Administration*; International Federation of Surveyors (FIG): Copenhagen, Denmark, 2014.
37. Zevenbergen, J.; De Vries, W.; Bennett, R.M. *Advances in Responsible Land Administration*; CRC Press: Boca Raton, FL, USA, 2015.
38. Ma, L.; Liu, Y.; Zhang, X.; Ye, Y.; Yin, G.; Johnson, B.A. Deep learning in remote sensing applications: A meta-analysis and review. *ISPRS J. Photogramm. Remote Sens.* **2019**, *152*, 166–177. [CrossRef]
39. Levin, N.; Ali, S.; Crandall, D. Utilizing remote sensing and big data to quantify conflict intensity: The Arab Spring as a case study. *Appl. Geogr.* **2018**, *94*, 1–17. [CrossRef]
40. Schoepfer, E.; Spröhnle, K.; Kranz, O.; Blaes, X.; Kolomaznik, J.; Hilgert, F.; Bartalos, T.; Kemper, T. Towards a multi-scale approach for an Earth observation-based assessment of natural resource exploitation in conflict regions. *Geocarto Int.* **2017**, *32*, 1139–1158. [CrossRef]
41. Gbanie, S.; Griffin, A.; Thornton, A. Impacts on the urban environment: Land cover change trajectories and landscape fragmentation in Post-War Western Area, Sierra Leone. *Remote Sens.* **2018**, *10*, 129. [CrossRef]
42. World Bank. Why Are Some Data Not Available? Available online: https://datahelpdesk.worldbank.org/knowledgebase/articles/191133-why-are-some-data-not-available (accessed on 21 September 2019).
43. Lee, C.; de Vries, W.T. A divided nation: Rethinking and rescaling land tenure in the Korean (re-)unification. *Land Use Policy* **2018**, *75*, 127–136. [CrossRef]
44. Lee, C.; de Vries, W.T.; Chigbu, U.E. Land Governance Re-Arrangements: The One-Country One-System (OCOS) Versus One-Country Two-System (OCTS) Approach. *Adm. Sci.* **2019**, *9*, 21. [CrossRef]
45. Gyeon, Y.; Lee, S. North Korea Forum Seminar: Collection and Management of North Korea Data and Statistics. *Plan. Policy* **2006**, *292*, 102–107.
46. United Nations. UN Entities in DPR Korea. Available online: https://dprkorea.un.org/about/un-entities-in-country (accessed on 23 September 2019).
47. He, J.; Xu, J. Is there decentralization in North Korea? Evidence and lessons from the sloping land management program 2004–2014. *Land Use Policy* **2017**, *61*, 113–125. [CrossRef]
48. Kim, N.; Lee, J.; Lee, H.; Seo, J. Accurate segmentation of land regions in historical cadastral maps. *J. Vis. Commun. Image Represent.* **2014**, *25*, 1262–1274. [CrossRef]
49. Lee, H.; Lee, S.; Kim, N.; Seo, J. JigsawMap: Connecting the past to the future by mapping historical textual cadasters. In Proceedings of the SIGCHI Conference on Human Factors in Computing Systems, Austin, TX, USA, 5–10 May 2012; pp. 463–472.
50. Ko, Y.; Yun, H. Analysis of Land Use in North Korea Using the Original of Cadastral Map. *J. Korean Soc. Cadastre* **2016**, *32*, 49–57.
51. Yu, J.; Lim, J.; Lee, K.-S. Investigation of drought-vulnerable regions in North Korea using remote sensing and cloud computing climate data. *Environ. Monit. Assess.* **2018**, *190*, 126. [CrossRef]
52. Um, D.-B.; Um, J.-S. Informed consent utilizing satellite imagery in forestry carbon trading with North Korea. *Int. Environ. Agreem. Politics Law Econ.* **2017**, *17*, 531–552. [CrossRef]
53. Lim, J.; Lee, K.-S. Flood Mapping Using Multi-Source Remotely Sensed Data and Logistic Regression in the Heterogeneous Mountainous Regions in North Korea. *Remote Sens.* **2018**, *10*, 1036. [CrossRef]
54. Lim, J.; Lee, K.-S. Investigating flood susceptible areas in inaccessible regions using remote sensing and geographic information systems. *Environ. Monit. Assess.* **2017**, *189*, 96. [CrossRef]
55. Ernst, M.; Jurowetzki, R. Satellite Data, Women Defectors and Black Markets in North Korea: A Quantitative Study of the North Korean Informal Sector Using Night-Time Lights Satellite Imagery. *North Korean Rev.* **2016**, *12*, 64–83.
56. Lim, H.-T.; Yang, S.-C.; Lee, S.-H.; Shin, D.-B.; Ahn, J.-W. *An Experimental Project on Land Information Establishment in North Korea Regarding Rapid Unification*; LX Spaital Information Research Institute: Seoul, Korea, 2015.
57. Lee, B.-M.; Kim, T.-J.; Kwak, B.-Y.; Lee, Y.-H.; Choi, J. Improvement of the Korean LADM country profile to build a 3D cadastre model. *Land Use Policy* **2015**, *49*, 660–667. [CrossRef]

58. Choi, J.-H.; Kim, B.-J.; Shim, J.-W. *A Study on Improvement of Land Characteristic Survey*; KAB Real Estate Research & Development Institute: Daegu, Korea, 2015.
59. LX Spatial Information Research Institute. *Study on Efficient Implementation of Cadastral Resurvey Project—Improvement of Land Category System by Land Use*; Ministry of Land, Transport and Maritime Affairs: Gwacheon, Korea, 2012.
60. Jensen, J.R.; Cowen, D.C. Remote sensing of urban suburban infrastructure and socio-economic attributes. *Photogramm. Eng. Remote Sens.* **1999**, *65*, 611–622.
61. Ramankutty, N.; Graumlich, L.; Achard, F.; Alves, D.; Chhabra, A.; DeFries, R.S.; Foley, J.A.; Geist, H.; Houghton, R.A.; Goldewijk, K.K.; et al. Global Land-Cover Change: Recent Progress, Remaining Challenges. In *Land-Use and Land-Cover Change: Local Processes and Global Impacts*; Lambin, E.F., Geist, H., Eds.; Springer: Berlin/Heidelberg, Germany, 2006; pp. 9–39. [CrossRef]
62. Wickham, J.; Stehman, S.V.; Homer, C.G. Spatial patterns of the United States National Land Cover Dataset (NLCD) land-cover change thematic accuracy (2001–2011). *Int. J. Remote Sens.* **2018**, *39*, 1729–1743. [CrossRef] [PubMed]
63. Lymburner, L.; Tan, P.; Mueller, N.; Thackway, R.; Lewis, A.; Thankappan, M.; Randall, L.; Islam, A.; Senarath, U. *The national dynamic land cover dataset - Technical Report*; Record 2011/031; Geoscience Australia: Canberra, Australia, 2011. Available online: http://pid.geoscience.gov.au/dataset/ga/71069 (accessed on 5 April 2019).
64. Büttner, G. CORINE Land Cover and Land Cover Change Products. In *Land Use and Land Cover Mapping in Europe: Practices & Trends*; Manakos, I., Braun, M., Eds.; Springer: Dordrecht, The Netherlands, 2014; pp. 55–74. [CrossRef]
65. Zhang, Z.; Wang, X.; Zhao, X.; Liu, B.; Yi, L.; Zuo, L.; Wen, Q.; Liu, F.; Xu, J.; Hu, S. A 2010 update of National Land Use/Cover Database of China at 1:100,000 scale using medium spatial resolution satellite images. *Remote Sens. Environ.* **2014**, *149*, 142–154. [CrossRef]
66. Bryman, A. *Social Research Methods*; Oxford University Press: Oxford, UK, 2016.
67. Fielding, N.G. Triangulation and Mixed Methods Designs: Data Integration with New Research Technologies. *J. Mixed Methods Res.* **2012**, *6*, 124–136. [CrossRef]
68. Larsson, G. *Land Management: Public Policy, Control and Participation*; Coronet Books: London, UK, 1997.
69. Payne, G. Land tenure and property rights: An introduction. *Habitat Int.* **2004**, *28*, 167–179. [CrossRef]
70. Fox, J.; Rindfuss, R.R.; Walsh, S.J.; Mishra, V. *People and the Environment: Approaches for Linking Household and Community Surveys to Remote Sensing and GIS*; Springer Science & Business Media: Berlin, Germany, 2003; Volume 1.
71. Ghaffarian, S.; Kerle, N.; Filatova, T. Remote Sensing-Based Proxies for Urban Disaster Risk Management and Resilience: A Review. *Remote Sens.* **2018**, *10*, 1760. [CrossRef]
72. Ghazouani, F.; Farah, I.R.; Solaiman, B. Semantic remote sensing scenes interpretation. In *Ontology in Information Science*; IntechOpen: London, UK, 2018.
73. Bosse, E.; Roy, J.; Wark, S. *Concepts, Models, and Tools for Information Fusion*; Artech House Inc.: Norwood, MA, USA, 2007.
74. Altaweel, M.R.; Alessa, L.N.; Kliskey, A.D.; Bone, C.E. Monitoring Land Use: Capturing Change through an Information Fusion Approach. *Sustainability* **2010**, *2*, 1182–1203. [CrossRef]
75. Bossé, É.; Solaiman, B. *Information Fusion and Analytics for Big Data and IoT*; Artech House: Norwood, MA, USA, 2016.
76. Kim, Y.-H. Changes in Agricultural Farms in North Korea—Focusing on Cooperative Farms. *Q. Agric. Trends North Korea* **2010**, *12*, 3–17, UCI: I410-ECN-0102-2018-500-004168983.
77. Kosztra, B.; Büttner, G.; Hazeu, G.; Arnold, S. *Updated CLC Illustrated Nomenclature Guidelines*; European Environmental Agency: Copenhagen, Denmark, 2017.
78. Choi, W.; Kang, S.; Choi, J.; Larsen, J.J.; Oh, C.; Na, Y.-G. Characteristics of deforestation in the Democratic People's Republic of Korea (North Korea) between the 1980s and 2000s. *Reg. Environ. Chang.* **2017**, *17*, 379–388. [CrossRef]
79. Kim, Y.-H.; Kwon, T.-J.; Lim, S.-K.; Choi, S.-K. *A Study on North Korea's Agricultural Reform Measures after the Economic Crisis*; Korea Rural Economic Institute: Jeollanam-do, Korea, 2013; pp. 1–146.
80. Kim, C.S. *DPR Korea 2008 Population Census National Report*; Central Bureau of Statistics: Pyongyang, DPR Korea, 2009.
81. Choi, S.-H.; Choi, D.-S.; Lee, J.-K.; Hong, S.-W. *North Korea Housing Survey and Analysis*; Land and Housing Institute: Deajeon, Korea, 2015.

82. Reenberg, A.; Lund, C. Land Use and Land Right Dynamics—Determinants for Resource Management Options in Eastern Burkina Faso. *Hum. Ecol.* **1998**, *26*, 599–620. [CrossRef]
83. Hettig, E.; Lay, J.; Sipangule, K. Drivers of Households' Land-Use Decisions: A Critical Review of Micro-Level Studies in Tropical Regions. *Land* **2016**, *5*, 32. [CrossRef]
84. Lim, S.H. *Coexistence of Planning and Markets*; Samsung Economic Research Institute: Seoul, Korea, 2008.
85. Stoter, J.E.; van Oosterom, P. *3D Cadastre in an International Context: Legal, Organizational, and Technological Aspects*; CRC Press: Boca Raton, FL, USA, 2006.
86. Li, J.; Huang, X.; Gong, J. Deep neural network for remote-sensing image interpretation: Status and perspectives. *Natl. Sci. Rev.* **2019**. [CrossRef]

remote sensing

MDPI

Article

Application of Deep Learning for Delineation of Visible Cadastral Boundaries from Remote Sensing Imagery

Sophie Crommelinck *, Mila Koeva, Michael Ying Yang and George Vosselman

Faculty of Geo-Information Science and Earth Observation (ITC), University of Twente, Hengelosestraat 99, 7514 AE Enschede, The Netherlands; m.n.koeva@utwente.nl (M.K.); michael.yang@utwente.nl (M.Y.Y.); george.vosselman@utwente.nl (G.V.)

* Correspondence: s.crommelinck@utwente.nl; Tel.: +31-53-489-5524

Received: 9 October 2019; Accepted: 23 October 2019; Published: 25 October 2019

Abstract: Cadastral boundaries are often demarcated by objects that are visible in remote sensing imagery. Indirect surveying relies on the delineation of visible parcel boundaries from such images. Despite advances in automated detection and localization of objects from images, indirect surveying is rarely automated and relies on manual on-screen delineation. We have previously introduced a boundary delineation workflow, comprising image segmentation, boundary classification and interactive delineation that we applied on Unmanned Aerial Vehicle (UAV) data to delineate roads. In this study, we improve each of these steps. For image segmentation, we remove the need to reduce the image resolution and we limit over-segmentation by reducing the number of segment lines by 80% through filtering. For boundary classification, we show how Convolutional Neural Networks (CNN) can be used for boundary line classification, thereby eliminating the previous need for Random Forest (RF) feature generation and thus achieving 71% accuracy. For interactive delineation, we develop additional and more intuitive delineation functionalities that cover more application cases. We test our approach on more varied and larger data sets by applying it to UAV and aerial imagery of 0.02–0.25 m resolution from Kenya, Rwanda and Ethiopia. We show that it is more effective in terms of clicks and time compared to manual delineation for parcels surrounded by visible boundaries. Strongest advantages are obtained for rural scenes delineated from aerial imagery, where the delineation effort per parcel requires 38% less time and 80% fewer clicks compared to manual delineation.

Keywords: cadastral mapping; indirect surveying; boundary extraction; boundary delineation; machine learning; deep learning; image analysis; CNN; RF

1. Introduction

Recording land rights provides land owners tenure security, a sustainable livelihood and increases financial opportunities. Estimates suggest that about 75% of the world population does not have access to a formal system to register and safeguard their land rights [1]. This lack of recorded land rights increases insecure land tenure and fosters existence-threatening conflicts, particularly in developing countries. Recording land rights spatially, i.e., cadastral mapping, is considered the most expensive part of a land administration system [2]. Recent developments in technology allow us to rethink contemporary cadastral mapping. The aim of this study is to make use of technological developments to create automated and thus more efficient approaches for cadastral mapping.

Automated cadastral boundary delineation based on remote sensing data has been rarely investigated, even though physical objects, which can be extracted using image analysis, often define visible cadastral boundaries [3].

Visible cadastral boundaries demarcated by physical objects such as hedges, fences, walls, bushes, roads, or rivers, can make up a large portion of all cadastral boundaries [4,5]. Indirect surveying relies on the delineation of such visible parcel boundaries from remote sensing imagery.

Automatically extracting visible cadastral boundaries combined with (legal) adjudication and an incorporation of local knowledge from human operators offers the potential to improve current cadastral mapping approaches in terms of time, cost, accuracy and acceptance [6]. High-resolution remote sensing imagery, increasingly captured with Unmanned Aerial Vehicles (UAVs) in land administration [7–11], provides the basis for such a semi-automated delineation of visible boundaries.

1.1. CNN Deep Learning for Cadastral Mapping

CNNs are one of the most popular and successful deep networks for image interpretation tasks. They are proven to work efficiently to identify various objects in remote sensing imagery [12–16]. Comprehensive overviews contextualizing the evolution of deep learning and CNNs in geoscience and remote sensing are provided by Bergen et al. and Zhu et al. [17,18]. In essence, CNNs are neural networks that incorporate the convolution and pooling operation as a layer. CNNs have been characterized by five concepts [19]:

- Convolution operation increases the network's simplicity, which makes training more efficient.
- Representation learning through filters requires the user to engineer the architecture rather than the features.
- Location invariance through pooling layers allows filters to detect features dissociated from their location.
- Hierarchy of layers allows the learning of abstract concepts based on simpler concepts.
- Feature extraction and classification are included in training, which eliminates the traditional machine learning need for hand-crafted features, and distinguishes CNN as a deep learning approach.

In deep learning, there are two approaches to train a CNN: From scratch or via transfer learning [20]. When trained from scratch, all features are learned from data to be provided, which demands large amounts of data and comes with a higher risk of overfitting. An over-fitted network can make accurate predictions for a certain dataset, but fails to generalize its learning capacity for another dataset. With transfer learning, part of the features are learned from a different, typically large dataset. These low-level features are more general and abstract. The network has proven excellence for a specific application. Its core architecture is kept and applied to a new application. Only the last convolution block is trained on specific data of the new application, resulting in specialized high-level features. Transfer learning requires learning fewer features, and thus fewer data. In our study, we investigate transfer learning as an existing CNN for cadastral mapping.

1.2. Study Objective

The main goal of our research is to develop an approach that simplifies image-based cadastral mapping to support the automated mapping of land tenure. We pursue this goal by developing an automated cadastral boundary delineation approach applicable to remote sensing imagery. In this study, we describe our approach in detail, optimize its components, apply it to more varied and larger remote sensing imagery from Kenya, Rwanda and Ethiopia, test its applicability to cadastral mapping, and assess its effectiveness compared to manual delineation.

We previously proposed a semi-automated indirect surveying approach for cadastral mapping from remote sensing data. To delineate roads from UAV imagery with that workflow, the number of clicks per 100 m compared to manual delineation was reduced by up to 86%, while obtaining a similar localization quality [21]. The workflow consists of: (i) Image segmentation to extract visible object outlines, (ii) boundary classification to predict boundary likelihoods for extracted segment lines, and (iii) interactive delineation to connect these lines based on the predicted boundary likelihood. In this study, we investigate improvements in all three steps (Figure 1). First, for step (i), we filter out small segments to reduce over-segmentation. Second, for step (ii), we replace hand-crafted features and line classification based on Random Forest (RF) by Convolutional Neural Networks (CNNs).

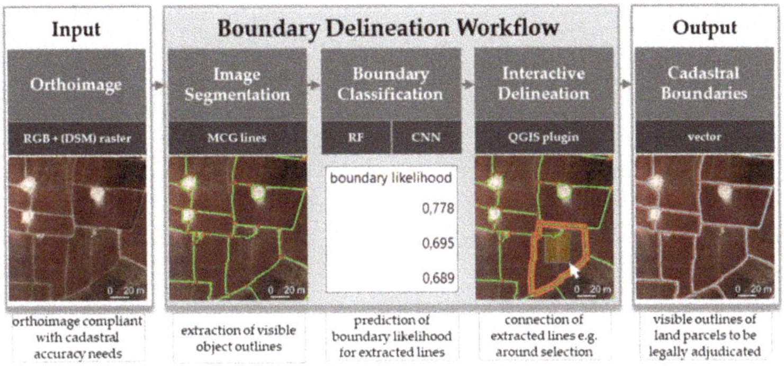

Figure 1. Boundary Delineation workflow proposed to improve indirect surveying. This study optimizes image segmentation, questions whether Random Forest (RF) or Convolutional Neural Networks (CNNs) are better suited to derive boundary likelihoods for visible object outlines, and introduces additional functionalities for the interactive delineation.

Finally, for step (iii), we introduce more intuitive and comprehensive delineation functionalities. While we tested the previous workflow on two UAV ortho-images in Germany, and delineated road outlines, we now test our workflow on imagery covering much larger extents and compare the results to cadastral boundaries. In this study, we test our workflow on UAV and aerial imagery of 0.02–0.25 m resolution from Kenya, Rwanda and Ethiopia covering 722 visible parcels.

Our new functionalities for the interactive delineation address cases for which the boundary classification fails, or is not necessary. Boundary classification comes into play in cases of over-segmentation, when many object outlines exist. Then, the delineator has to choose which lines demarcate the cadastral boundary. Support comes from the lines' boundary likelihood predicted by RF or CNN. In this study, we introduce functionalities that allow connecting image segmentation lines to cadastral boundaries, regardless of their boundary likelihood.

2. Materials and Methods

2.1. Image Data

An aerial image of 0.25 m Ground Sample Distance (GSD) of a rural scene in Ethiopia is used (Figure 2a,b). The local agricultural practice consists mostly of smallholder farming. The image was captured during the dry season around March. The crops within our study area consist mostly of millet, corn, and a local grain called teff. Since the crops are in the beginning of an agricultural cycle, they do not cover the visible cover parcel boundaries. The cadastral reference data cover 33 km^2 containing 9,454 plots with a median size of 2500 m^2. The cadastral reference data is derived through on-screen manual delineation from the aerial image. In case of uncertainty or invisible boundaries, the boundary is captured together with land owners in the field using high-resolution total stations. For a later assessment, in which we compare our approach to on-screen manual delineation, additional Unmanned Aerial Vehicle (UAV) data from Kenya and Rwanda is used (Figure 2c,d). The UAV data in Rwanda have a GSD of 0.02 m, and were captured with a rotary-wing Inspire 2 (SZ DJI Technology Co., Ltd., Shenzhen, Guangdong, People's Republic of China). The UAV data in Kenya have a GSD of 0.06 and were captured with a fixed-wing DT18 (Delair-Tech, Delair, Toulouse, France).

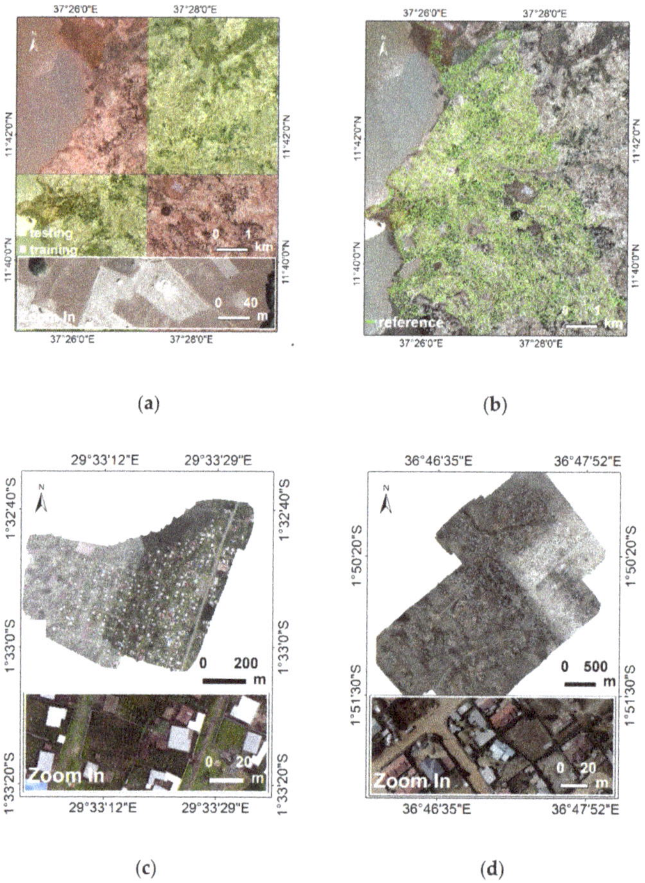

Figure 2. (**a**) Aerial image of 0.25 m Ground Sample Distance (GSD) for a rural scene in Ethiopia, divided into areas for training and testing our approach before comparing results to (**b**) the cadastral reference. Unmanned Aerial Vehicle (UAV) images for peri-urban scenes in (**c**) Rwanda (0.02 m GSD), and (**d**) Kenya (0.06 m GSD) to compare automated to manual delineation.

2.2. Boundary Mapping Approach

The boundary mapping approach refers to the one we previously described [22]. In the following, modifications and the data-dependent implementation of the three workflow steps are described. The source code is publically available [23].

Image segmentation is based on Multiresolution Combinatorial Grouping (MCG) [24], which delivers closed contours, capturing the outlines of visible objects. To run the original MCG implementation, the Ethiopian aerial image is tiled to 20 tiles of 8000 × 8000 pixels. The parameter k regulating over- and under-segmentation is set to produce over-segmentation (k = 0.1). This setting creates outlines around the majority of visible objects. Tests with parameters (k = 0.3 and k = 0.5) resulting in less over-segmentation show that visible object outlines are partly missed, while irrelevant lines around small objects are still produced. To reduce the number of irrelevant lines produced through over-segmentation, the lines are simplified through filtering (Figure 3): Lines around areas smaller than 30 m^2 are merged to the neighboring segments, which reduces the line count by 80% to 600,000 lines.

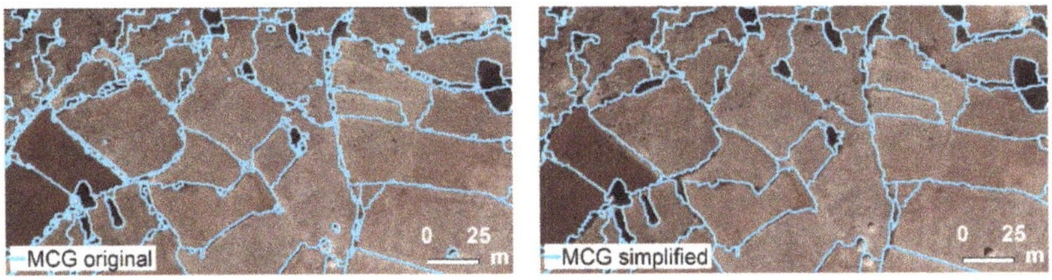

(a) (b)

Figure 3. Multiresolution Combinatorial Grouping (MCG) image segmentation lines around visible objects (**a**) before and (**b**) after simplification reducing the line count by 80%.

According to our visual inspections, this post-processing removes artefacts in the segmentation results and keeps outlines of large objects being more relevant for cadastral mapping. For the high-resolution data from Rwanda and Kenya, we proceed similarly by tiling the data and setting k = 0.4 and k = 0.3, respectively.

Boundary classification is applied to the post-processed 600,000 MCG lines. We investigate two machine learning approaches to derive the boundary likelihood per MCG line: Random Forest (RF) and Convolutional Neural Networks (CNN). Both require the labeling of training data as 'boundary' and 'not boundary'. The training data for RF consist of lines, that for CNN of image tiles. For both approaches, the cadastral reference is buffered by a radius of 0.4 m. This size accounts for inaccuracies in the cadastral reference and the ortho-image, enlarges the number of 'boundary' samples, and is identical to the one applied to derive hand-crafted RF features. For both approaches, the ratio between training and testing data is set to 50%. The number of 'boundary' and 'not boundary' training samples is balanced to 1:1 by randomly under-sampling 'not boundary' tiles (Table 1). The areas for training and testing are randomly selected and large to minimize the number of lines at the borders of each area that are clipped and of limited use for further analysis (Figure 2). The boundary likelihood predicted by both approaches represents the probability ($\hat{y}$) of a line being 'boundary':

$$boundary\ likelihood\ [0;1] = \hat{y}_{boundary} \quad (1)$$

Table 1. Distribution of training and testing data for boundary classification based on Random Forest (RF) and Convolutional Neural Networks (CNN).

	RF Classification		**CNN Classification**	
	Number of Lines		**Number of Tiles**	
Label	**Training**	**Testing**	**Training**	**Testing**
'boundary'	12,280 (50%)	9,742 (3%)	35,643 (50%)	34,721 (4%)
'not boundary'	12,280 (50%)	280,108 (97%)	34,665 (50%)	746,349 (96%)
$\sum$	24,560	289,850	70,308	781,070

RF classification is applied as shown in Figure 4 [22]. Instead of manually labeling lines for training, a line is now automatically labeled as 'boundary' when it overlaps with the cadastral reference buffer of 0.4 m by more than 50%. This value aligns with the threshold at which a CNN-derived result is labeled as 'boundary' or 'not boundary'. Since no DSM information is available for the study area, the feature dsm grad is not calculated.

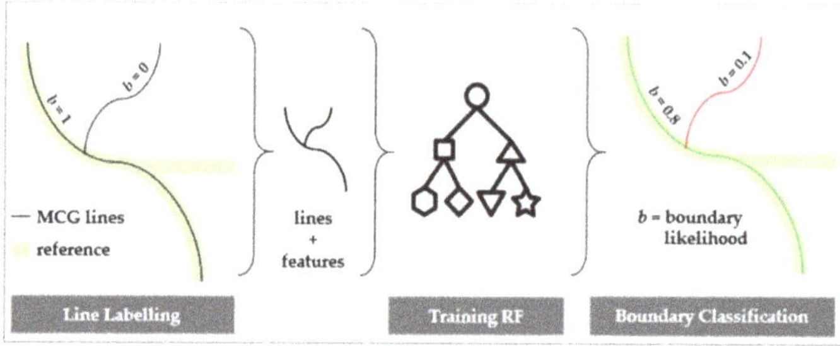

Figure 4. Boundary line classification based on Random Forest (RF) to derive boundary likelihoods for MCG lines.

CNN classification is investigated by training state-of-the-art tile-based CNNs (Figure 5). We reformulate our problem of generating boundary likelihoods for MCG lines to be solvable by a tile-based CNN as follows: At first, image tiles of 224 × 224 pixels centered on an MCG line are cropped from the ortho-image. 224 × 224 × 3 is the standard size of images required by the used CNN. A tile is labeled as 'boundary' if the center pixel covering an MCG line overlaps with the cadastral reference buffer. A tile is created every 5 m along an MCG line. Decreasing this distance would increase the overlap, and thus the redundancy, of the image content per tile. Increasing this distance would reduce the number of tiles and thus the number of training data. With these settings, we generate 1.5 million tiles surrounding MCG pixels of which 5% are labeled as 'boundary' and 95% as 'not boundary'. After training, the CNN predicts boundary likelihoods for unseen testing areas (Figure 2a). The likelihoods of all tiles per MCG line are averaged based on the 97th percentile. This value aligns with the distribution of 'boundary' and 'not boundary' lines in the training data (Table 1). We use a pre-trained CNN architecture. We apply transfer learning by adding additional trainable layers: A global spatial average pooling layer, a fully connected layer with rectified linear unit (ReLU) activation, a dropout layer and a logistic layer with softmax activation. Only these last layers are trainable. We investigate using different pre-trained CNN architectures, including the Visual Geometry Group (VGG) [25], ResNet [26], Inception [27], Xception [28], MobileNet [29] and DenseNet [30], as well as different hyper-parameter settings on the learning optimizer, the depth of the fully connected layer and the dropout rate.

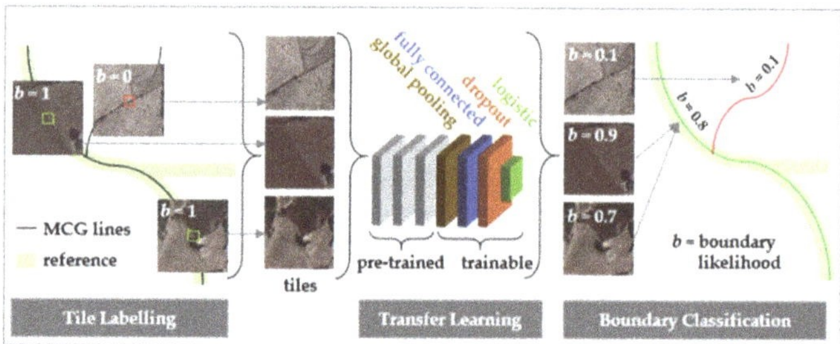

Figure 5. Boundary line classification based on Convolutional Neural Networks (CNNs) to derive boundary likelihoods for MCG lines.

Interactive delineation supports the creation of final cadastral boundaries. In comparison to our previous study [22], we now include more functionalities to delineate parcels (Table 2) and redesigned the Graphical User Interface (GUI). The interactive delineation is implemented in the open source geographic information system QGIS [31] as BoundaryDelineation plugin [32]:

Table 2. Delineation functionalities of BoundaryDelineation QGIS plugin.

Functionality	Description
Connect around selection	Connect lines surrounding a click or selection of lines (Figure 6a,b)
Connect lines' endpoints	Connect endpoints of selected lines to a polygon, regardless of MCG lines (Figure 6c)
Connect along optimal path	Connect vertices along least-cost-path based on a selected attribute, e.g., boundary likelihood (Figure 6d)
Connect manual clicks	Manual delineation with the option to snap to input lines and vertices
Update edits	Update input lines based on manual edits
Polygonize results	Convert created boundary lines to polygons

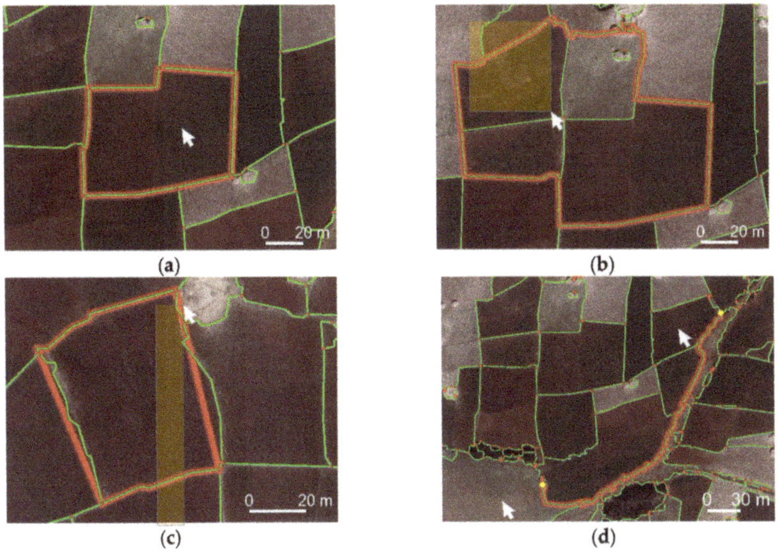

Figure 6. Interactive delineation functionalities: (**a**) Connect lines surrounding a click, or (**b**) a selection of lines. (**c**) Close endpoints of selected lines to a polygon. (**d**) Connect lines along least-cost-path.

2.3. Accuracy Assessment

The accuracy assessment investigates multiple aspects of our workflow, each requiring a different analysis:

CNN Architecture: This analysis aims to optimize the CNN architecture by considering loss and accuracy for training and validation data per epoch. The curves for training loss and validation loss, as well as for training accuracy and validation accuracy, are expected to converge with incremental epochs. Loss is the summation of errors made for each example in training, and should be minimized. We use cross-entropy loss that increases as the predicted probability ($\hat{y}_i$) diverges from the actual label (y_i):

$$cross - entropy\ loss = -(y_i \log(\hat{y}_i) + (1 - y_i) \log(1 - \hat{y}_i)) \tag{2}$$

All predictions < 0.5 are considered as 'not boundary', those ≥ 0.5 as 'boundary'. This results in a confusion matrix showing the number of tiles being False Positive (FP), True Positive (TP), False Negative (FN) and True Negative (TN). From this matrix, the accuracy is derived as the sum of correctly classified tiles divided by all tiles:

$$accuracy\ [0; 1] = \frac{\mathrm{TP} + \mathrm{TN}}{\mathrm{TP} + \mathrm{FP} + \mathrm{FN} + \mathrm{TN}} \tag{3}$$

RF vs. CNN Classification: This analysis compares the boundary likelihood obtained through RF and CNN to the percentage to which an MCG line overlaps with the cadastral reference. Both are buffered with a radius of 0.4 m. The area of their overlap in relation to the entire MCG buffer area represents the percentage of overlap:

$$overlap\ [0;1] = \frac{area_{MCG-buffer} \cap area_{cadastral-buffer}}{area_{MCG-buffer}} \quad (4)$$

We investigate whether lines that should get a boundary likelihood > 0, i.e., those that fall within the cadastral reference buffer, are assigned a boundary likelihood > 0:

$$recall\ [0;1] = \frac{\mathrm{TP}}{\mathrm{TP} + \mathrm{FN}} \quad (5)$$

Then, we check whether the assigned boundary likelihood is valid, i.e., whether it is equal to the line's overlap with the cadastral reference buffer. This is indicated by the precision that captures the ratio of lines having a boundary likelihood that aligns with overlap to the sum of lines having a correct or too positive boundary likelihood:

$$precision\ [0;1] = \frac{\mathrm{TP}}{\mathrm{TP} + \mathrm{FP}} \quad (6)$$

Since the boundary likelihood captures the probability of a line being a 'boundary' line, a high boundary likelihood should go along with a high overlap between the MCG and cadastral reference buffer:

$$overlap\ [0;1] boundary\ likelihood\ [0;1] \quad (7)$$

Both values are not expected to be identical, which can be influenced by altering the buffer size. Our focus is on comparing RF to CNN, and secondarily on the boundary likelihood itself. Results are considered only in areas for testing in which we have cadastral reference data (Figure 2).

Manual vs. Automated Delineation: This analysis compares the time and number of clicks required to delineate visible boundaries, once manually, and once with the automated approach. Manual delineation refers to delineating parcels based on the ortho-image without further guidance. Automated delineation refers to our approach, including RF or CNN classification depending on which approach shows superior results in this study. All delineations should fall within the cadastral reference buffer of the 0.4 m radius. The buffer size represents the local accepted accuracy for cadastral delineation and falls within the 2.4 m proposed for rural areas by the International Association of Assessing Officers (IAAO) [33].

The comparison is conducted for a rural area in Ethiopia and two peri-urban areas in Rwanda and Kenya (Figure 2). No urban area is selected, as indirect surveying relies on the existence of visible boundaries, which are rare in densely populated areas. Furthermore, indirect surveying in urban areas saves less logistics for field surveys, due to smaller parcel sizes. Only parcels for which all boundaries are visible, and thus detectable from the ortho-image, are kept for this analysis. Since no digital up-to-date cadastral reference exists for our areas in Kenya and Rwanda, cadastral reference data are created based on local knowledge in alignment with visible boundaries.

3. Results

3.1. CNN Architecture

We first tested different pre-trained base CNNs (VGG, ResNet, Inception, Xception, MobileNet and DenseNet) to which we added trainable layers. The combined CNN model was trained with a batch size of 32 for 100 epochs. In the case of no learning, the training stopped earlier. Out of the 50% of balanced data used for training, we used 10% for validation. These data were not seen by the network, but were used only to calculate loss and accuracy per epoch. These metrics and their curves looked most promising for VGG19 [25]. Then we applied the trained network to the remaining 50% of testing data. VGG19 is a 19 layer deep CNN developed by the Visual Geometry Group (VGG) from the University of Oxford (Oxford, Oxfordshire, UK). It is trained to classify images into 1000 object categories, such as keyboard, mouse pencil and many animals. It has learned high-level features for a wide range of images from ImageNet [34]. ImageNet is a dataset of over 15 million labeled high-resolution images with around 22,000 categories. Compared to other CNNs, VGG has shown to generalize well, compared to more complex and less deep CNN architectures [25].

We used VGG19 layers pre-trained for 20,024,384 parameters as a base model. Next, we modified hyper-parameters for VGG19 on the learning optimizer, the depth of the fully connected layer, and the dropout rate to optimize accuracy and loss. We used softmax as an activation function to retrieve predictions for tiles being 'not boundary' in the range [0, 1]. These values represent the weights for the later least-cost-path calculation. Sigmoid activation, which is a type of softmax for a binary classification problem, provided similar results in terms of accuracy and loss. However, it required more post-processing, as the resulting value in the range [0, 1] cannot be understood as described for softmax activation.

The aim was to maximize the accuracy for training and validation data, while minimizing loss. To avoid over-fitting, the curves for training and validation accuracy should not diverge, which was achieved by increasing the dropout rate from 0.5 to 0.8. To avoid under-fitting, the curve for training accuracy should not be below that of the validation accuracy, which was avoided by increasing the depth of the fully connected layer from 16 to 1,024. To avoid oscillations in loss, the learning rate was lowered from 0.01 to 0.001. Learning was stopped once the validation accuracy did not further improve. Results and observations derived from different hyper-parameter settings and different pre-trained base CNNs are provided in the Appendix A (Table A1).

We achieved the best results after training 8,242 parameters on four trainable layers added to 22 pre-trained VGG19 layers (Table 3). This led to a validation accuracy of 71% and a validation loss of 0.598 after 200 epochs (Figure 7). The accuracy could be increased by 1% after 300 epochs, with validation loss restarting to increase to 0.623. We conclude that optimal results are achieved after 200 epochs. 100 epochs halve the training time to 11 hours, whilst obtaining 1% less accuracy and a loss of 0.588. The implementation relies on the open source library Keras [35], and this is publically available [23]. All experiments are conducted on a machine having a NVIDIA GM200 (GeForce GTX TITAN X) GPU with 128 GB RAM (Nvidia Corporation, Santa Clara, CA, US).

Table 3. Settings for our fine-tuned CNN based on Visual Geometry Group 19 (VGG19).

	Settings	Parameters
untrainable layers	VGG19 pre-trained on ImageNet	exclusion of final pooling and fully connected layer
trainable layers	pooling layer	global average pooling 2D
	fully connected layer	Depth = 1024, activation = ReLu
	dropout layer	dropout rate = 0.8
	logistic layer	Activation = softmax
learning optimizer	stochastic gradient descent (SGD) optimizer	learning rate = 0.001 momentum = 0.9 decay = learning rate/epochs
training	shuffled training tiles and un-shuffled validation tiles	Epochs = max. 200 batch size = 32

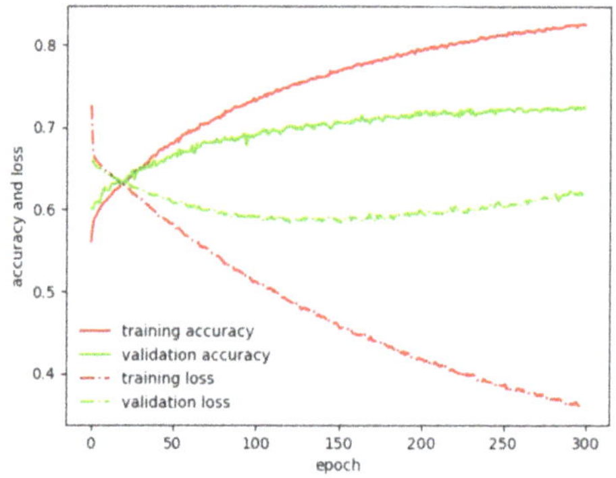

Figure 7. Accuracy and loss for our fine-tuned VGG19.

3.2. *RF vs. CNN Classification*

Of those lines that should get a boundary likelihood > 0, i.e., those that fall within the cadastral reference buffer, 100% for RF and 98% for CNN are assigned a boundary likelihood > 0 (Table 4). This means that both classifiers predict a boundary likelihood in the range [0, 1] when there is some overlap with the cadastral reference buffer.

Table 4. Is the boundary likelihood predicted for the correct lines?

			Overlap			
			0	]0; 1]	Σ	Σ %
boundary likelihood	RF	0	535	265	800	0
		]0; 1]	150,583	59,123	209,706	100
		Σ	151,118	59,388	210,506	
		Σ %	72	28		100
	CNN	0	7,560	1,794	9,354	4
		]0; 1]	145,558	57,594	201,152	96
		Σ	151,118	59,388	210,506	
		Σ %	72	28		100

Next, we looked at how valid the boundary likelihood is, i.e., whether its value is equal to the line's overlap with the cadastral reference buffer. For this we excluded lines having no overlap with the cadastral reference buffer, i.e., those having an overlap = 0. We grouped the remaining lines to compare boundary likelihood and overlap values (Table 5). For RF-derived boundary likelihoods, we obtained an accuracy of 41% and a precision of 49%. For CNN-derived boundary likelihoods, we obtained an accuracy of 52% and a precision of 76%. The percentage of lines per value interval of 0.25 for the same boundary likelihood and overlap value deviated on average by 15% for RF and by 7% for CNN (Table 5).

Overall, CNN-derived boundary likelihoods obtained a similar recall, a higher accuracy, and a higher precision (Table 4). The percentage of lines for different ranges of boundary likelihoods represented the distribution of overlap values more accurately (Table 5). Even though the values of overlap and boundary likelihood do not express the same, they provide a valid comparison between RF- and CNN-derived boundary likelihoods. We consider CNN-derived boundary likelihoods a better input for the interactive delineation, and continue the accuracy assessment for a boundary classification based on CNN.

Table 5. How correct is the predicted boundary likelihood?

			overlap						
			0	]0; 0.25]	]0.25; 0.5]	]0.5; 0.75]	]0.75; 1]	Σ	Σ %
boundary likelihood	RF	]0; 0.25]	151,118	15,176	3,633	481	95	19,385	32
		]0.25; 0.5]		11,553	5633	2178	730	20,094	34
		]0.5; 0.75]		6827	4849	3120	1617	16,413	28
		]0.75; 1]		973	1002	813	708	3496	6
		Σ		34,529	15,117	6592	3150	59,388	
		Σ %	/	58	26	11	5		100
	CNN	]0; 0.25]	151,118	26,546	10,472	4305	1981	43,304	73
		]0.25; 0.5]		5974	3307	1534	765	11,580	19
		]0.5; 0.75]		1751	1177	655	328	3,911	7
		]0.75; 1]		258	161	97	77	593	1
		Σ		34,529	15,117	6591	3151	59,388	
		Σ %	/	58	26	11	5		100

3.3. Manual vs. Automated Delineation

Indirect surveying, comprising of manual or automated delineation, both rely on visible boundaries. Before comparing manual to automated delineation, we filtered the cadastral reference data for Ethiopia (Figure 2b) to contain visible parcels only. We kept only those parcels for which all boundary parts were visually demarcated. As in Kohli et al. [4], we consider only fully closed polygons that are entirely visible in the image. From the original cadastral reference data, we kept 38% of all parcels for which all boundaries were visible. In Kohli et al. [4], the portion of fully visible parcels has been reported to average around 71% of all cadastral parcels in rural Ethiopian areas. We can confirm 71% for parts of our study area that cover smallholder farms. Cadastral data for Rwanda and Kenya were delineated based on local knowledge in alignment with visible boundaries. As for Ethiopia, only fully closed and visible parcels were considered. The mean size of our visible parcels amounts to 2,725 m^2 for Ethiopia, 656 m^2 for Rwanda, and 730 m^2 for Kenya.

When manually delineating visible boundaries, we observed how tiring a task this manual delineation is: The delineator has to continuously scan the image for visible boundaries to then click precisely and repeatedly along the boundary to be delineated. Apart from the visual observation of the ortho-image, the delineator has no further guidance on where to click. Each parcel is delineated the same way, which makes it a highly repetitive task that exhausts the eyes and fingers in no time.

When comparing manual to automated delineation, this impression changes: The delineator now has lines and vertices to choose from, which can be connected automatically using multiple functionalities (Table 2, Figure 6). Complex, as well as simple, parcels require fewer clicking when delineating with the automated approach: To follow a curved outline, manual delineation requires frequent and accurate clicking while zooming in and out. Automated delineation requires clicking on vertices covering the start and endpoint once, before they are automatically connected precisely following object outlines (Figure 6d). Similarly, the automated delineation is superior for simple rectangular parcels: While manual delineation requires accurate clicking on each of the at least four corners of a rectangle, automated delineation allows clicking once somewhere inside the rectangle to retrieve its outline (Figure 8a).

However, choosing the optimal functionality can be time-consuming, especially in cases of fragmented MCG lines obtained from high-resolution UAV data. We assume that the time for automated delineation can be reduced through increased familiarity with all functionalities and by further developing their usability, e.g., by keyboard shortcuts.

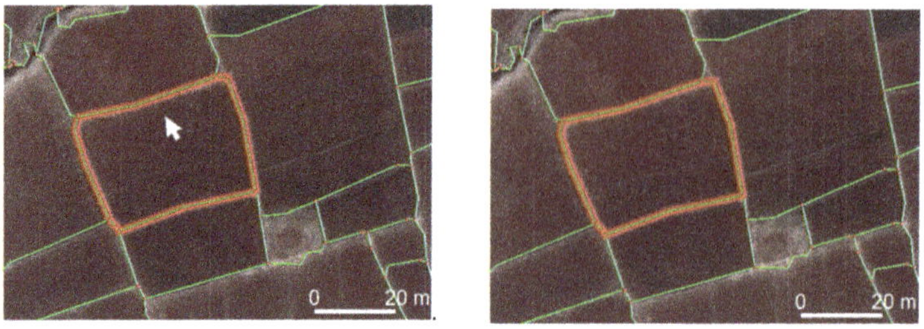

(**a**) (**b**)

Figure 8. (**a**) Automated delineation requires clicking once somewhere in the parcel, while manual delineation requires precise clicking at least four times on each corner. (**b**) Boundaries partly covered or delineated by vegetation impede indirect surveying and limit the effectiveness of our automated delineation compared to manual delineation.

Automated delineation required fewer clicks for our rural and peri-urban study areas (Table 6). Only those parcels for which one of our functionalities was more effective than manual delineation are considered for the automated delineation, amounting to 40–58% of all visible parcels. The effectiveness of manual delineation is considered for all 100% of the visible parcels. By maximizing the number of delineated parcels, we aimed to minimize the effect of unusual parcels that required much effort to delineate manually. We expect the measures that we obtained for the manual delineation to be similar for the 40–58% of parcels considered for the automated delineation. For the remaining parcels, MCG lines were either not available, or not aligning enough with the reference data. Manually delineating these parcels with the plugin requires the same number of clicks and time as conventional manual delineation, but is partly less tiring, as the delineation can be snapped to the MCG lines and vertices.

Table 6. Does automated delineation cost less effort?

	Manual Delineation			Automated Dlineation		
	Parcel Count	$\frac{time}{parcel}$ [s]	$\frac{clicks}{parcel}$	Parcel Count	$\frac{time}{parcel}$ [s]	$\frac{clicks}{parcel}$
Ethiopia (rural)	350	13	10	181 (52%)	8	2
Rwanda (peri-urban)	100	12	7	40 (40%)	25	5
Kenya (peri-urban)	272	11	5	157 (58%)	10	4

Nevertheless, the lines and vertices can also impede the visibility: For our data from Rwanda and Kenya, the boundaries are not continuously visible. The partly vegetation-covered boundaries result in zigzagged and fragmented MCG lines (Figure 8b). Additionally, visible boundaries with low contrast were partly missed by MCG image segmentation. In both cases, the advantages of automated delineation are limited.

We claimed that the least-cost-path based on the boundary likelihood is beneficial to delineate long and curved outlines [21,22]. For the Ethiopian data, we now barely made use of the boundary likelihood: For the often small and rectangular parcels, connecting all lines surrounding a click or a selection of lines was more efficient. For areas with few fragmented, long or curved outlines, the workflow is assumed to be of similar effectiveness when leaving out the boundary classification. To include the boundary classification is beneficial when boundaries are demarcated, e.g., by long and curved boundaries, such as roads, waterbodies, or vegetation.

For our data from Kenya and Rwanda, we omitted the boundary classification, since we hardly used it for the Ethiopian data. The least-cost-path, for which a weight attribute can be selected in the plugin interface, used line length instead of boundary likelihood. Since the boundaries differ from the boundaries in the Ethiopian scene, the CNN would need to be retrained or fine-tuned for the new boundary types. Retrieving CNN-derived boundary likelihoods for these UAV data would require further experiments on whether and how to rescale tiles to 224 × 224 pixels, while providing context comparable to our aerial tiles (Figure 5).

Overall, the automated delineation provided diverse functionalities for different boundary types (Table 7), which made delineation less tiring and more effective (Table 6). Improvements to manual delineation were the strongest for parcels fully surrounded by MCG lines. Such parcels were mostly found in the Ethiopian rural scene, where boundaries aligned with agricultural fields. In the Rwandan scene, automated delineation was time-consuming, since the boundaries were not demarcated consistently. Selecting and joining fragmented MCG lines required more careful visual inspection compared to the rural Ethiopian scene. In the Kenyan scene, the boundaries were less often covered by vegetation, and thus were in general better visible. Compared to the rural Ethiopian scene, the automated delineation still required more zooming, as boundaries were demarcated by more diverse objects.

Table 7. Which plugin functionality to use for which boundary type?

Functionality	Boundary Type	Boundary ≙ Segmentation	Example Boundary
Connect around selection	complex or rectangular	yes	agricultural field
Connect lines' endpoints	small or rectangular	partly	vegetation-covered
Connect along optimal path	long or curved	yes	curved river
Connect manual clicks	fragmented or partly invisible	no or partly	low-contrast

4. Discussion

4.1. Working Hypothesis: Improving Boundary Mapping Approach

Compared to our previous workflow [21], we improved each of the three workflow steps. For image segmentation, we remove the previous need to reduce the image resolution for images larger than 1000 × 1000 pixels, and we introduce a filtering step that allows us to limit over-segmentation by reducing the number of segment lines by 80%. For boundary classification, we implement Convolutional Neural Networks (CNNs) and thereby eliminate the previous need for Random Forest (RF) hand-crafted feature generation. For interactive delineation, we develop two additional delineation functionalities ('Connect around selection', 'Connect lines' endpoints'), we develop an attribute selection for the least-cost-path functionality ('Connect along optimal path') and redesign the GUI to be more intuitive. While we previously tested our approach on road outlines only, we now show advantages compared to manual delineation for cadastral mapping, which includes various object types. The number of clicks per 100 m compared to manual delineation was previously reduced by 76% and 86%, respectively, when delineating roads from two UAV images. Now we applied our approach to delineate 378 visible cadastral boundaries from UAV and aerial imagery of larger extents, while requiring on average 80% fewer clicks compared to manual delineation.

4.2. Working Hypothesis: CNN vs. RF

By reformulating our problem to be solvable by a CNN, we have investigated integrating a more state-of-the-art approach in our previously proposed boundary delineation workflow [21]. A deep learning CNN was assumed to be superior to a machine learning RF, as CNNs require no hand-crafted features, and can be trained incrementally. This starting hypothesis holds true: Even though pre-trained on images from computer vision, transfer-learning a CNN on remote sensing data provided more accurate predictions for boundary likelihoods compared to RF. Our successful integration reduces the effect of possibly meaningless or biased hand-crafted features, and increases the degree of automation of our approach. However, when conducting the final workflow step, i.e., interactive delineation, we found that we seldom made use of the boundary likelihood. We reduced over-segmentation, due to post-processing the image segmentation. This, in combination with new interactive delineation functionalities, is more effective than manual delineation for regular-shaped parcels surrounded by visible boundaries. The delineation functionality that uses boundary likelihood is beneficial for long or curved boundaries, which was rare in our study areas.

4.3. Limitations & Future Work

When training a network to predict boundary likelihoods for visible object outlines, our training data based on cadastral reference are beneficial, as it is available without further processing. The data have little bias, as no human annotator with domain knowledge is required [36]. However, the data could be improved: Cadastral data contain invisible boundaries not detectable by MCG. To limit training data to visible boundaries would match better with what the network is expected to learn, and thus increase achievable accuracy metrics. When deciding whether to use RF or CNN for boundary classification, one needs to balance feature extraction for RF [37] against training data preparation and computational requirements for CNN [18]. In cases of limited training data for CNN, our CNN-based boundary classification may be adopted by data augmentation and re-balancing class weights. One advantage of our RF-based boundary classification is that it contains a feature capturing 3D information from a Digital Surface Model (DSM) [21]. 3D information still needs to be included in the CNN-based boundary classification. Compared to computer vision, the amount and size of benchmark image data are marginal: Existing benchmarks cover aerial data for urban object classification [38] and building extraction [39], satellite imagery for road extraction, building extraction and land cover classification [40], as well as satellite and aerial imagery for road extraction [41]. Such benchmarks in combination with open data initiatives for governmental cadastral data [42], aerial imagery [43] and crowdsourced labeling [44–46] may propel deep learning frameworks for cadastral boundary delineation, i.e., cadastral intelligence. Instead of using a VGG pre-trained on ImageNet, our approach could then be trained on diverse remote sensing and cadastral data, resulting in a possibly more effective and scalable network.

Despite the shown advances, automating cadastral boundary delineation is not at its end. Identifying areas in which a large portion of cadastral boundaries is visible, and for which high-resolution remote sensing and up-to-date cadastral data are available in digital form, still impedes methodological development. Future work could investigate the approach's applicability for invisible boundaries, that are marked before UAV data capture, e.g., with paint or other temporary boundary markers. In this context, the degree to which the approach can support participatory mapping could also be investigated. Furthermore, research needs to be done on how to align innovative approaches with existing technical, social, legal and institutional frameworks in land administration [47–49]. We are currently pursuing this by developing documentation and testing material [50] that enables surveyors and policy makers in land administration to easily understand, test and adapt our approach.

4.4. Comparison to Previous Studies

How we reformulated our problem to be solvable by a tile-based CNN has been similarly proposed in biomedical optics [51]. Fang et al. crop tiles centered on retinal boundary pixels and train a CNN to predict nine different boundary labels. Correspondingly labeled pixels are connected with a graph-based approach. To transfer the latter to our case, we may investigate whether connecting tiles of similar boundary likelihood can omit the need for an initial MCG image segmentation: By using Fully Convolutional Networks (FCNs) [52] each pixel of the input image would be assigned a boundary likelihood, which can be connected using Ultrametric Contour Maps (UCMs) [53] included in MCG [54]. Connecting pixels of corresponding boundary likelihoods could also be realized by using MCG-based contour closure [55], line integral convolution [56], or template matching [57].

Alternatively, the topology of MCG lines can be used to sort out false boundary likelihoods before aggregating them per line: This could be realized by not shuffling training data, and thus maintaining more context information per batch, or by using graph-based approaches such as active contour models [58] suggested for road detection [59,60], or region-growing models suggested for RF-based identification of linear vegetation [61].

Predicting the optimal MCG parameter k per image may also be achieved with CNNs. Depending on whether an area is, e.g., rural or urban, cadastral parcels vary in size and shape. Larger parcels demand less over-segmentation and a higher k. Similarly, our high-resolution UAV data required a higher k, i.e., 0.3 and 0.4 as compared to 0.1 for the aerial data. Challenges to be addressed are training with data from multiple sensors, varying parcel sizes in training and automatically labeling data with the optimal segmentation parameter k.

5. Conclusions

We have introduced a workflow that simplifies the image-based delineation of visible boundaries to support the automated mapping of land tenure from various sources' remote sensing imagery. In comparison to our previous work [21], the approach is now more automated and more accurate due to the integration of CNN deep learning, compared to RF machine learning. For RF-derived boundary likelihoods, we obtained an accuracy of 41% and a precision of 49%. For CNN-derived boundary likelihoods, we obtained an accuracy of 52% and a precision of 76%. CNNs eliminate the need to generate hand-crafted features required for RF. Furthermore, our approach has proven to be less tiring and more effective compared to manual delineation, due to the decreased over-segmentation and our new delineation functionalities. We limit over-segmentation by reducing the number of segment lines by 80% through filtering. Through the new delineation functionalities, the delineation effort per parcel requires 38% less time and 80% fewer clicks compared to manual delineation. The approach works on data from different sensors (aerial and UAV) of different resolutions (0.02–0.25 m). Advantages are strongest when delineating in rural areas due to the continuous visibility of monotonic boundaries. Manual delineation remains superior in cases where the boundary is not fully visible, i.e., covered by shadow or vegetation. While our approach has been developed for cadastral mapping, it can also be used to delineate objects in other application fields, such as land use mapping, agricultural monitoring, topographical mapping, road tracking, or building extraction.

Author Contributions: Conceptualization, S.C., M.K., M.Y.Y. and G.V.; methodology, S.C., M.K., M.Y.Y. and G.V.; software, S.C.; validation, S.C., M.K., M.Y.Y. and G.V.; formal analysis, S.C.; investigation, S.C.; resources, S.C.; data curation, S.C.; writing—original draft preparation, S.C.; writing—review and editing, S.C., M.K., M.Y.Y. and G.V.; visualization, S.C.; supervision, M.K., M.Y.Y. and G.V.; project administration, M.K.; funding acquisition, G.V.

Funding: This research was funded by Horizon 2020 program of the European Union (project number 687828).

Acknowledgments: We are grateful for the support of our African project partners INES Ruhengeri (Rwanda), Bahir Dar University (Ethiopia), Technical University of Kenya, and Esri Rwanda to support the data capture, which was guided and further processed by Claudia Stöcker from University of Twente. Berhanu Kefale Alemie from Bahir Dar University provided aerial data and reference data, as well as knowledge about local land administration situation.

Conflicts of Interest: The authors declare no conflict of interest.

Appendix A

Table A1. Results obtained on validation data for different fine-tuned CNNs. The one used for further analysis in our study is outlined in green. The legend text corresponds to that of Figure 7.

Parameter	Value	Acc.	Loss	Plot
base model	VGG19	0.607	0.654	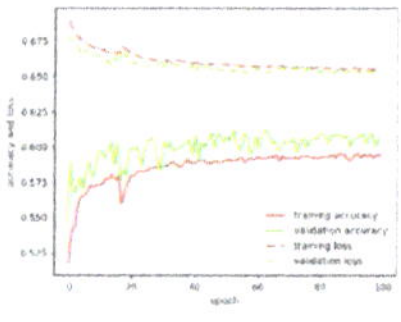
dense layer depth	/			
dense layer depth	16			
dropout rate	0.5			
learning rate	0.01			
base model	VGG19	0.705	0.66	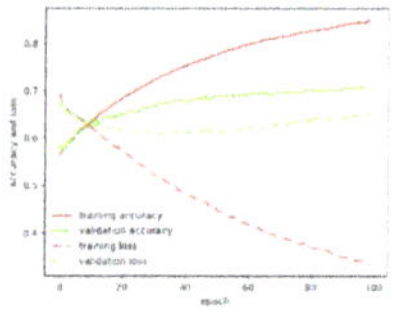
dense layer depth	/			
dense layer depth	1024			
dropout rate	0.5			
learning rate	0.01			
base model	VGG19	0.693	1.632	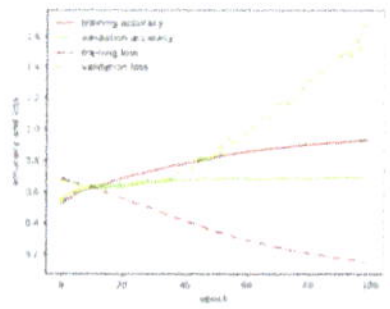
dense layer depth	512			
dense layer depth	16			
dropout rate	0.5			
learning rate	0.01			
base model	VGG19	0.613	0.643	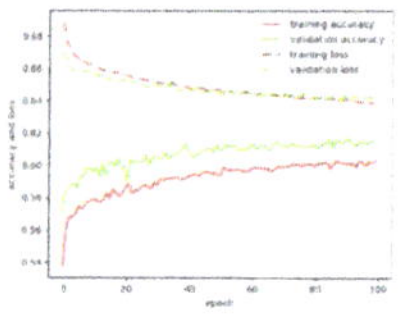
dense layer depth	/			
dense layer depth	16			
dropout rate	0.5			
learning rate	0.001			
base model	VGG19	0.615	0.646	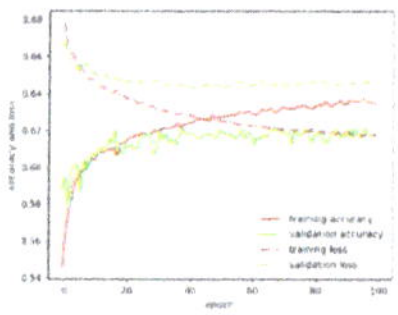
dense layer depth	/			
dense layer depth	16			
dropout rate	0.2			
learning rate	0.01			
base model	VGG19	0.6	0.656	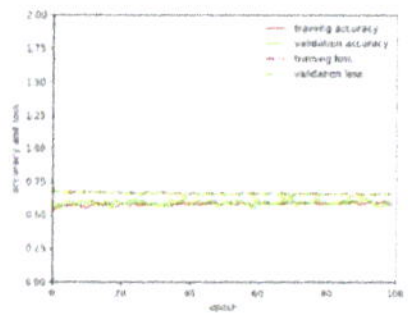
dense layer depth	16			
dense layer depth	16			
dropout rate	0.8			
learning rate	0.001			
base model	VGG16	0.667	0.608	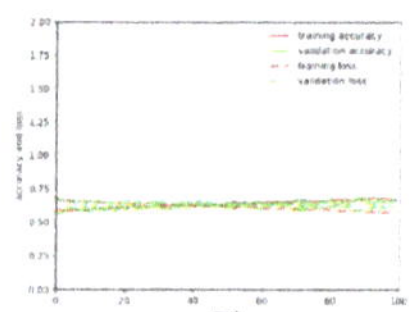
dense layer depth	/			
dense layer depth	1024			
dropout rate	0.8			
learning rate	0.001			

Table A1. *Cont.*

Parameter	Value	Acc.	Loss	Plot
base model	VGG19	0.692	0.586	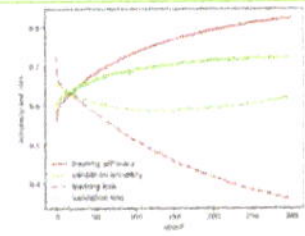
dense layer depth	/			
dense layer depth	1024			
dropout rate	0.8			
learning rate	0.001			
base model	VGG19	0.733	1.205	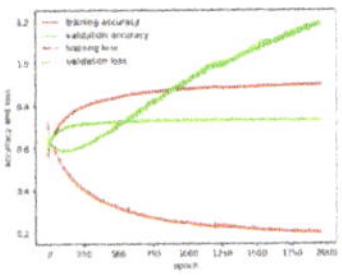
dense layer depth	/			
dense layer depth	1024			
dropout rate	0.8			
learning rate	0.001			
base model	ResNet50	0.571	0.742	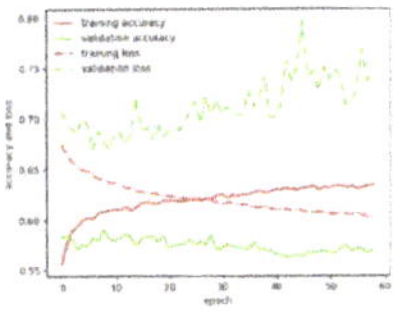
dense layer depth	/			
dense layer depth	16			
dropout rate	0.5			
learning rate	0.01			
base model	ResNet50	0.561	2.367	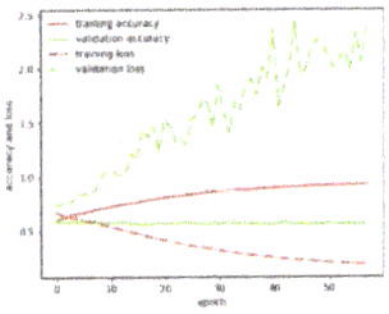
dense layer depth	/			
dense layer depth	1024			
dropout rate	0.5			
learning rate	0.01			
base model	ResNet50	0.546	3.86	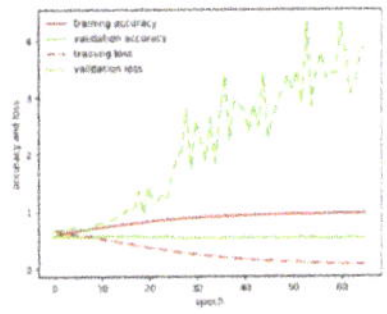
dense layer depth	512			
dense layer depth	16			
dropout rate	0.5			
learning rate	0.01			
base model	ResNet50	0.577	0.787	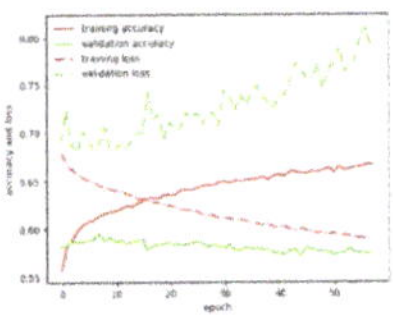
dense layer depth	/			
dense layer depth	16			
dropout rate	0.5			
learning rate	0.001			
base model	ResNet50	0.578	0.838	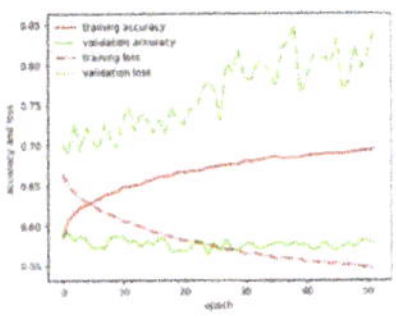
dense layer depth	/			
dense layer depth	16			
dropout rate	0.2			
learning rate	0.01			
base model	InceptionV3	0.543	0.792	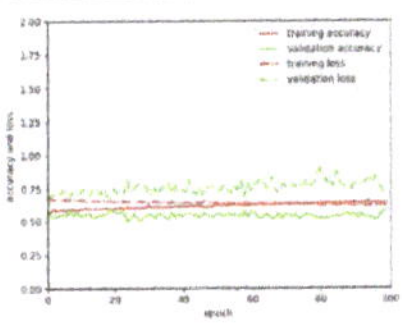
dense layer depth	/			
dense layer depth	1024			
dropout rate	0.8			
learning rate	0.001			

Table A1. *Cont.*

Parameter	Value	Acc.	Loss	Plot
base model	Xception	0.559	0.777	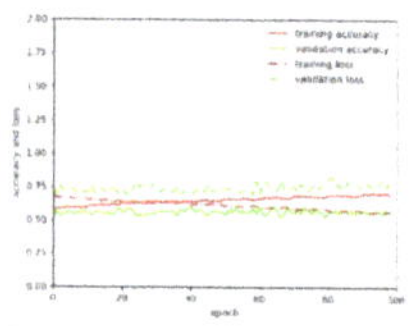
dense layer depth	/			
dense layer depth	1024			
dropout rate	0.8			
learning rate	0.001			
base model	MobileNet	0.612	0.775	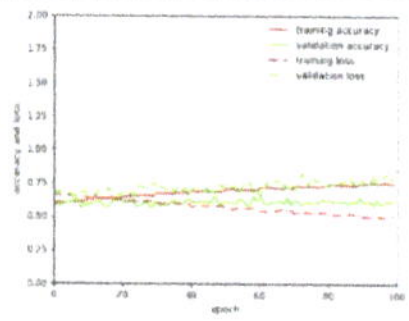
dense layer depth	/			
dense layer depth	1024			
dropout rate	0.8			
learning rate	0.001			
base model	DenseNet201	0.569	0.895	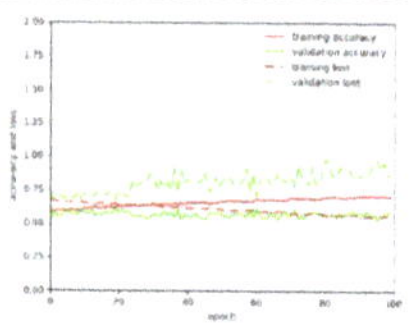
dense layer depth	/			
dense layer depth	1024			
dropout rate	0.8			
learning rate	0.001			

References

1. Enemark, S.; Bell, K.C.; Lemmen, C.; McLaren, R. *Fit-for-Purpose Land Administration*; International Federation of Surveyors: Frederiksberg, Denmark, 2014; p. 42.
2. Williamson, I.; Enemark, S.; Wallace, J.; Rajabifard, A. *Land Administration for Sustainable Development*; ESRI Press Academic: Redlands, CA, USA, 2010; p. 472.
3. Zevenbergen, J.; Bennett, R. The Visible Boundary: More Than Just a Line Between Coordinates. In Proceedings of the GeoTechRwanda, Kigali, Rwanda, 8–20 November 2015; pp. 1–4.
4. Kohli, D.; Bennett, R.; Lemmen, C.; Asiama, K.; Zevenbergen, J. A Quantitative Comparison of Completely Visible Cadastral Parcels Using Satellite Images: A Step towards Automation. In Proceedings of the FIG Working Week 2017, Helsinki, Finland, 29 May–2 June 2017; pp. 1–14.
5. Luo, X.; Bennett, R.; Koeva, M.; Lemmen, C.; Quadros, N. Quantifying the overlap between cadastral and visual boundaries: A case study from Vanuatu. *Urban Sci.* **2017**, *1*, 32. [CrossRef]
6. Crommelinck, S.; Bennett, R.; Gerke, M.; Nex, F.; Yang, M.Y.; Vosselman, G. Review of automatic feature extraction from high-resolution optical sensor data for UAV-based cadastral mapping. *Remote Sens.* **2016**, *8*, 689. [CrossRef]
7. Manyoky, M.; Theiler, P.; Steudler, D.; Eisenbeiss, H. Unmanned aerial vehicle in cadastral applications. In Proceedings of the International Archives of the Photogrammetry, Remote Sensing and Spatial Information Sciences, Zurich, Switzerland, 14–16 September 2011; pp. 1–6.
8. Jazayeri, I.; Rajabifard, A.; Kalantari, M. A geometric and semantic evaluation of 3D data sourcing methods for land and property information. *Land Use Policy* **2014**, *36*, 219–230. [CrossRef]
9. Koeva, M.; Muneza, M.; Gevaert, C.; Gerke, M.; Nex, F. Using UAVs for map creation and updating. A case study in Rwanda. *Surv. Rev.* **2018**, *50*, 312–325. [CrossRef]
10. Gevaert, C.M. Unmanned Aerial Vehicle Mapping for Settlement Upgrading. Ph.D. Thesis, University of Twente, Enskede, The Netherlands, 2018.
11. Maurice, M.J.; Koeva, M.N.; Gerke, M.; Nex, F.; Gevaert, C. A photogrammetric approach for map updating using UAV in Rwanda. In Proceedings of the GeoTechRwanda 2015, Kigali, Rwanda, 18–20 November 2015; pp. 1–8.
12. He, H.; Zhou, J.; Chen, M.; Chen, T.; Li, D.; Cheng, P. Building extraction from UAV images jointly using 6D-SLIC and multiscale Siamese convolutional networks. *Remote Sens.* **2019**, *11*, 1040. [CrossRef]

13. Marmanis, D.; Wegner, J.D.; Galliani, S.; Schindler, K.; Datcu, M.; Stilla, U. Semantic segmentation of aerial images with an ensemble of CNNs. *ISPRS Ann. Photogramm. Remote Sens. Spat. Inf. Sci.* **2016**, *3*, 473–480. [CrossRef]
14. Yang, M.Y.; Liao, W.; Li, X.; Rosenhahn, B. Deep learning for vehicle detection in aerial images. In Proceedings of the IEEE International Conference on Image Processing (ICIP), Athens, Greece, 7–10 October 2018; pp. 3079–3083.
15. Long, Y.; Gong, Y.; Xiao, Z.; Liu, Q. Accurate object localization in remote sensing images based on convolutional neural networks. *IEEE Trans. Geosci. Remote Sens.* **2017**, *55*, 2486–2498. [CrossRef]
16. Persello, C.; Tolpekin, V.A.; Bergado, J.R.; de By, R.A. Delineation of agricultural fields in smallholder farms from satellite images using fully convolutional networks and combinatorial grouping. *Remote Sens. Environ.* **2019**, *231*, 111253. [CrossRef]
17. Bergen, K.J.; Johnson, P.A.; Maarten, V.; Beroza, G.C. Machine learning for data-driven discovery in solid earth geoscience. *Science* **2019**, *363*, 0323. [CrossRef]
18. Zhu, X.X.; Tuia, D.; Mou, L.; Xia, G.; Zhang, L.; Xu, F.; Fraundorfer, F. Deep learning in remote sensing: A comprehensive review and list of resources. *IEEE Geosci. Remote Sens. Mag.* **2017**, *5*, 8–36. [CrossRef]
19. Ketkar, N. *Deep Learning with Python*; Apress: New York, NY, USA, 2017; p. 169.
20. Garcia-Gasulla, D.; Parés, F.; Vilalta, A.; Moreno, J.; Ayguadé, E.; Labarta, J.; Cortés, U.; Suzumura, T. On the behavior of convolutional nets for feature extraction. *J. Artif. Intell. Res.* **2018**, *61*, 563–592. [CrossRef]
21. Crommelinck, S.; Höfle, B.; Koeva, M.; Yang, M.Y.; Vosselman, G. Interactive Boundary Delineation from UAV data. In Proceedings of the ISPRS Annals of the Photogrammetry, Remote Sensing and Spatial Information Sciences, Riva del Garda, Italy, 4–7 June 2018; pp. 81–88.
22. Crommelinck, S.; Koeva, M.; Yang, M.Y.; Vosselman, G. Robust object extraction from remote sensing data. *arXiv* **2019**, arXiv:1904.12586.
23. Crommelinck, S. Delineation-Tool GitHub. Available online: https://github.com/its4land/delineation-tool (accessed on 10 October 2019).
24. Pont-Tuset, J.; Arbeláez, P.; Barron, J.T.; Marques, F.; Malik, J. Multiscale combinatorial grouping for image segmentation and object proposal generation. *IEEE Trans. Pattern Anal. Mach. Intell.* **2017**, *39*, 128–140. [CrossRef] [PubMed]
25. Simonyan, K.; Zisserman, A. Very deep convolutional networks for large-scale image recognition. *arXiv* **2014**, arXiv:1409.1556.
26. He, K.; Zhang, X.; Ren, S.; Sun, J. Deep residual learning for image recognition. In Proceedings of the IEEE Conference on Computer Vision and Pattern Recognition (CVPR), Las Vegas, NV, USA, 26 June–1 July 2016; pp. 770–778.
27. Szegedy, C.; Vanhoucke, V.; Ioffe, S.; Shlens, J.; Wojna, Z. Rethinking the inception architecture for computer vision. In Proceedings of the IEEE Conference on Computer Vision and Pattern Recognition (CVPR), Las Vegas, NV, USA, 26 June–1 July 2016; pp. 2818–2826.
28. Chollet, F. Xception: Deep learning with depthwise separable convolutions. In Proceedings of the IEEE Conference on Computer Vision and Pattern Recognition (CVPR), Honolulu, HI, USA, 21–26 July 2017; pp. 1251–1258.
29. Howard, A.G.; Zhu, M.; Chen, B.; Kalenichenko, D.; Wang, W.; Weyand, T.; Andreetto, M.; Adam, H. Mobilenets: Efficient convolutional neural networks for mobile vision applications. *arXiv* **2017**, arXiv:1704.04861.
30. Iandola, F.; Moskewicz, M.; Karayev, S.; Girshick, R.; Darrell, T.; Keutzer, K. Densenet: Implementing efficient convnet descriptor pyramids. *arXiv* **2014**, arXiv:1404.1869.
31. QGIS Development Team. Qgis Geographic Information System, Open Source Geospatial Foundation. Available online: https://www.qgis.org (accessed on 10 October 2019).
32. Crommelinck, S. QGIS Plugin Repository: BoundaryDelineation. Available online: http://plugins.qgis.org/plugins/BoundaryDelineation/ (accessed on 10 October 2019).
33. International Association of Assessing Officers (IAAO). *Standard on Digital Cadastral Maps and Parcel Identifiers*; International Association of Assessing Officers (IAAO): Kansas City, MO, USA, 2015; p. 24.
34. Deng, J.; Dong, W.; Socher, R.; Li, L.J.; Li, K.; Li, F.F. Imagenet: A large-scale hierarchical image database. In Proceedings of the IEEE Conference on Computer Vision and Pattern Recognition (CVPR), Miami Beach, FL, USA, 20–25 June 2009; pp. 248–255.
35. Chollet, F. Keras. Available online: https://keras.io (accessed on 10 October 2019).

36. Nyandwi, E.; Koeva, M.; Kohli, D.; Bennett, R. Comparing Human versus Machine-Driven Cadastral Boundary Feature Extraction. *Preprints* **2019**. [CrossRef]
37. Warner, T.A.; Fang, F. Implementation of machine-learning classification in remote sensing: An applied review. *Int. J. Remote Sens.* **2018**, *39*, 2784–2817.
38. Rottensteiner, F.; Sohn, G.; Jung, J.; Gerke, M.; Baillard, C.; Benitez, S.; Breitkopf, U. The ISPRS benchmark on urban object classification and 3D building reconstruction. *ISPRS Ann. Photogramm. Remote Sens. Spat. Inf. Sci.* **2012**, *1*, 293–298. [CrossRef]
39. Chen, Q.; Wang, L.; Wu, Y.; Wu, G.; Guo, Z.; Waslander, S.L. Aerial imagery for roof segmentation: A large-scale dataset towards automatic mapping of buildings. *ISPRS J. Photogramm. Remote Sens.* **2018**, *147*, 42–55. [CrossRef]
40. Demir, I.; Koperski, K.; Lindenbaum, D.; Pang, G.; Huang, J.; Basu, S.; Hughes, F.; Tuia, D.; Raskar, R. Deepglobe 2018: A challenge to parse the earth through satellite images. In Proceedings of the IEEE/CVF Conference on Computer Vision and Pattern Recognition Workshops (CVPRW), Salt Lake City, UT, USA, 18–22 June 2018; pp. 172–181.
41. Cardim, G.; Silva, E.; Dias, M.; Bravo, I.; Gardel, A. Statistical evaluation and analysis of road extraction methodologies using a unique dataset from remote sensing. *Remote Sens.* **2018**, *10*, 620. [CrossRef]
42. University of Auckland. Land Information New Zeeland Data Service. Available online: https://data.linz.govt.nz/ (accessed on 10 October 2019).
43. Humanitarian OpenStreetMap Team. OpenAerialMap. Available online: https://openaerialmap.org/ (accessed on 10 October 2019).
44. Debats, S.R.; Estes, L.D.; Thompson, D.R.; Caylor, K.K. Integrating active learning and crowdsourcing into large-scale supervised landcover mapping algorithms. *PeerJ Prepr.* **2017**. [CrossRef]
45. Keenja, E.; De Vries, W.; Bennett, R.; Laarakker, P. Crowd sourcing for land administration: Perceptions within Netherlands kadaster. In Proceedings of the FIG Working Week, Rome, Italy, 6–10 May 2012; pp. 1–12.
46. Basiouka, S.; Potsiou, C. VGI in cadastre: A Greek experiment to investigate the potential of crowd sourcing techniques in cadastral mapping. *Surv. Rev.* **2012**, *44*, 153–161. [CrossRef]
47. Moreri, K.; Fairbairn, D.; James, P. Issues in developing a fit for purpose system for incorporating VGI in land administration in Botswana. *Land Use Policy* **2018**, *77*, 402–411. [CrossRef]
48. Stöcker, C.; Ho, S.; Nkerabigwi, P.; Schmidt, C.; Koeva, M.; Bennett, R.; Zevenbergen, J. Unmanned Aerial System Imagery, Land Data and User Needs: A Socio-Technical Assessment in Rwanda. *Remote Sens.* **2019**, *11*, 1035. [CrossRef]
49. Spatial Collective. Mapping: (No) Big Deal. Available online: http://mappingnobigdeal.com/ (accessed on 10 October 2019).
50. Crommelinck, S. Delineation-Tool Wiki. Available online: https://github.com/its4land/delineation-tool/wiki (accessed on 10 October 2019).
51. Fang, L.; Cunefare, D.; Wang, C.; Guymer, R.H.; Li, S.; Farsiu, S. Automatic segmentation of nine retinal layer boundaries in OCT images of non-exudative AMD patients using deep learning and graph search. *Biomed. Opt. Express* **2017**, *8*, 2732–2744. [CrossRef]
52. Long, J.; Shelhamer, E.; Darrell, T. Fully convolutional networks for semantic segmentation. In Proceedings of the IEEE Conference on Computer Vision and Pattern Recognition (CVPR), Boston, MA, USA, 7–12 June 2015; pp. 3431–3440.
53. Arbeláez, P. Boundary extraction in natural images using ultrametric contour maps. In Proceedings of the Conference on Computer Vision and Pattern Recognition Workshop (CVPRW), New York, NY, USA, 17–22 June 2006; pp. 1–8.
54. Volpi, M.; Tuia, D. Deep multi-task learning for a geographically-regularized semantic segmentation of aerial images. *ISPRS J. Photogramm. Remote Sens.* **2018**, *144*, 48–60. [CrossRef]
55. Yang, J.; Price, B.; Cohen, S.; Lee, H.; Yang, M.H. Object contour detection with a fully convolutional encoder-decoder network. In Proceedings of the IEEE Conference on Computer Vision and Pattern Recognition (CVPR), Las Vegas, NV, USA, 26 June–1 July 2016; pp. 193–202.
56. Li, P.; Zang, Y.; Wang, C.; Li, J.; Cheng, M.; Luo, L.; Yu, Y. Road network extraction via deep learning and line integral convolution. In Proceedings of the IEEE International Geoscience and Remote Sensing Symposium (IGARSS), Beijing, China, 10–15 July 2016; pp. 1599–1602.

57. Zhou, W.; Wu, C.; Yi, Y.; Du, W. Automatic detection of exudates in digital color fundus images using superpixel multi-feature classification. *IEEE Access* **2017**, *5*, 17077–17088. [CrossRef]
58. Kass, M.; Witkin, A.; Terzopoulos, D. Snakes: Active contour models. *Int. J. Comput. Vis.* **1988**, *1*, 321–331. [CrossRef]
59. Butenuth, M.; Heipke, C. Network snakes: Graph-based object delineation with active contour models. *Mach. Vis. Appl.* **2012**, *23*, 91–109. [CrossRef]
60. Gerke, M.; Butenuth, M.; Heipke, C.; Willrich, F. Graph-supported verification of road databases. *ISPRS J. Photogramm. Remote Sens.* **2004**, *58*, 152–165. [CrossRef]
61. Lucas, C.; Bouten, W.; Koma, Z.; Kissling, W.D.; Seijmonsbergen, A.C. Identification of linear vegetation elements in a rural landscape using LiDAR point clouds. *Remote Sens.* **2019**, *11*, 292. [CrossRef]

remote sensing

MDPI

Article

Towards 3D Indoor Cadastre Based on Change Detection from Point Clouds

Mila Koeva *, Shayan Nikoohemat, Sander Oude Elberink, Javier Morales, Christiaan Lemmen and Jaap Zevenbergen

Faculty of Geo-Information Science and Earth Observation (ITC), University of Twente, 7514 AE Enschede, The Netherlands
* Correspondence: m.n.koeva@utwente.nl; Tel.: +31-534874410

Received: 12 June 2019; Accepted: 12 August 2019; Published: 21 August 2019

Abstract: 3D Cadastre models capture both the complex interrelations between physical objects and their corresponding legal rights, restrictions, and responsibilities. Most of the ongoing research on 3D Cadastre worldwide is focused on interrelations at the level of buildings and infrastructures. So far, the analysis of such interrelations in terms of indoor spaces, considering the time aspect, has not been explored yet. In The Netherlands, there are many examples of changes in the functionality of buildings over time. Tracking these changes is challenging, especially when the geometry of the spaces changes as well; for example, a change in functionality, from administrative to residential use of the space or a change in the geometry when merging two spaces in a building without modifying the functionality. To record the changes, a common practice is to use 2D plans for subdivisions and assign new rights, restrictions, and responsibilities to the changed spaces in a building. In the meantime, with the advances of 3D data collection techniques, the benefits of 3D models in various forms are increasingly being researched. This work explores the opportunities for using 3D point clouds to establish a platform for 3D Cadastre studies in indoor environments. We investigate the changes in time of the geometry of the building that can be automatically detected from point clouds, and how they can be linked with a Land Administration Model (LADM) and included in a 3D spatial database, to update the 3D indoor Cadastre. The results we have obtained are promising. The permanent changes (e.g., walls, rooms) are automatically distinguished from dynamic changes (e.g., human, furniture) and are linked to the space subdivisions.

Keywords: point clouds; indoor change detection; laser scanning; 3D indoor modelling; Cadastre

1. Introduction

With the increasing complexity of the buildings in highly urban areas since the late 90s, 3D Cadastre has been a subject of interest. 3D Cadastre is beneficial for land registries, architects, surveyors, urban planners, engineers, real estate agencies, etc. [1]. On one hand, it shows the spatial extent of the ownership and, on the other, it facilitates 3D property rights, restrictions, and responsibilities [2,3]. However, for realization of the 3D Cadastre concept, there is no one single solution. User needs, the national political and legal situation, and technical possibilities should be taken into account. This was also clear from the International Federation of Surveyors (FIG) questionnaire completed by many countries in 2010 [4,5]. In recent years, many 3D Cadastre activities have been initiated worldwide, since 3D information is essential for efficient land and property management [6–9]. An investigation into the legal foundation has been done for 15 countries covering Europe, North and Latin America, the Middle East, and Australia [10], not only overground, but also underground [11]. However, there is still no fully implemented 3D Cadastre in the world [4] due to a lack of integration between legal, institutional, and technical parties involved. With the technical developments, physical and legal representation for the purposes of 3D Cadastre are being actively researched; however, considering the dynamics of the complex relationships between people and their properties, we must take into account the time aspect, which needs more attention [12]. Most of the ongoing research on 3D Cadastre worldwide is focused on interrelations at the level of buildings and infrastructures. So far, the analysis of such interrelations in terms of indoor spaces, considering the time aspect, has not yet been explored. Therefore, the current paper aims to investigate the opportunities provided by automatic techniques for detecting changes based on point clouds in support of 3D indoor Cadastre. When using the term "automatic", we mean that the process of change detection and separation of permanent changes from temporary changes are automatic. However, setting relevant parameters for each step is required by an expert. Moreover, Cadastral expert intervention is required to connect the land administration database, if it exists, to the physical space subdivision extracted from point clouds. The remainder of this section includes the relevance of our research, showing a real example in The Nederlands and related scientific work in the field.

In recent years, many examples can be found of changes in the functionality of buildings. According to the statistics shared by Rijksolverheid [13] in The Netherlands, 17% of the commercial real estate is empty. The Ministry of Interior and Kingdom Relations (BZK) and the Association of Dutch Municipalities (VNG) set up an expert team to support municipalities in the transformation of empty buildings from commercial to residential use. One of the examples is a nursing home located in the city of Hoorn (Figure 1a), 40% of which was owned by housing associations and 60% by health care organizations and was changed in 2015 into student accommodation and privately owned apartments (Figure 1b).

(a)

(b)

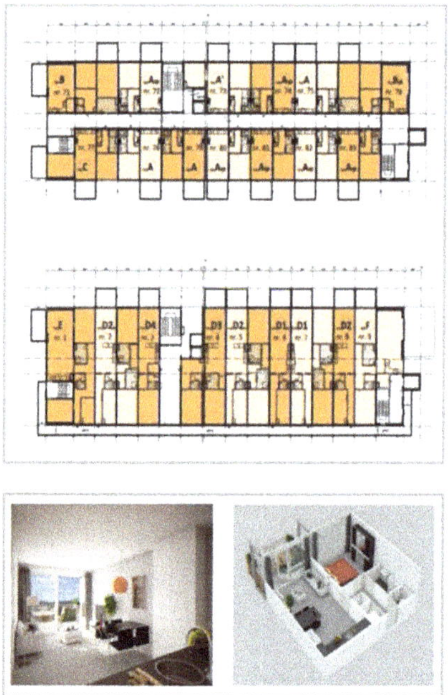

Figure 1. (**a**) Changing from a nursing house (**top**) to student accommodation (**middle**) and (**b**) privately owned apartments (**bottom**) [13].

From the recent research in the field, it was observed that point clouds are a valuable source for decision makers in the domain of urban planning and land administration. Laser scanner data acquired with aerial laser scanners (ALS), mobile laser scanners (MLS), and terrestrial laser scanners

(TLS) have been used for reconstruction of 3D cities, building facades, roof reconstruction [14–16], and damage assessment of the buildings before and after a disaster [17]. In the domain of forestry, point clouds are used for monitoring the growth of trees and changes in the forest canopy. Xiao et al. [18] used point clouds to monitor the changes of trees in urban canopies. Regarding buildings, some methods combine images with laser scanner data for facade reconstruction [19–21]. There has been incredible progress in recent years in the automation of 3D modeling based on point clouds [22–24] and more specifically in subdividing the space to semantic subdivisions, such as offices, corridors, staircases, and so forth [25–27]. Challenges for detecting changes for updating 3D Cadastre in an urban environment using ALS and image-based point clouds for 3D Cadastre were also explored [28]. Regarding indoor spaces, geometric changes during the lifetime of a building were analyzed for the Technical University of Munich (TUM) [29], as shown in Figure 2; however, they were not related to Cadastre. This fact motivates us to use point clouds and monitor changes for updating 3D Cadastre.

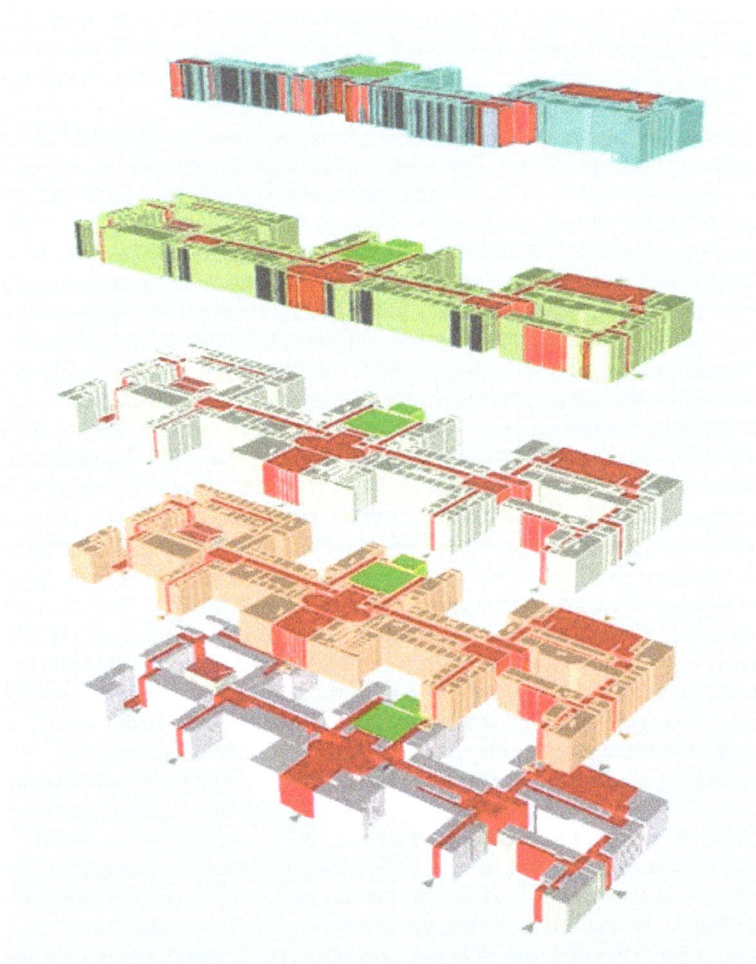

Figure 2. Geometric changes of indoor spaces during the lifetime of a building for complex buildings for the Technical University of Munich (TUM). Campus 3D map-rendered corridors [29].

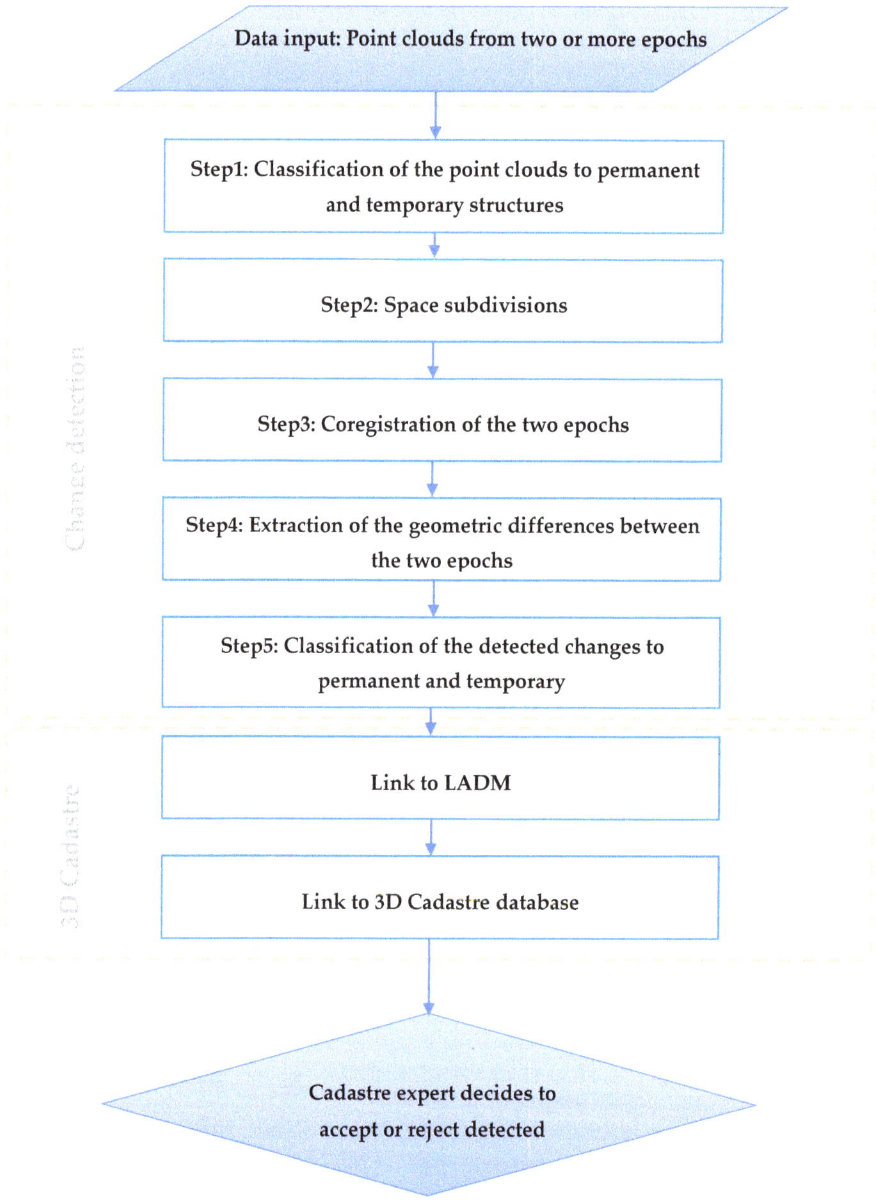

Figure 3. Methodology.

2.1. *Case Studies*

For the current research, two case study examples are used. The first case study is the building of the Technical University in Braunschweig (TUB) and the second is the University of Twente Faculty of Geo-Information Science and Earth Observation (ITC) building. The floor plans of these buildings are shown in Figure 4. In Figure 4a, the highlighted area shows that a wall was removed and rooms were merged into one, and Figure 4b shows the two rooms before removing the walls.

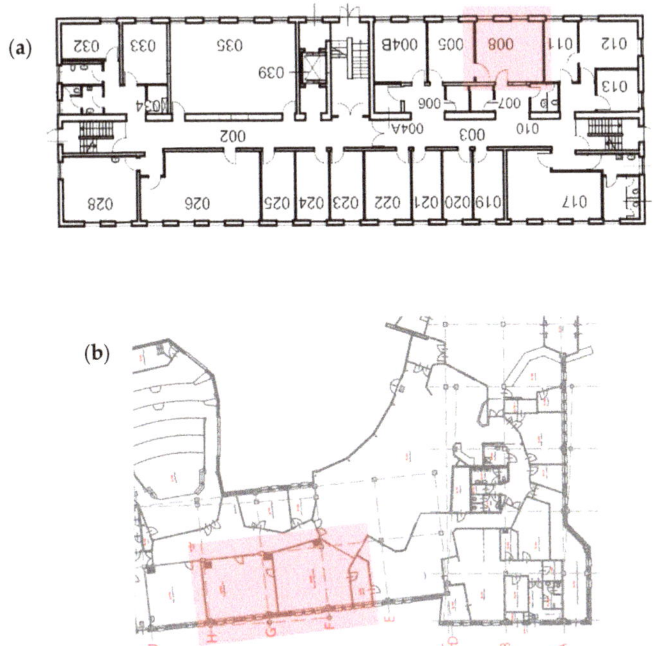

Figure 4. The floor plans for our two case studies. (**a**) (**top**) TU Braunschweig after the change (floor2), (**b**) (**bottom**) University of Twente ITC building (floor 1) before the change. The highlighted areas show the rooms where the changes happened.

Point cloud data for the two case studies were collected with different scanners (Figure 5). The data for the Braunschweig building were collected with an ITC Indoor Mobile Mapping System (ITC-IMMS) (epoch1) [43] and a Zeb-Revo (epoch2) [44]. For the ITC building, we used the Riegl [45] terrestrial laser scanning system and a Viametris device. The accuracy of the point clouds varied from 0.01 to 0.06 m depending on the laser scanner system. While the noise in mobile mapping systems was louder than the terrestrial laser scanner (TLS), the scene coverage of a mobile mapping system was more than a TLS. The noise in the data could have been caused by sensors, data acquisition algorithms, and the reflective surfaces. For more information on the comparison of scanning systems, refer to the study by Lehtola et al. [46].

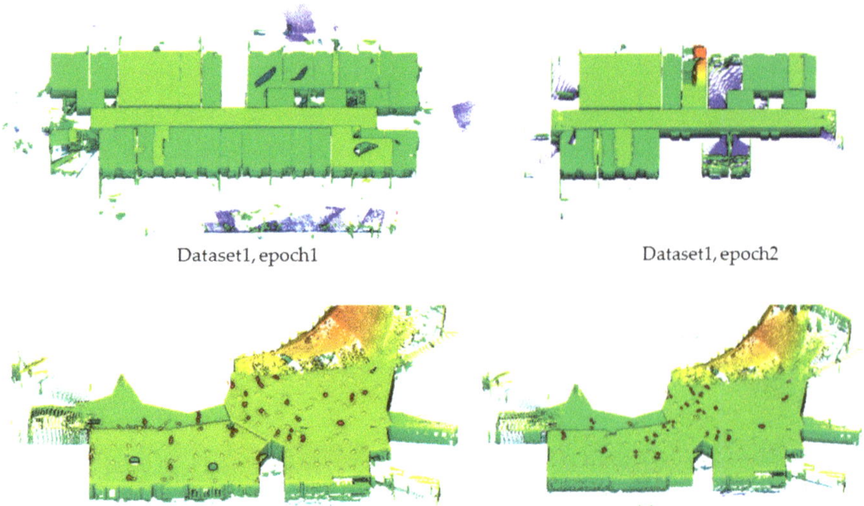

Figure 5. The datasets for two different epochs. The first row is the dataset which belongs to the Braunschweig building and the second row is from the ITC building and is a more complex dataset, with furniture and large glass windows.

In the following subsections, the detailed methodology is explained based on the first case study.

2.2. Indoor Change Detection from Point Clouds

Differences in two epochs of point clouds inside the buildings can be categorized as:

1. Changes in the dynamic objects (e.g., furniture);
2. Changes in the permanent structure (walls, floors, rooms).

There are some other differences between two epochs of point clouds that are interpreted as:

3. Differences because of the acquisition coverage;
4. Differences because of the difference in the sensors.

In our approach, categories number 1 and 2 were dealt with as important changes for 3D Cadastre, and categories 3 and 4 were just inevitable differences in two epochs that occurred because of data acquisition systems and were not relevant to the 3D Cadastre. We acquired two point clouds of two time periods with two different laser scanners, one a Zeb-Revo [44] handheld MLS and the other an ITC-IMMS [43]. The motivation to use different sensors is to explore all realistic possible causes of differences between epochs. The process of change detection starts with the co-registration of two point clouds (Step 3 from Figure 3). The co-registration of two point cloud datasets was a straightforward approach, such as using the iterative closest point (ICP) [47–49]. After the registration, two point clouds were compared based on the distance threshold to detect the differences caused by the registration error and sensors differences (4th category; Step 4 from Figure 3). The distance threshold was chosen by summing the registration error and sensor noise. The registration error and sensor noise already introduced some differences between the two datasets. The registration errors were the residuals of each co-registration process (less than 10 cm). The sensor noise was specified in the specification of the systems. This threshold, d, described points from two datasets with the distance less than the threshold. They were not considered as changes and they were in the 4th category because of the differences in the sensors. Points that had distances more than the threshold were in one of the other three categories. In our experiment, we defined the distance threshold of less than 0.10 m.

Let the point clouds (PC) from epoch one (acquired by a backpack) be PC1 and the point clouds from the second epoch (acquired by Zeb-Revo) be PC2. The point to point comparison was based on the reconstruction of a Kd-tree [50,51] and a comparison of the distance of the points in PC1 from PC2 and was stored in PC1. Using this method, the differences caused by the acquisition system and registration errors were excluded from the real changes.

In the next Step 5 from Figure 3, the differences were further analyzed to detect and exclude the acquisition coverage (3rd category). Our change detection method was based on analyzing two geometric differences between two point clouds. This was done in two steps: (1) The distinction was made between object changes and coverage differences and (2) the object changes were separated into changes on permanent structures and dynamic objects, such as persons and furniture (Section 2.2).

The geometric differences were calculated by determining the nearest 2D point and the nearest 3D point in the other epoch. The first nearest point was based on the X, Y coordinates and the second on X, Y, Z coordinates. Figure 6 shows both geometric distances as a point attribute categorized in three colors: Green <20 cm, yellow >20 cm and <50 cm, red >50 cm to the nearest.

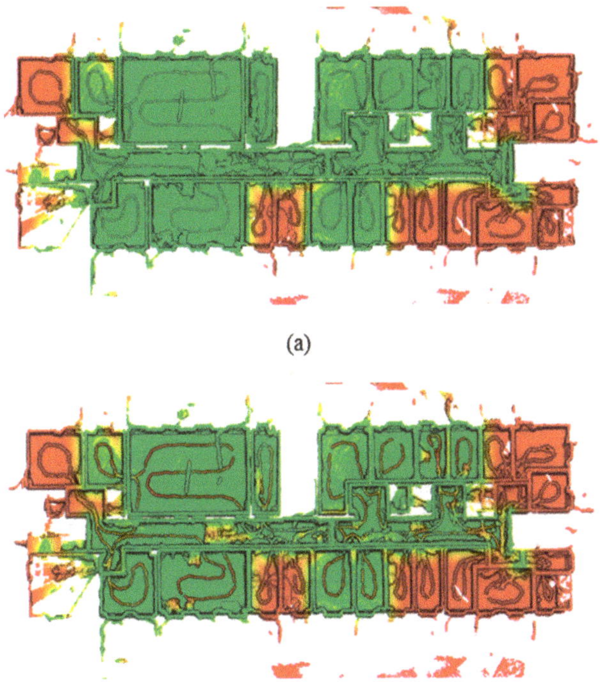

(a)

(b)

Figure 6. The distance (green <20 cm, yellow <50 cm, red >50 cm) to the nearest point in (**a**) 2D and (**b**) 3D.

For both object changes and coverage differences, it was expected that the nearest 3D point was further than a certain threshold. However, the nearest 2D point may have been close to a changed object, but not in case of coverage differences. Points were temporarily labeled as part of changed objects if the distance to the nearest point in 3D was larger than 20 cm, but the nearest point in 2D was less than 20 cm. Threshold values were chosen such that they were larger than the expected registration errors but small enough to detect changes larger than 20 cm. Next, the whole point cloud was segmented into planar segments and only the vertical segments with a majority (more than 50%)

of points labeled as potentially changed were considered to be changed. The planar segmentation was performed by a region growing algorithm presented by Vosselman et al., [52]. Note that, in this way, the points on a newly built wall near the ground or ceiling, with a small 3D distance to the nearest point in the other epoch, were included in the changed objects as they belonged to a segment with more than 50% points with a large perpendicular distance to the plane in the other epoch. By using planar segments and calculating perpendicular distances from a point in one epoch to a plane in the other epoch, we avoided the influence of differences in point densities between the point clouds. The vertical segments labeled as changed objects included permanent structures, such as walls, but also dynamic objects, such as persons. In the second step, the aim was to separate permanent from temporary changes by looking at a method described in [23] and [42].

2.2.1. Classify Changes to Permanent and Non-Permanent

The next step was to separate the changes that were part of the permanent structure from dynamic objects. This involved classifying the point clouds in each epoch to a permanent structure (e.g., walls, floors, ceilings) and a non-permanent structure (e.g., furniture, clutter and outliers).

In Figure 7c, the blue color represents the areas captured by Zeb-Revo and the red areas show the differences in the coverage where PC1 is not covered by PC2. In Figure 7d, the point clouds of epoch1 after the comparison with the epoch2 are shown and the blue points show the points in which their distance differences are less than the threshold and are not changed. The green points show the changes, because of coverage or furniture, or a permanent change, and the ceiling is removed for better visualization. We applied a method from [23] to classify the permanent structures in each epoch (see Figure 8). Four main classes were important for our change detection process. Walls, floors, and ceilings were three classes that belonged to the permanent structures. The non-permanent structures were, for example, furniture, outliers, and unknown points, which were classified as the clutter. The classification started with surface growing segmentation and generating an adjacency graph from the connected segments. By analyzing the adjacency graph, it was possible to separate permanent structures, such as walls, because of their connection to the floor and ceiling. The normal angle of the planes was important in this decision because walls in most indoor environments have an angle of more than 45 degrees with the positive direction of the z-axis. Figure 8c shows that the permanent structure (walls and floor) was separated from the clutter.

After the classification of points in each epoch, by comparing the changes with the semantic labels (walls, floors, and ceilings), it is possible to distinguish relevant changes for 3D Cadastre. Each point in the set of changes is a possible change for 3D Cadastre if is labeled as a wall, floor, or ceiling, otherwise it is a change only because of furniture or dynamic objects or outliers. Table 1 shows how we identified changes with labels per point, respecting the permanent structure. According to the table, points with label 1 are important for change detection in 3D Cadastre because they represent a permanent change in the building. Figure 5 represents the changes with different colors according to their label.

Table 1. The table shows how the point clouds are labeled regarding the changes and their role in the building structure. The points with label 1 are interesting for change detection of the 3D Cadastre.

Labels of Points in the Data	Non-Permanent Structure	Permanent Structure
Change	0	1
No change	2	3

Figure 7. (**a**) Point clouds from a backpack system from the first epoch. (**b**) Point clouds from a Zeb-Revo system from the second epoch. (**c**) Co-registered point clouds. (**d**) Point clouds of epoch1 after the comparison with epoch2.

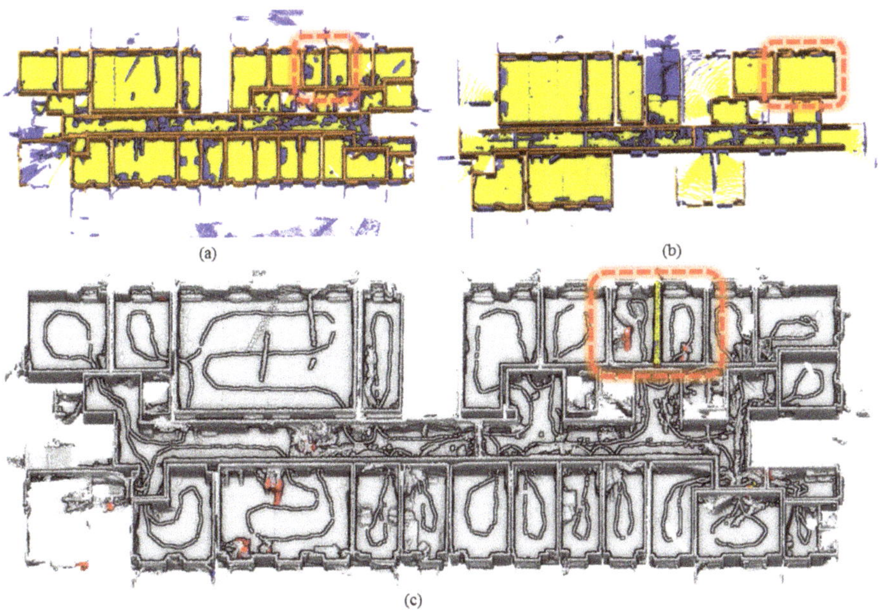

Figure 8. (**a**) PC1 acquired by a backpack and (**b**) PC2 is acquired by Zeb-Revo, walls (orange), floor (yellow), clutter (blue). (**c**) The changes are detected in PC1 and classified to permanent structure changes (yellow) and non-permanent changes (red). The red rectangle shows the wall that is showing a permanent change.

2.2.2. Changes in Relation to Indoor Space Subdivisions

The process of detection of permanent changes is continued by linking detected changes to the volumetric space or space subdivisions. Space subdivisions represent the semantic space in an indoor environment, such as offices, corridors, parking areas, staircases, and so forth. Each space subdivision is connected to space in a spatial unit in the 3D Cadastre model and all laser points in the space subdivisions carry the attributes of the corresponding Cadastre administration. In this step, we explain how these space subdivisions were extracted from the point clouds and linked to the previously detected changes. Note that an apartment may consist of one or more spatial units, a spatial unit may consist of one or more spaces, and a spatial unit may have invisible boundaries and needs to be checked by a Cadastre expert.

Following the method in [23], after the extraction of the permanent structures in each epoch, a voxel grid was reconstructed from the point clouds, including walls, floors, and ceilings. Using a 3D morphology operation on the voxel grid, space was then subdivided into rooms and corridors. Each space subdivision was represented with the center of voxels as a point cloud segment. To find out which changes occured in which space subdivisions, we intersected the space subdivisions of each epoch with the permanent changes detected earlier (see Figure 9). For example, in Figure 8, we can see that in the second epoch (Figure 8b) a wall was removed and two spaces were merged. Since this wall was detected as a change during the previous step (Figure 8c) by the intersection of changed objects with subdivisions, the changes in the two epochs were extracted (Figure 9). These changes were linked to a space subdivision, and each space subdivision or a group of them (e.g., a building level) may represent a spatial unit in the 3D Cadastre model.

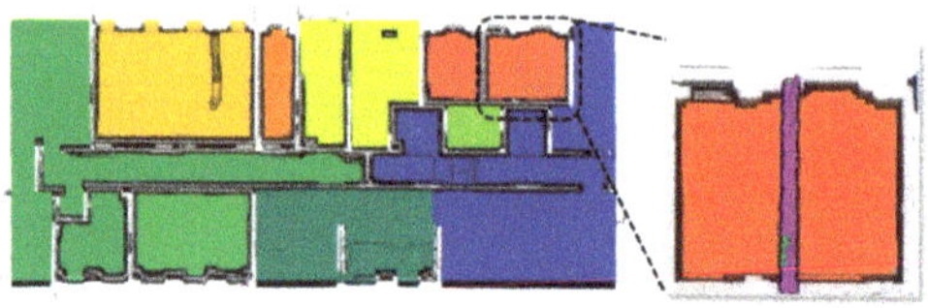

Figure 9. The space subdivisions of PC2 (second epoch) after the change. The purple wall in the right image shows the intersection of a detected change with a space subdivision.

2.2.3. Changes in Relation to the 3D Cadastre Model

To link the Cadastre to the detected changes, we assumed that every space subdivision in the point clouds was represented in the object description of the spatial unit in the LADM, considering that an interactive refinement on the space subdivision from the previous step was necessary to group some of the subdivisions, according to the 3D Cadastre legal spatial units. For example, a group of offices that belonged to the same owner had an invisible boundary that should be interactively corrected. LADM represents legal spaces in spatial units. Spatial units were refined into two specializations [38].

(1) Building units, as instances of class:

LA_LegalSpaceBuildingUnit. A building unit concerns the legal space, which does not necessarily coincide with the physical space of a building. A building unit is a component of the building (the legal, recorded, or informal space of the physical entity). A building unit may be used for different purposes (e.g., living or commercial) or it can be under construction. An example of a building unit is a space in a building, an apartment, a garage, a parking space, or a laundry space.

(2) Utility networks, as instances of a class:

LA_LegalSpaceUtilityNetwork. A utility network concerns legal space, which does not necessarily coincide with the physical space of a utility network.

The LADM class LA_BAUnit (Figure 10) allowed the association of one right to a combination of spatial units (e.g., an apartment and a parking place).

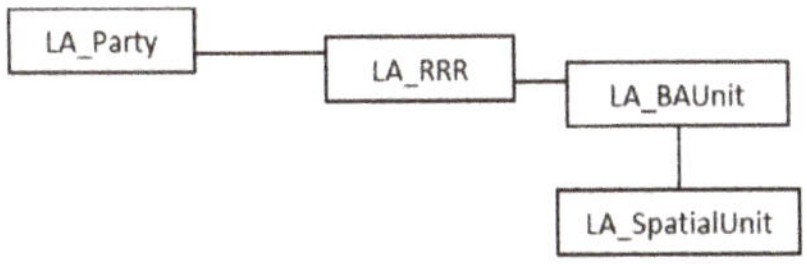

Figure 10. Basic classes of the LADM [38].

A basic administrative unit (LA_BAUnit) in LADM is an administrative entity, subject to registration, consisting of 0 or more spatial units, against which (one or more) homogeneous and unique rights (e.g., ownership right or land use right), responsibilities, or restrictions are associated to the whole entity, as included in a land administration system. In LADM, each space is represented as a spatial unit and then uses a LADM class LA_BAUnit to associate those spatial units to a legal unit. The type of building units were individual or shared. An individual building unit is an apartment and represents a legal space. A building contains individual units (apartments), a shared unit with a common threshold (entrance), and a ground parcel. Each unit owner holds a share in the shared unit and the ground parcel.

Every spatial unit in LADM was modelled with GM_MultiSurface. 2D parcels were modelled by boundary face string (LA_BoundaryFace). The representation of 3D spatial units was done by boundary face (LA_BoundaryFace), and for the storage a GM_Surface was used (see Figure 11). However, in our approach, we are aiming to keep the point clouds until the last step for spatial analysis. Therefore, we just used the calculated features, such as volume, area, and neighboring units, to insert them as classes in the LADM. All spatial attributes and legal issues, such as rights, restrictions, and responsibilities, could be associated between point clouds and LADM. The measured spaces were important because, apart from the floor space, the volumes are also known. This is relevant for valuation purposes of the individual spaces in apartments.

Figure 12 illustrates the LADM representation of an apartment—in this case, owned by a party (right holder) named Frank. This party has an individual space and a share (1/100) in the common or shared space. Individual and shared spaces (including the ground parcel) compose the building as a whole.

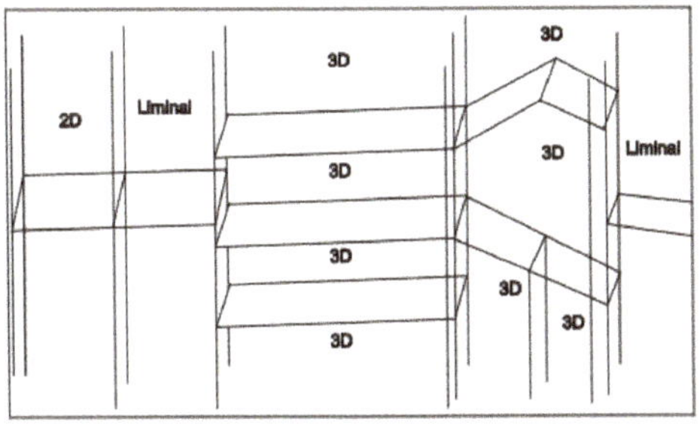

Figure 11. Mixed use of boundary face strings and boundary faces defining both bounded and unbounded 3D volumes: Annex B in [38].

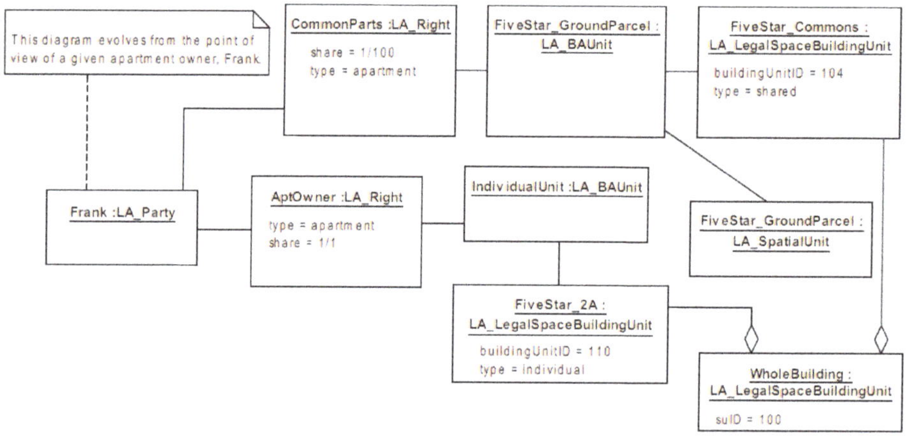

Figure 12. An apartment building in Land Administration Model (LADM) and its legal space [38].

The limiting factor of associating detected space changes to the LADM is that the LADM only provides an abstract representation of 3D objects with no direct mapping to an implementation. There are also specialized data structures, such as CityGML or IndoorGML, which can be used to store 3D data as specified in the LADM model. The issue here is that these data structures are primarily designed for visualization and indoor navigation and not for the management of rights of legal spaces. This becomes more apparent when looking at the definition of primal and dual spaces. The primal space is used to represent semantic subdivisions (e.g., a room, a corridor) and the dual space is used to represent the navigability of the primal space. For the proper management of legal space in a database and to properly determine which changes in the layout of a building affect legal spaces, additional information is needed to be stored in the database, namely, a direct relationship between visible and invisible subdivisions of space and the legal objects in the 3D Cadastre.

Given today's database technology, the available option for the implementation of 3D legal spaces and their corresponding topological relationships is a GM_PolyhedralSurface [53]. A PolyhedralSurface datatype is defined as a collection of polygons connected by edges which may enclose a solid. When using such a data structure, it is possible to define a subdivision in a building as the primal space and a legal object as the dual space. This way, properties can be assigned to, for example, walls to define whether it corresponds to a legal boundary or not (or where in the wall the boundary is). Similarly, properties can be assigned to invisible space subdivisions that define a change in the rights of the spaces. In this scenario, the dual of an edge is a face and the dual of a face is solid, which will represent a LA_BAUnit. A database implementation of the topological relationships of a PolyhedralSurface as required by a 3D Cadastre can be based on dual half-edges [54,55]. With this approach, each face is stored as an array of half-edges and can be associated with a set of attributes. These attributes can be defined as a result of the face detection from the point-cloud analysis. Since each face is associated to a legal object, it is possible to support the update of the 3D Cadastre by directly updating changes detected in the latest point cloud epoch on the database structure of the 3D Cadastre. This has to be followed by an update on the rights of the legal objects which will require the intervention of the cadastral expert responsible for mutations and transaction in the land administration system.

3. Results and Discussion

The proposed method is tested on two datasets. One dataset has a smaller amount of clutter and the shape of the building has a regular structure. Therefore, the separation of walls is easier. To challenge the robustness of our method with a complex structure and more furniture, a dataset with arbitrary wall layout and glass surfaces is selected (ITC restaurant, Figure 12). The details of the datasets for each epoch are in Table 2.

Table 2. The details of the datasets and two case studies. The first and second rows belong to the first case study. The table shows the number of points and scanning device per dataset. The fourth column shows the number of changed rooms before and after the renovation of the building. The fifth column shows the items which are identified as changes.

Dataset	# Points	Scanner	# of Changed Rooms	Changed Items	Figures
TU Braunschweig (epoch 1)	1.7 M	ITC-IMMS	2	Clutter, walls	4, 5, 6, 8, 8, 14
TU Braunschweig (epoch 2)	1.8 M	Zeb-Revo	1	Wall is removed.	4, 5, 6, 8, 9, 10, 14, 16
ITC Restaurant (epoch 1)	2.8 M	Viametris, Riegl	3	Clutter, walls, curtains	4, 5, 13, 15
ITC Restaurant (epoch 2)	1.0 M	Viametris	1	Two walls are removed.	4, 5, 13, 15

First, the datasets from two different epochs were co-registered using the iterative closest point ICP algorithm (Figure 13). Then the changes between two epochs were identified in 2D and 3D, as explained in the methodology (Section 2.2). The classification algorithm separated the permanent changes from non-permanent changes and then we intersected the permanent changes with the reconstructed spaces from two epochs (Figures 14 and 15). In this way, the changes in the rooms in the second epoch of both datasets can automatically be identified. To identify the relation of physical changes with the 3D Cadastre, a user adds the ownership of the spaces as an attribute to each space. For example, the spaces which have the same rights and ownership obtain the same label and form a new physical space (Figure 16). Then it is possible to connect them to the basic class of the LA_Spatial Unit in the LADM and update the spatial unit class in the LADM.

In dataset 2 (ITC restaurant), part of the curtain was identified as the permanent change because the curtains were covering the walls and they were detected as a permanent structure. However, this can be the inaccuracy of the classification method, for identifying the changes in the space is not problematic because it has a slight change in the space partitioning.

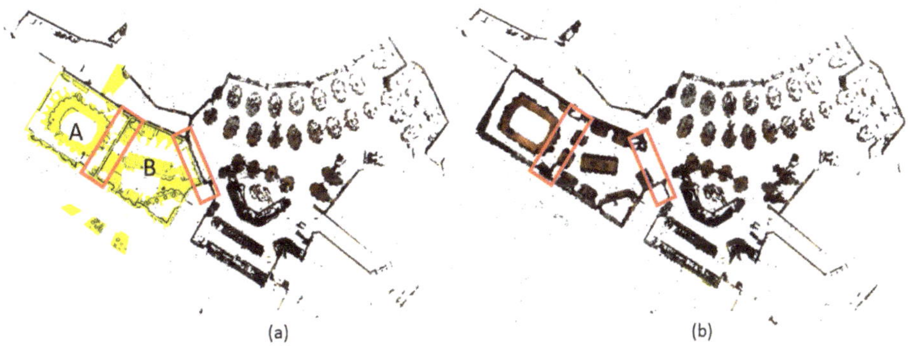

Figure 13. The figure shows the top view of two epochs of our use case. The floor and ceiling are removed for a clear visualization. (**a**) The data is collected by a Riegl terrestrial laser scanners (TLS) [45] (rooms A and B in yellow) and is co-registered with the data collected by the Viametris system [56]. (**b**) The second epoch is also collected by the Viametris system and the walls in the red rectangles are removed.

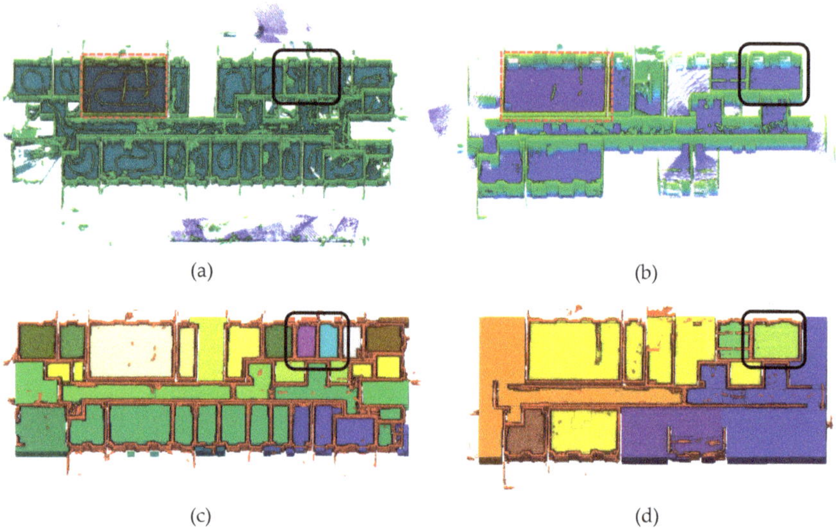

Figure 14. The figure represents the changes in the detected permanent structure and then the spaces. (**a**) and (**b**) show the changes in the walls (black rectangles). The red transparent rectangle is for the orientation between two images. (**c**) and (**d**) show the detected walls in orange and space partitions in random colors. The black rectangles show how the room changed after removing a wall.

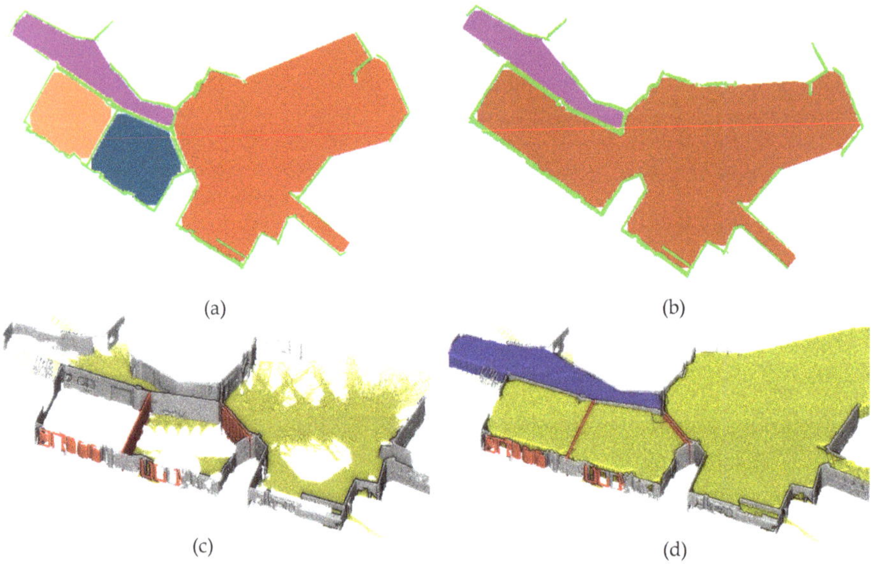

Figure 15. The top view of the spaces and permanent changes. (**a**) Epoch one, walls are in green and four spaces in random colors. (**b**) After removing walls, two rooms in epoch one are merged with the large space in brown color, and, in total, it forms two spaces with the rest of the interiors. (**c**) Detected permanent changes are shown in red. (**d**) The spaces from the second epoch are intersected with the permeant changes to identify the changes in the space.

The important parameter for the detection of changes is the distance threshold (d) to identify the changes from the differences caused by noise and registration errors. We set this parameter slightly larger than the sum up of the sensor noise coming from the scanning device and the residuals coming from the ICP algorithm (less than 10 cm). In our experiments, we set this threshold on 20 cm, which implies that we cannot detect changes which are smaller than 20 cm. For planar segmentation of the point clouds, the smoothness parameter for a surface growing algorithm is important, which depends on the noise and point spacing in the data. We set the smoothness threshold to 8 cm because the noise from MLS systems (Viametris and Zeb-Revo) is around 5 cm. The smoothness parameter was set slightly larger than the sensor noise and point spacing. The point spacing was 5 cm, which meant we could subsample point clouds to reach 5 cm point spacing. The parameters for detecting the permanent structure were chosen according to [23]. Segments with more than 500 supporting points were selected for creating the adjacency graph and smaller segments were discarded. The voxel size for space partitioning was 10 cm, which is an apppriate voxel size to have enough precision to identify changes and avoid expensive computations.

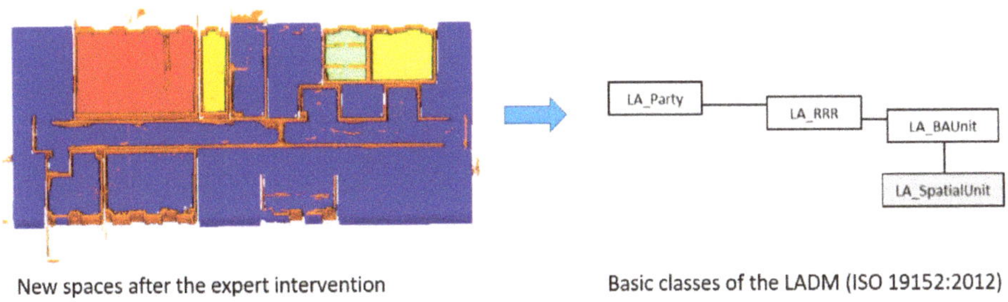

Figure 16. New spaces with the same rights and ownership obtain the same label and color and form a new physical space that can be linked to the LA_SpatialUnit.

The running time for surface growing segmentation, identifying the permanent structure, and detecting the changes for the first dataset with 1.7 million points took 2.4 min, 5.6 min, and 7 min, respectively. The space partitioning was computationally more expensive than other processes and it took 10 min for dataset 1 with the voxel size of 10 cm, and it depended on the volume of the building. Larger volumes required more voxels for morphological space partitioning.

In our workflow, the challenge was detecting the permanent changes from the dynamic changes, which were not important for the Cadastre. According to [23], this process can have an average accuracy of 93% for permanent structures and 90% for spaces [57]. Furthermore, the extraction of spaces are really crucial in the process, because the volume and area is calculated from the space subdivision result. Therefore, an expert should check the results of space subdivision and merge or split some of the spaces that are extracted from the point clouds. The interactive corrections are less than 10% of the whole process and, for a building of three floors as large as our case study, it does not take more than 10 min.

The process of linking the spatial units to the 3D Cadastre model was not automated in our approach. This was because of the lack of possibilities for representation and visualization of 3D objects in the 3D Cadastre models. Therefore, our method was limited when it comes to the storage of 3D spatial objects in the Cadastre databases. As future work, linking the 3D objects and 3D Cadastre models, one solution we intend to investigate is using the point clouds as external classes and trying to keep the 3D objects as point clouds for all steps. The extraction of vector boundaries for the Cadastre models can be done with functions from the point clouds.

4. Conclusions

In this paper, we have shown that permanent changes in buildings can be found automatically using multi-epoch mobile laser data. The detection is based on the selection of planar segments with a majority (i.e., more than 50%) of points in one epoch with a distance larger than 20 cm to the nearest points in the other epoch. In our approach, changes are detected as dynamic changes (e.g., human, furniture) and permanent changes (e.g., walls, rooms). The permanent changes are then linked to the space subdivisions, which are extracted from the point clouds of each epoch (Section 2.2). A Cadastre expert will need to interactively group some of the space subdivision according to their legal attributes. The spaces that are changed and identified during the process will then be further analyzed to extract spatial attributes, such as boundary, area, and volume. This process can be done on point clouds where changes have occurred. Extracted spatial attributes can be exchanged between a Cadastre model, such as LADM, and the point clouds. A Cadastre expert should make decisions on updating the model according to the spatial changes. In the future, we plan to investigate the link between designed space by the architect or civil engineer and the real constructed space as measured with point clouds. This measurement is relevant for the composition of legal space in LADM, but also for building and other permits (e.g., for shops, companies, etc.). It was proved that it is also relevant for crisis management using smart indoor models in 3D [58]. Moreover, the latest updates in 3D mapping using multi-acquisition capabilities, virtual reality, and augmented reality in combination with precise architectural plans and BIM provide immense opportunities. Apart from the technical advances, our future research will be aligned with the second edition of LADM, which is currently under development and includes extensions incorporating the usage of point clouds, BIM, etc. [59].

The process of representing and linking 3D objects to the 3D Cadastre, especially for indoor use, is ongoing research. The authors of this paper hope that this work will introduce a new research avenue regarding the connection between point clouds and indoor Cadastre models.

Author Contributions: Conceptualization, M.K., S.N., S.O.E., J.M., C.L., and J.Z.; investigation, M.N. and S.N.; visualization, M.K., S.N., and S.O.E.; writing—original draft, M.K., S.N., S.O.E., and J.M.; review and editing, M.K., S.N., S.O.E., J.M., C.L., and J.Z.

Funding: The work of Shayan Nikoohemat is part of the TTW Maps4Society project Smart 3D indoor models to support crisis management in large public buildings (13742), which are (partly) financed by the Netherlands Organization for Scientific Research (NWO).

Acknowledgments: The authors would like to thank Yolla Al Asmar for capturing the data for the ITC building. We acknowledge the Institute of Geodesy and Photogrammetry of University of Braunschweig, Markus Gerke and his team, and Isabelle Dikhoff and Yahya Ghassoun for providing their MLS device to collect the data.

Conflicts of Interest: The authors declare no conflict of interest.

References

1. Aien, A.; Kalantari, M.; Rajabifard, A.; Williamson, I.; Wallace, J. Towards integration of 3D legal and physical objects in cadastral data models. *Land Use Policy* **2013**, *35*, 140–154. [CrossRef]
2. Lemmen, C. A Domain Model for Land Administration. Ph.D. Thesis, Delft University of Technology, Delft, The Netherlands, 2012.
3. Zevenbergen, J.; De Vries, W.; Bennett, R. *Advances in Responsible Land Administration*; CRC Press: Padstow, UK, 2015.
4. Van Oosterom, P.; Stoter, J.; Ploeger, H. World-wide inventory of the status of 3D cadastres in 2010 and expectations for 2014. In Proceedings of the FIG Working Week 2011, Marrakech, Morocco, 18–22 May 2011; pp. 1–21.
5. Van Oosterom, P. Research and development in 3D cadastres. *Comput. Environ. Urban Syst.* **2013**, *40*, 1–6. [CrossRef]
6. Dimopoulou, E.; Karki, S.; Roić, M.; de Almeida, P.D.; Griffith-Charles, C.; Thompson, R.; Ying, S.; Paasch, J.; van Oosterom, P. Initial registration of 3D parcels. In *Best Practices 3D Cadastres*; van Oosterom, P., Ed.; International Federation of Surveyors: Copenhagen, Denmark, 2018; pp. 67–94.
7. Van Oosterom, P.; Lemmen, C.; Thompson, R.; Janečka, K.; Zlatanova, S.; Kalantari, M. 3D Cadastral Information Modelling. In *Best Practices 3D Cadastres*; van Oosterom, P., Ed.; International Federation of Surveyors: Copenhagen, Denmark, 2018; pp. 95–133.
8. Rajabifard, A.; Kalantari, M.; Williamson, I. Land and property information in 3D. In Proceedings of the FIG Working Week, Rome, Italy, 6–10 May 2012.

9. Nikoohemat, S.; Koeva, M.; Oude Elberink, S.J.; Lemmen, C.H.J. Change detection from point clouds to support indoor 3D cadastre. In *International Archives of the Photogrammetry, Remote Sensing and Spatial Information Sciences*; ISPRS: Hannover, Germany, 2018; Volume 42.
10. Kitsakis, D.; Paasch, J.; Paulsson, J.; Navratil, G.; Vucic, H.; Karabin, M.; El-Mekawy, M.; Koeva, M.; Janecka, K.; Erba, D.; et al. Legal Foundations. In *Best Practices 3D Cadastres*; van Oosterom, P., Ed.; International Federation of Surveyors (FIG): Copenhagen, Denmark, 2018; pp. 1–66.
11. Kitsakis, D.; Paasch, J.; Paulsson, J.; Navratil, G.; Vucic, N.; Lisec, A.; Koeva, M.N.; Janecka, K.; Karabin, M. Layer Approach to Ownership in 3D cadastre—A subway case. In Proceedings of the 6th International FIG 3D Cadastre Workshop, Delft, The Netherlands, 2–4 October 2018.
12. Jing, Y.; Bennett, R.M.; Zevenbergen, J.A. Up-to-dateness in land administration: Setting the record straight. *Coordinates* **2014**, *10*, 37–42.
13. MBZK. Rapportage Transformatie Zorgvastgoed Naar Wonen—Rapport. Available online: https://www.rijksoverheid.nl/documenten/rapporten/2015/06/30/transformatie-zorgvastgoed-tien-praktijkvoorbeelden (accessed on 23 July 2018).
14. Delbecq, A.L.; Van de Ven, A.H. A group process model for problem identification and program planning. *J. Appl. Behav. Sci.* **1971**, *7*, 466–492. [CrossRef]
15. Elberink, S.O.; Vosselman, G. Building reconstruction by target based graph matching on incomplete laser data: Analysis and limitations. *Sensors* **2009**, *9*, 6101–6118. [CrossRef] [PubMed]
16. Pu, S.; Vosselman, G. Knowledge based reconstruction of building models from terrestrial laser scanning data. *ISPRS J. Photogramm. Remote Sens.* **2009**, *64*, 575–584. [CrossRef]
17. Vetrivel, A.; Gerke, M.; Kerle, N.; Vosselman, G. Identification of damage in buildings based on gaps in 3D point clouds from very high resolution oblique airborne images. *ISPRS J. Photogramm. Remote Sens.* **2015**, *105*, 61–78. [CrossRef]
18. Xiao, W.; Xu, S.; Oude Elberink, S. Change detection of trees in urban areas using multi-temporal airborne lidar point clouds. In Proceedings of the SPIE Remote Sensing of the Ocean, Sea Ice, Coastal Waters, and Large Water Regions, Edinburgh, UK, 24–27 September 2012; Volume 8532.
19. Müller, P.; Zeng, G.; Wonka, P.; Van Gool, L. Image-based procedural modeling of facades. *ACM Trans. Graph.* **2007**, *26*, 85. [CrossRef]
20. Pu, S.; Vosselman, G. Building facade reconstruction by fusing terrestrial laser points and images. *Sensors* **2009**, *9*, 4525–4542. [CrossRef]
21. Teboul, O.; Simon, L.; Koutsourakis, P.; Paragios, N. Segmentation of building facades using procedural shape priors. In Proceedings of the 2010 IEEE Computer Society Conference on Computer Vision and Pattern Recognition, San Francisco, CA, USA, 13–18 June 2010; pp. 3105–3112.
22. Mura, C.; Mattausch, O.; Pajarola, R. Piecewise-planar Reconstruction of Multi-room Interiors with Arbitrary Wall Arrangements. *Comput. Graph. Forum* **2016**, *35*, 179–188. [CrossRef]
23. Nikoohemat, S.; Peter, M.; Elberink, S.O.; Vosselman, G. Exploiting Indoor Mobile Laser Scanner Trajectories for Semantic Interpretation of Point Clouds. In Proceedings of the ISPRS Geospatial Week 2017, Wuhan, China, 18–22 September 2017.
24. Ochmann, S.; Vock, R.; Wessel, R.; Klein, R. Automatic reconstruction of parametric building models from indoor point clouds. *Comput. Graph.* **2016**, *54*, 94–103. [CrossRef]
25. Bobkov, D.; Kiechle, M.; Hilsenbeck, S. Room segmentation in 3D point clouds using anisotropic potential fields. In Proceedings of the 2017 IEEE International Conference on Multimedia and Expo, Hong Kong, China, 10–14 July 2017; pp. 727–732.
26. Jung, J.; Stachniss, C.; Kim, C. Automatic Room Segmentation of 3D Laser Data Using Morphological Processing. *ISPRS Int. J. Geo-Inf.* **2017**, *6*, 206. [CrossRef]
27. Elseicy, A.; Nikoohemat, S.; Peter, M.; Elberink, S.O.; Elseicy, A.; Nikoohemat, S.; Peter, M.; Elberink, S.O. Space Subdivision of Indoor Mobile Laser Scanning Data Based on the Scanner Trajectory. *Remote Sens.* **2018**, *10*, 1815. [CrossRef]
28. Koeva, M.; Oude Elberink, S. Challenges for Updating 3D Cadastral Objects using LiDAR and Image-based Point Clouds. In Proceedings of the 5th International FIG Workshop on 3D Cadastre, Athens, Greece, 18–20 October 2016; pp. 169–182.
29. Nikoohemat, S. Smart Campus Map. Master's Thesis, Technical University of Munich, Munich, Germany, 2013.

30. Van Oosterom, P. *Best Practices 3D Cadastre*; International Federation of Surveyors (FIG): Copenhagen, Denmark, 2018.
31. ISO. Available online: https://www.iso.org/standard/51206.html (accessed on 12 July 2018).
32. Lemmen, C. *Land Information Modelling: Inaugural Address to Mark the Occasion of the Appointment of Prof. Dr. Christiaan Lemmen as Professor of Land Information Modeling at the Faculty of Geo-Information Science and Earth Observation, ITC, at the University of Twente on Thursday 25 October 2018*; University of Twente: Enschede, The Netherlands, 2018.
33. Oldfield, J.; van Oosterom, P.; Beetz, J.; Krijnen, T. Working with Open BIM Standards to Source Legal Spaces for a 3D Cadastre. *ISPRS Int. J. Geo-Inf.* **2017**, *6*, 351. [CrossRef]
34. Alattas, A.; Zlatanova, S.; Oosterom, P.; Chatzinikolaou, E.; Lemmen, C.; Li, K. Supporting indoor navigation using access rights to spaces based on combined use of IndoorGML and LADM models. *ISPRS Int. J. Geo-Inf.* **2017**, *6*, 384. [CrossRef]
35. Kalantari, M.; Yip, K.; Atazadeh, B.; ISA, D. An LADM-based Approach for Developing and Implementing a National 3D Cadastre–A Case Study of Malaysia. In Proceedings of the 7th International FIG Workshop on the Land Administration Domain Model, Zagreb, Croatia, 11–13 April 2018.
36. Kalogianni, E.; Dimopoulou, E.; van Osterom, P. 3D Cadastre and LADM—Needs and Expectations towards LADM Revision. In Proceedings of the 7th Land Administration Domain Model Workshop, Zagreb, Croatia, 11–13 April 2018; pp. 1–22.
37. Thompson, R.; van Oosterom, P.; Cemellini, B. Developing an LADM Compliant Dissemination and Visualization System for 3D Spatial Units. In Proceedings of the 7th International FIG Workshop on the Land Administration Domain Modeln Model, Zagreb, Croatia, 11–13 April 2018.
38. Atazadeh, B.; Kalantari, M. Connecting LADM and IFC Standards–Pathways towards an Integrated Legal-Physical Model. In Proceedings of the 7th International FIG Workshop on the Land Administration Domain Model, Zagreb, Croatia, 11–13 April 2018.
39. Cadastral Data Content Standard—Federal Geographic Data Committee. Available online: https://www.fgdc.gov/standards/projects/FGDC-standards-projects/cadastral/index_html (accessed on 17 July 2018).
40. Steudler, D. Swiss cadastral core data model—experiences of the last 15 years. *Comput. Environ. Urban Syst.* **2006**, *30*, 600–613. [CrossRef]
41. Soltanieh, S.M.K. Cadastral Data Modelling—A Tool for e-Land Administration. Ph.D. Thesis, The University of Melbourne, Victoria, Australia, 2008.
42. Nikoohemat, S.; Peter, M.; Oude Elberink, S.; Vosselman, G. Semantic Interpretation of Mobile Laser Scanner Point Clouds in Indoor Scenes Using Trajectories. *Remote Sens.* **2018**, *10*, 1754. [CrossRef]
43. Karam, S.; Vosselman, G.; Peter, M.; Hosseinyalamdary, S.; Lehtola, V. Design, Calibration, and Evaluation of a Backpack Indoor Mobile Mapping System. *Remote Sens.* **2019**, *11*, 905. [CrossRef]
44. GeoSLAM ZEB-REVO Handheld Mapping and Real-Time Data Processing. Available online: https://geoslam.com/zeb-revo-rt/ (accessed on 18 July 2019).
45. Riegl VZ-400 V-Line Google Search. Available online: https://www.google.nl/search?source=hp&ei=3YwwXebGHsjLwQLVgJmYBw&q=Riegl+VZ-400+V-Line&oq=Riegl+VZ-400+V-Line&gs_l=psy-ab.3..33i160.566.566..1139...0.0..0.108.164.1j1......0....2j1..gws-wiz.....0.otat6JCr2uA&ved=0ahUKEwimqbuL4b7jAhXIZVAKHVVABnMQ4dUDCAU&ua (accessed on 18 July 2019).
46. Lehtola, V.; Kaartinen, H.; Nüchter, A.; Kaijaluoto, R.; Kukko, A.; Litkey, P.; Honkavaara, E.; Rosnell, T.; Vaaja, M.; Virtanen, J.P.; et al. Comparison of the Selected State-Of-The-Art 3D Indoor Scanning and Point Cloud Generation Methods. *Remote Sens.* **2017**, *9*, 796. [CrossRef]
47. Besl, P.; McKay, N.; Schenker, P. Method for registration of 3-D shapes. *Sens. Fusion IV Control Paradig. Data Struct.* **1992**, *1611*, 586–606. [CrossRef]
48. Makadia, A.; Patterson, A.; Daniilidis, K. Fully automatic registration of 3D point clouds. In Proceedings of the 2006 IEEE Computer Society Conference on Computer Vision and Pattern Recognition, New York, NY, USA, 17–22 June 2006.
49. Rabbani, T.; Dijkman, S.; van den Heuvel, F.; Vosselman, G. An integrated approach for modelling and global registration of point clouds. *ISPRS J. Photogramm. Remote Sens.* **2007**, *61*, 355–370. [CrossRef]
50. Friedman, J.; Bentley, J.; Finkel, R. An algorithm for finding best matches in logarithmic time. *C. Trans. Math. Softw.* **1977**, *3*, 209–226. [CrossRef]

51. Greenspan, M.; Yurick, M. Approximate kd tree search for efficient ICP. In Proceedings of the Fourth International Conference on 3-D Digital Imaging and Modeling, Banff, AB, Canada, 6–10 October 2003; pp. 442–448.
52. Vosselman, G.; Gorte, B.; Sithole, G. Recognising structure in laser scanner point clouds. In *International Archives of the Photogrammetry, Remote Sensing and Spatial Information Sciences*; University of Freiburg: Freiburg Germany, 2004; Volume 46, pp. 33–38.
53. ISO. 19107:2003—Geographic Information—Spatial Schema. Available online: https://www.iso.org/standard/26012.html (accessed on 4 June 2019).
54. Boguslawski, P. Modelling and Analysing 3D Building Interiors with the Dual Half-Edge Data Structure. Ph.D. Thesis, University of Glamorgan, Pontypridd, UK, 2011.
55. Boguslawski, P.; Gold, C. Euler Operators and Navigation of Multi-shell Building Models. In *Developments in 3D Geo-Information Sciences, Lecture Notes in Geoinformation and Cartography*; Springer: Berlin/Heidelberg, Germany, 2010; pp. 1–16.
56. Viametris Continuous Indoor Mobile Scanner iMS3D. Available online: https://www.viametris.com/ims3d (accessed on 18 July 2019).
57. Nikoohemat, S.; Diakité, A.; Zlatanova, S.; Vosselman, G. Indoor 3d modeling and flexible space subdivision from point clouds. In Proceedings of the ISPRS Geospatial Week 2019, Enschede, The Netherlands, 10–14 June 2019.
58. Smart Indoor Models in 3D (SIMs3D) Easy. Available online: https://easy.dans.knaw.nl/ui/datasets/id/easy-dataset:115892 (accessed on 6 June 2019).
59. LADM. The 8th FIG Land Administration Domain Model Workshop EuroSDR. Available online: http://www.eurosdr.net/workshops/ladm-2019-8th-fig-land-administration-domain-model-workshop (accessed on 7 June 2019).

remote sensing

MDPI

Article

Towards an Underground Utilities 3D Data Model for Land Administration

Jingya Yan [1,*], Siow Wei Jaw [1,2,3], Kean Huat Soon [4], Andreas Wieser [5] and Gerhard Schrotter [6]

1 ETH Zürich, Future Cities Laboratory, Singapore-ETH Centre, Singapore 138602, Singapore
2 Geoscience & Digital Eaeth Centre (INSTeG), Research Institute for Sustainable Environment, Universiti Teknologi Malaysia, Johor Bahru 81310, Malaysia
3 Department of Geoinformation, Faculty of Built Environment and Surveying, Universiti Teknologi Malaysia, Johor Bahru 81310, Malaysia
4 Singapore Land Authority, Singapore 307987, Singapore
5 ETH Zürich, Institute of Geodesy and Photogrammetry, 8093 Zürich, Switzerland
6 Geomatik + Vermessung Stadt Zürich, 8004 Zürich, Switzerland
* Correspondence: Jingya.yan@arch.ethz.ch

Received: 6 July 2019; Accepted: 15 August 2019; Published: 21 August 2019

Abstract: With the pressure of the increasing density of urban areas, some public infrastructures are moving to the underground to free up space above, such as utility lines, rail lines and roads. In the big data era, the three-dimensional (3D) data can be beneficial to understand the complex urban area. Comparing to spatial data and information of the above ground, we lack the precise and detailed information about underground infrastructures, such as the spatial information of underground infrastructure, the ownership of underground objects and the interdependence of infrastructures in the above and below ground. How can we map reliable 3D underground utility networks and use them in the land administration? First, to explain the importance of this work and find a possible solution, this paper observes the current issues of the existing underground utility database in Singapore. A framework for utility data governance is proposed to manage the work process from the underground utility data capture to data usage. This is the backbone to support the coordination of different roles in the utility data governance and usage. Then, an initial design of the 3D underground utility data model is introduced to describe the 3D geometric and spatial information about underground utility data and connect it to the cadastral parcel for land administration. In the case study, the newly collected data from mobile Ground Penetrating Radar is integrated with the existing utility data for 3D modelling. It is expected to explore the integration of new collected 3D data, the existing 2D data and cadastral information for land administration of underground utilities.

Keywords: 3D data model; data governance; underground utility networks; underground mapping; utility cadastre; land administration

1. Introduction

Rapid urbanization creates a strong need to optimize land use in densely populated cities. Attention is thus shifting from the very limited available space above ground to generation and increased use of underground spaces. Comparing to the above ground, underground is an unseen space. The trench for the building and maintenance of underground infrastructure costs a lot of money, as well as faces high risks. A prerequisite for including the underground in urban planning is the availability of sufficiently complete, accurate and up-to-date 3D maps of the underground. However, such maps are not yet widely available, if at all, and the required data acquisition is much more challenging than for spaces above ground.

With a population of more than five million living in an area of 720 square kilometres, Singapore has revealed a plan for placing infrastructure underground [1]. Currently, a data sharing platform, which is called GeoSpace, is maintained by the Singapore Land Authority (SLA) and used by government agencies (e.g., utility owners, land developers, and land owners) to establish a 2D map of Singapore underground including utility services. Figure 1 shows an example in the Marina Bay region of Singapore includes water supply, sewage, drainage, telecommunication and power grid networks. All the existing data are the 2D format. The 2D visualization with overlap pipelines has limitation to provide accurate and reliable information about underground utilities to various applications.

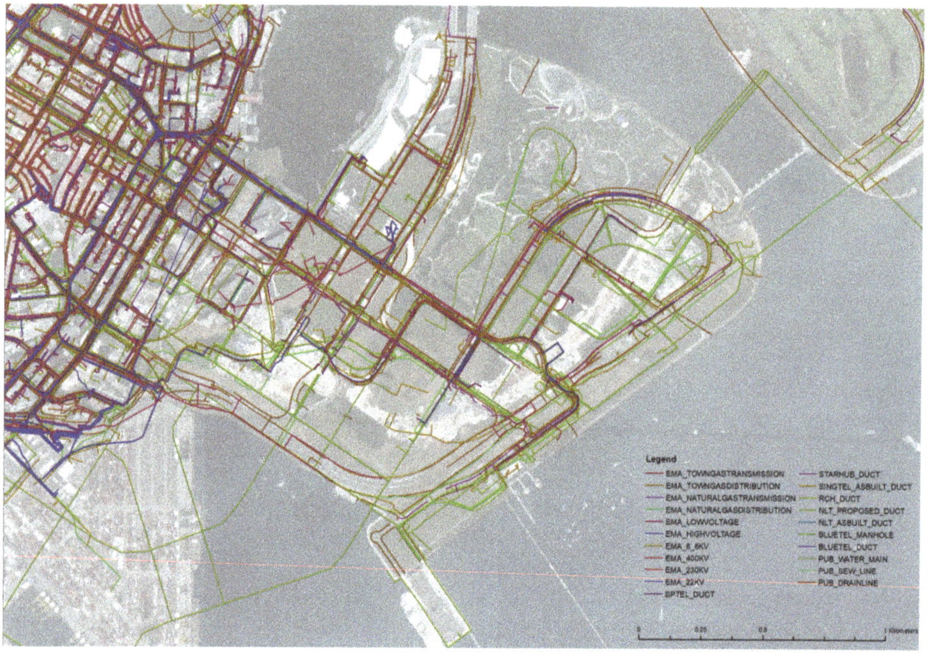

Figure 1. An example of utility data in the Marina Bay region of Singapore (source: Singapore Land Authority, 2018).

To observe the existing data, we zoom in to a corner of the Marina Bay region. Figure 2a presents five layers of different power grid networks. In the real world, the five different power grid networks may be located at the same place and different depths. However, these data have the same x, y value in the database, which makes them impossible to identify in the vertical space and distinguish them in 2D. All of the existing data are as-build data. We can not trust them to present the real situation of underground utility networks. From Figure 2b, the limited attributes are provided from the current database. Only the main water pipes have a diameter. Most of them have 2D geospatial information. In addition, data owners have more details of existing utility data. However, most of them are 2D data as well. Depending on the requirement of the application, some data owners try to collect 3D data. There are some issues during the data capture to usage. Without the utility survey standard, some of them only use the traditional survey method to get the 3D points data of pipelines and overlay on the existing data. Nobody can guarantee the quality of these data. Meanwhile, because of the limitation of the existing data model, it is difficult to integrate 3D data with the existing 2D data. Update cycles were observed to be infrequent and slow, which is once per six months. We not only need time information in the data model to maintain utility database frequently, but also should improve data governance procedures for updating. In general, some issues prevent these data from being sufficient for urban planning, land administration, and on-site work. In fact, many existing databases, not only the ones in Singapore, contribute incompletely to the spatial understanding of the underground because of similar restrictions. In particular:

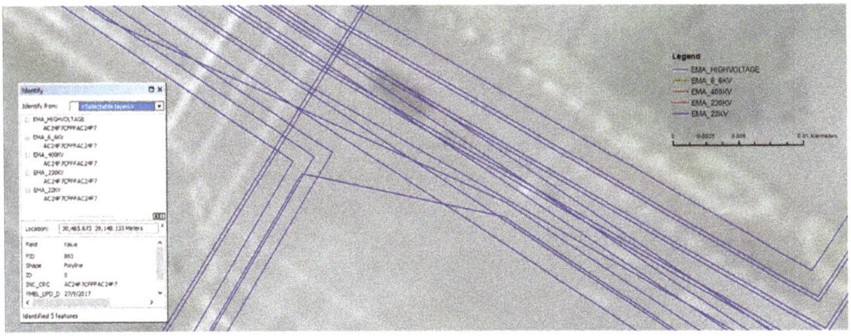

(**a**) An example of power grid data

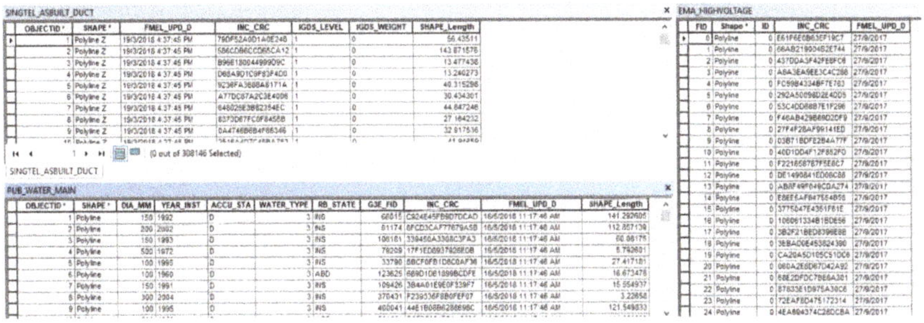

(**b**) The attributes of existing utility data

Figure 2. The issues of existing utility data (Source: Singapore Land Authority, 2018).

- The data are often only 2D i.e., lacking depth information entirely, or 2.5D (i.e., featuring depth as an attribute to a horizontal position rather than as an independent coordinate. Furthermore, the depth information may be sparse with depths measured at few locations only, e.g., at accessible manholes, and it may be ambiguous because it is not always clear whether the values represent depth relative to a specific surface with unknown elevation or height relative to an established height datum.
- It is unknown whether the data represent the current situation, the possibly different as-built state, or just the as-designed state. Furthermore, the geometric accuracy and the completeness of the area often unknown.
- Much of the attribute information (e.g., diameter, material, installation date) required to support specific applications is not available or does not represent the appropriate level of detail.
- There is a lack of standards for organizing the data and semantic information of underground utilities, impairing data sharing and use of the shared data.

Overall, the reliable and accurate 3D data of utility networks are sorely demanded. Therefore, the Singapore-ETH Centre together with the SLA and the Geomatics Department of the City of Zürich have started a related project under the name "Digital Underground" [2]. The initial goals of this project are to develop a road map, a data model and a concept for deriving a unified and complete 3D map of the relevant underground structures (in particular of utilities and spaces like corridors or tunnels). Collecting best practices for underground utility mapping is a special focus within the project. Figure 3 describes the workflow of data governance for 3D underground utility mapping. In the data capture, different types of survey techniques (e.g., Ground Penetrating Radar (GPR), Gyro-based system) are explored and compared to find the optimal underground utility survey approach. After the data processing, the newly collected data should be integrated into the existing database aiming to improve the information of underground utility. As the backbone of the 3D underground utility map, the 3D consolidated database of underground utilities should be developed for data storage. This is a loop workflow. The data capture could improve and update the database. At the same time, the underground utility database should provide information to support data capture. In order to organize these four steps, we need two main components in the data governance. One is the framework to manage different roles and communication between them in data governance. The other is the underground utility data model, which is a conceptual model to describe the structure and content of geodata independent from the used hard- and software systems. It will provide the standard for the presentation of geometrical information, data quality management and various applications. This paper focuses on the design of the framework of data governance and underground utility data model. To ensure legal compliance, efficiency, and resilience of these utility networks, the reliable 3D underground utility data could shed light on their ownership and operation [3]. Then, the underground utility data can be used in various applications. To provide sufficiently and consistently accurate information about underground utilities, it is necessary to fill the gap between engineering practices and mapping disciplines. Meanwhile, we need to find the solution for how to use the existing data and integrate it with newly collected data.

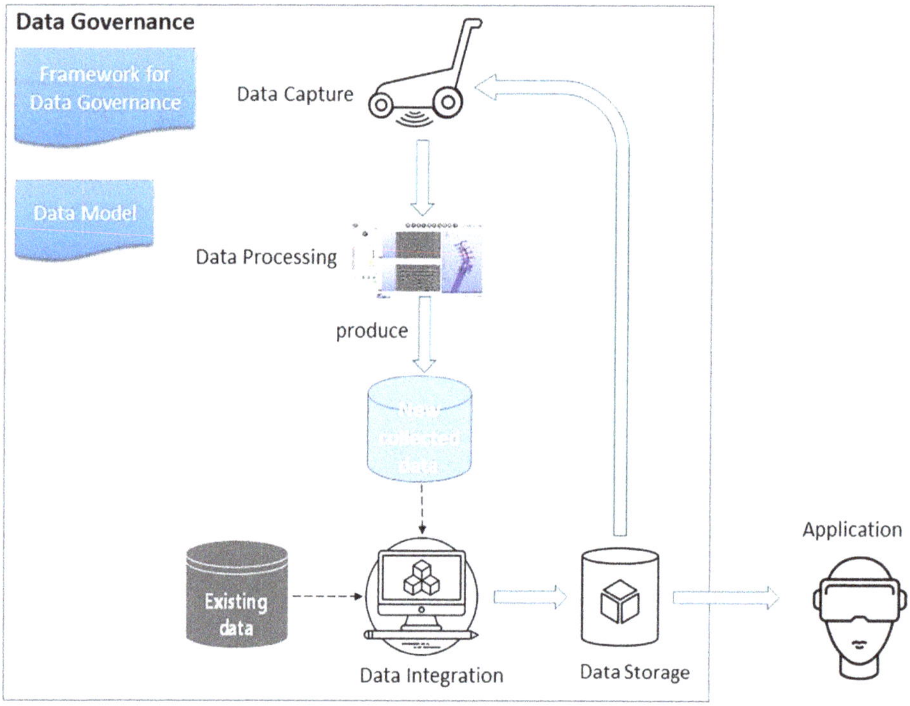

Figure 3. Workflow of underground utility mapping.

Here, we focus on underground utilities, ignoring other underground structures that eventually need to be represented in the same 3D database as the utilities. This work aims at bridging the gap between underground utility surveying and data governance for land administration. Our proposal addresses the following:

- The organization of different phases and roles from data capture to usage. It is necessary to make a clear definition of different roles. During this work process, the communication between different roles (e.g., data producers, owners and users) is very important.
- Different roles have different rights to access, change, delete or add data. These permissions must be defined and maintained administratively.
- Building and updating the 3D map of the underground requires integration of datasets of a different type, quality and source. Data may originate from recent surveying e.g., using GPR or self-contained sensors tracking their movement through a pipe. Data for building a map may also be derived from other databases. This integration requires handling various data formats, and quantifying and properly taking into account the respective data quality.
- The underground data need to be convertible into the data formats required by a variety of different applications and end users without loss of relevant information.

Subsequently, we first introduce related works on 3D underground utility data acquisition and review the underground utility data governance for land administration in some countries or regions. In Section 3, we propose a framework to resolve the above issues about data governance and explain the design of a 3D underground utility data model. In Section 4, we briefly summarize a Singapore case study covering the work process from large scale GPR-based data acquisition to 3D visualization. We conclude with a summary and an outlook on future work.

2. Related Works

2.1. The technologies for 3D underground utility data acquisition

Information about the buried utility networks can be retrieved without any excavation underground utility mapping using non-destructive technologies. However, this is more challenging than above ground mapping. Established approaches for surveying (e.g., photogrammetry, laser scanning, total station measurements or global positioning system) require clear line-of-sight between the instrument and the points to be measured, or between these points and the satellites. They are applicable to (parts of) utility networks while those are exposed in an open pit, e.g., during construction. In some special cases, and with considerable effort, it may even be possible to use such technologies inside buried utilities. However, underground utility mapping comprising detection, location and identification of buried utilities requires approaches without excavation [4,5]. Subsurface geophysical technologies [6,7], such as Ground Penetrating Radar or Electromagnetic Locators, can be used for this purpose [5,8]. In addition, a gyroscope-based system [9] is available for measuring the trajectory of certain utilities (newly laid pipelines with a suitable radius through which the measurement system can travel). Table 1 lists the technologies used for utility mapping with a general review of their accuracy. As positioning using a GPR requires manual processing, manufacturers typically do not mention any type of horizontal or depth measurement accuracy. However, surveying standards such as PAS128 [10] provide some accuracy indications for GPR. According to PAS128, a horizontal accuracy of 250 mm or + 40% of detected depth (whichever is greater) can be achieved when using one of Pipe and Cable Locator (PCL) and GPR, and a horizontal accuracy of 150 mm or + 15% of detected depth (whichever is greater) when using both. Additionally, PAS128 indicates that a depth measurement accuracy of 40% of buried depth can be achieved when using one of PCL and GPR, and 15% of buried depth when both are used. However, PAS128 does not elaborate on how these numbers have been established. In this paper, we focus on GPR due to its popularity in underground utility mapping [5] and on a gyroscope-based system as it is not limited by the depth of the pipeline, by other utilities nearby or by electromagnetic disturbances [9].

Table 1. Data capture methods for underground utility services.

Method	Technology	Use Case	Typical (Primary) Data	Accuracy
Conventional surveying	Total Station	Open pit	Sparse point trajectory	Measure distance up to 1500 m; accuracy about 1.5 mm + 2 parts per million.
	Laser scanning photogrammetry	Open pit	Dense point cloud	Handheld: depends on the distance to the subject and the quality of scan reconstruction; Desktop: consistent accuracy within the constrained scan volume.
Geophysical	GPR	Buried	Radargram	Horizontal: 250 mm or +40% of detected depth
Electromagnetic	PCL	Buried	Set of points	3% for <3 m; 5% for 3–5 m
Gyro-based	IMU-based system	Buried newly laid	Dense point trajectory	XY - 0.25%; Z - 0.1%
Electromagnetic RFID	Marker tagging	Buried	Sparse set of points	X, Y, Z axes: ±10 cm; max. depth: 1.5 m

2.1.1. Ground Penetrating Radar (GPR)

GPR is a widely used technology for characterizing structures in the underground. It is based on recording the delay and power of electromagnetic (EM) signals scattered and reflected at discontinuities of the permittivity. Such discontinuities are associated with differences in materials or differences in material properties allowing for detecting, e.g., man-made objects, holes, and layers of different composition or water content in the underground [11,12]. GPR is used for a variety of applications, among them geophysical exploration, archaeology, and inspection of buried utility networks [13,14]. Depending on the type of transmitted signals, impulse radar systems and continuous wave radar systems that are distinguished, with the former being more common [15]. The penetration depth, i.e., the maximum depth at which discontinuities can be detected using GPR is on the order of a few centimeters to a few tens of meters, depending on the soil characteristics, transmission power, signal stacking time and the frequency, which typically ranges from 10 MHz to 4 GHz. Lower frequencies require bigger antennas but facilitate higher penetration depths. Higher frequencies, on the other hand, yield better spatial resolution and thus allow for correctly locating smaller objects or distinguishing objects at smaller distances [13]. 3D information is obtained by moving the radar antennas along the ground surface, recording data quasi-continuously, and subsequently analyzing the data tomographically. Figure 4a shows two examples of GPR instruments, one being integrated with a mobile mapping trailer, and the other one a manually pushed cart.

Although GPR measurement can be very accurate, the responses may vary according to the measurement. A so-called B-scan (i.e., a 2D distance-depth representation of the underground) (see Figure 4b for an example) can be very challenging and is normally done by an experienced radargram analyst. The experience can be generated from a series of signal traces along a trajectory. B-scan normally represented by black and white colours indicative of the different signal strengths and polarities of the objects. These signals are analyzed for anomalous responses. If the positions of these anomalies form a linear line, it is interpreted as a utility feature. The interpretation of B-scan is subjected to the expertise of the radargram analyst or GPR specialist. Such interpretation experience can be gained from a regularly used system of proper training provided by the manufacturer or consultant.

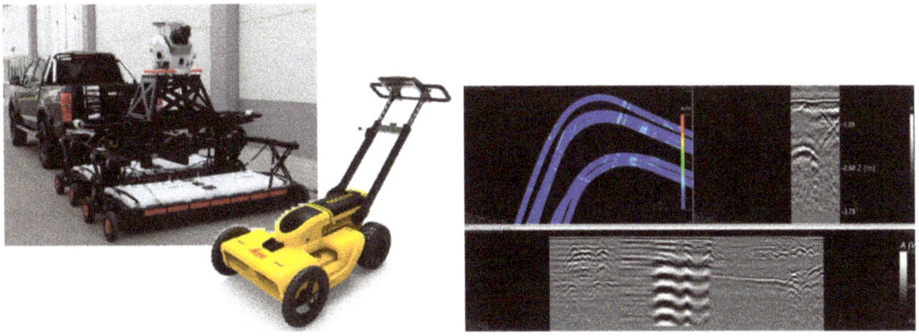

(**a**) (**b**)

Figure 4. Examples of **GPR** instruments (**a**) and **GPR** data (**b**); the data show a radargram of a longitudinal cross section of the top-most about 2.85 m along an asphalt paved road (bottom), a perpendicular cross section of one lane (top right) and the top view of the scanning tracks covered by GPR measurements (top left).

2.1.2. Gyroscope-Based Systems

Utilities with a diameter of more than about 5 cm through which a probe can travel may be accessible to mapping with an inertial measurement unit (IMU). The IMU measures the 3-axis acceleration and 3-axis rotation rates that can be integrated over time yielding position and orientation changes of the unit. If the unit is mounted within a probe and the probe travels through the utility (typically a pipe), it can record the trajectory of the probe—and thus the 3D coordinates of points along the axis of the utility [9].

The potential benefits of such a measurement system are that (i) it can acquire the as-built information of the suitable utilities even if they are buried at a depth exceeding the penetration depth of GPR, (ii) the location can be geometrically more accurate than using above-ground measuring technologies for the location of underground structures, (iii) it can acquire data irrespective of the properties of the surrounding underground (e.g., soil composition, water content) and of electromagnetic fields, and (iv) that the probe can be equipped with additional sensors capturing more information than just the coordinates (e.g., diameter, the radius of curvature, corrosion). Major disadvantages are that (i) only pipes with sufficient diameter, sufficient minimum radius or curvature and accessibility can be measured, (ii) depending on the measurement system, the pipe needs to be empty during the measurement i.e., the service of the utility is interrupted, (iii) the accuracy of the 3D coordinates degrades rapidly with time such that only short parts of the utility, with known coordinates of the start and end point, can be measured if high accuracy is needed, and (iv) additional provisions may be required, e.g., short periods through which the probe remains stationary while moving fast at others. Figure 5b shows an example of such a probe and a 3D map of utilities mapped using it.

(**a**) (**b**)

Figure 5. An example of a gyroscope-based pipeline measurement system (**a**) and the 3D map of the measured pipes (**b**).

At present, GPR seems to be of paramount importance for mapping the underground utilities. However, there are others' current technology that overcomes the shortcoming of GPR available on the market, such as laser scanning or gyro-based system. No single detection technique can detect the entire type of utilities in every location. Hence, GPR is not the only solution for underground utility mapping, as using more technologies increases the detection capability, coverage, efficiency and accuracy. Irrespective of the data acquisition technologies chosen, the information extracted from the measurements, in particular 3D locations, needs to be integrated with attributes of the respective utilities, e.g., type and dimension, in a geospatial database to support 3D visualization, urban planning and other applications.

2.2. *The Review of Underground Utility Data Governance*

Some utility data models have been developed for storage, visualization, exchange, and analysis in the geospatial domain. Obviously, the general data model is not enough to reach all the requirements from different users. In order to develop the 3D data model for the land administration of underground utilities, this work reviews the underground utility data governance in land administration from some countries and the existing data models that are related to underground utility networks and land administration.

2.2.1. Underground Utility Data Governance for Land Administration

The rapid urbanization and increasing complexity of urban spaces worldwide present an urgent need to provide much more and precise information for land usage. Obviously, 2D cadastral information and visualization are not enough for current land administration. During the past decade, a number of works have been conducted to study on the 3D cadastre from various aspects, such as legal, organization and technique [16–18]. The Land Administration Domain Model (LADM) [19] is an important legal framework to define and integrate concepts and terminology of Land Administration for 3D representations. As an international standard, the LADM provides a flexible conceptual schema from three main aspects: organizations, rights and spatial in formations [17]. The integration of 2D and 3D information in the LADM can provide solutions for 3D cadastre. The LADM has two classes (*LA_LegalSpaceutilityNetworke* and *ExPhysicalUtilityNetwork*) specifically describe information about the underground utility, which is not enough to define the 3D geometric and topological characteristics and support to land administration of underground utility.

In recent years, some researchers or government agencies have begun to consider the cadastre for underground infrastructures. To analyze the impact of 4D cadastres in the registration of underground utilities, Döner et al. [20] compared the physical and legal registration of utilities in three countries (Turkey, the Netherlands and Queensland, Australia). Obviously, all of them are supported by a 4D cadastral registration. Pouliot and Girard [18] provided a discussion about the integration of underground utility networks in the land administration system. Based on the case study of Quebec, they discussed three key questions in the following:

- Do we need to register underground objects?
- Should underground networks be registered in the Land Register, with the same specifications as land parcels?
- Which information should be part of the registration process?

Some countries and institutions have implemented or at least conceptualized the 3D mapping of underground utility network and their management in a related cadastral system. Until now, a few countries have utility data with cadastral information and related legislation, includes Switzerland, The Netherlands, Turkey, United Kingdom, Serbia, Sweden, Croatia [21,22].

In Switzerland, the Canton of Zürich started to establish a comprehensive Canton-wide utility cadastre map based on the Cantonal Act on Geoinformation of 2011 [23], derived from the Federal Act on Geoinformation of 2007 [24] and the Cantonal Regulation on Utility Cadastre of

2012 [25]. The regulation sets a deadline for each municipality to deliver and maintain a digital utility map latest until 2021. The City of Zürich has its own utility cadastre since 1999 and set up a governance framework with the corresponding utility providers [26]. Figure 6 shows an example of the utility map of the City of Zürich. The utility cadastre is a subset of the utility documentation of the utility owners. The most important media are included: gas, water, sewage, district heating, power, and telecommunications. SIA 405 [27] is a well-defined standard by the SIA (Swiss society of engineers and architects) for the exchange and publication of utility data. The data model LKMap, part of SIA 405, was introduced to define a unified visualisation/presentation of the utility map. The data are automatically delivered through well defined interfaces at least once a week by the utility owners to the cadastre operator (GeoZ) (central data storage). The utility owners are surveying and using partly 3D coordinates. During the exchange of information between owners and the operator, the information is not yet considered.

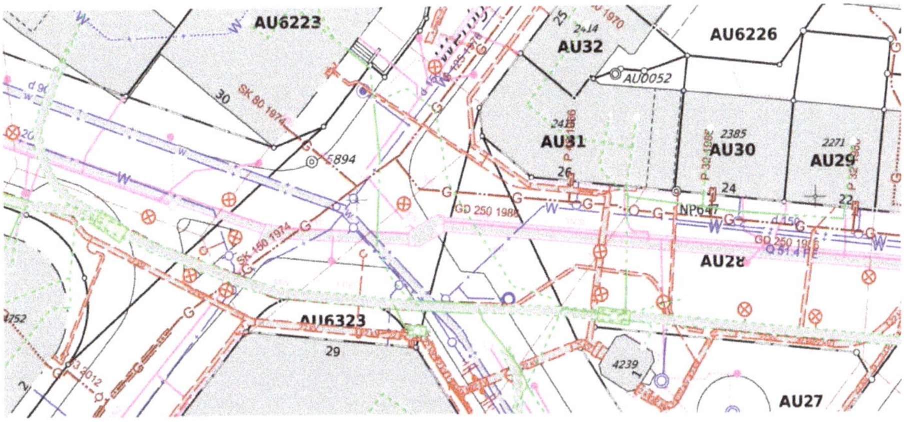

Figure 6. Utility map of City of Zürich (Source: Geomatik + Vermessung Stadt Zürich).

A number of laws related to the exchange of information on utility location exist in the Netherlands. In 2018, the law for storage and exchange of underground utility information was amended. To accommodate the changes introduced by that law as well as the EU INSPIRE guidelines, the KLIC-WIN program was launched. KLIC-WIN is a program (initiated by the digging sector in the Netherlands) that guides, develops and implements changes triggered by the introduction of both the WIBON, which is the law on information exchange of above ground and underground networks, and the new EU INSPIRE guidelines for utility network information retrieval. KLIC-WIN aims to introduce some changes that are required to comply with the new WIBON law and INSPIRE guidelines:

- Representation of utility information according to a new information model,
- The ability to (optionally) centrally store utility information at Kadaster,
- The gradual change of utility data formats for delivery to end users (from raster now to vector data in 2019 and/or beyond).

Furthermore, Serbia extends its LADM based country profile to include utility information for utility network cadastre [28]. Based on this data model, they will develop a system to register and maintain the ownership of the underground utility network. The United Kingdom began the registry of underground utilities and created a national underground assets mapping platform in 2018. The register aims to show where electricity and telecom cables, and gas and water pipes are buried and is intended to prevent both accidents and disruption to the economy. In Croatia, the utility cadastre information contains the type, purpose, basic technical features, and location of built utility lines, and lists the names and addresses of their managers [29]. The Croatia changed the law to organize the physical registration of utilities at a national level since 2016 [30]. Moreover, Canada has developed 3D maps of underground utility networks as well [18,31]. In general, some countries have 2D visualization of utility networks on cadastral map, legal document about utility data governance, registration of legal ownership of utility networks by law. Most of them begin to develop the 3D/4D utility cadastre. All of the current work is just beginning and ongoing. This has been a new challenging topic in recent years.

2.2.2. The 3D Data Model for Underground Utility Networks

The CityGML utility network Application Domain Extension (ADE) [32] focuses mainly on three aspects: (i) the general 3D geometric of network components; (ii) the 3D topographical structure of the entire utility network; and (iii) the functional information of different types of the network [32,33]. Based on the general concepts of the utility network, different types of utility networks can be implemented with their specific function [32]. Moreover, the interdependence between utility network features and city objects can be presented in 3D space [34]. Because this data model is an extension of CityGML [35], which is the popular standard for 3D city modelling (e.g., building), it is beneficial to integrate information of utility networks to the infrastructures to support urban planning and the other city studies. However, it does not consider the accuracy of the data. Some works begin to extend the existing data model to consider many more details about utility networks, such as [36], represent geographical uncertainties of utility locations based on CityGML Utility Network ADE.

The Industry Foundation Classes (IFC) utility model [37] is an ISO standard for data exchange of buildings in the architecture and civil engineering domain [32]. In the utility part, it describes 2D and 3D geometries of utility elements. Meanwhile, two different ways of connection are defined to describe the relationship between supply service components within the building, which is a logical and physical connection. In addition, it has a comprehensive semantic definition of utility network objects. However, this standard only focuses on the building level and lacks spatial information.

The INSPIRE Data Specification on Utility and Government Services—Technical Guidelines [38] organize the basic information of utility networks and administrative services of utility networks in a city or country range. It is a part of INSPIRE, which is a standard of the European Union to describe the spatial information of infrastructures. However, the INSPIRE Utility networks are lacking a definition of 3D geometric information of utility networks.

ESRI Utility Network model [39] provides a GIS-based utility solution to represent the basic logical and physical structure of all types of utility networks, which is composed of edges and junctions. This model is a general utility data model to represent the 2D geometric information and connections of the utility networks.

Until now, there has not been an international standard that has been widely used for 3D modelling of underground utility [40]. Although some existing standardized data models have been developed to integrate multi-utility networks, they can not guarantee the information to be reliable [3]. In order to develop a comprehensive utility database, we have the challenge to integrate different types of utility datasets from multiple surveying techniques, as well as the existing 2D data. Table 2 compares four popular utility data models relevant to the objectives of this work. Obviously, most of the existing utility data models are to focus on the 3D representation, and include 3D geometric and topological information. The existing data models provide a good reference to describe the geometric and spatial information of utility networks in 3D. Nevertheless, none of them considers the accuracy of data of underground utility networks. On the one side, the survey technique directly impacts the data accuracy. However, industry service providers are not usually aware of these extensive standards [3]. On the other side, different applications might use data at different levels of accuracy. Hence, we need an ideally 3D utility data model to support mapping procedures and control accuracy of underground utility network data.

Table 2. Comparison of model characteristics.

	CityGML Utility Network ADE	ArcGIS Utility Networks	INSPIRE Utility Networks	IFC
3D representation modelling				
-3D geometries	+	-	-	++
-Topological aspects	++	++	++	++
-Hierarchical modelling	++	.	++	+
Land Administration	-	-	-	-
Data quality management	-	-	-	-

-: no support, .: basic support, +: sophisticated support, ++: comprehensive support.

On the basis of their discussion and the situation of Singapore, it is necessary to register the utility segments as the legal objects in the land administration system, which helps to identify the ownership of underground utility. An integral approach needs to be developed based on legislative and technology solutions. It is essential to establish a degree of reliability and consistency between data produced by different service providers. It is essential to standardize the practices regarding the use of those techniques and various information management. In the underground utility data model, land parcel, as an important role in the land administration, should be connected to the underground utility networks [18,21].

3. Design of the 3D Data Model for Underground Utility Networks

3.1. A Framework for Utility Data Governance

From data capture to usage, the whole work process includes several participants in different stages. Hence, in order to improve the communication between different organizations at each phase, our previous work [3] proposed a framework for underground utility data governance. After observing the current work process in Singapore, this framework has been improved to organize the entire work process (Figure 7). This framework consists of five roles that are listed in the following:

- The data producer is the surveying constructor and/or surveyor in the data regulatory bodies' organization. In the utility survey phase, the data producer captures data in the field work and submits data to the utility network database.
- The data owner manages their collected data. This role could be companies or data regulatory bodies.
- Data regulatory bodies are government agencies, such as SLA or Public Utilities Board (PUB) of Singapore. They manage their utility data based on their utility network data model. The data regulatory bodies should provide clear permission for data integrator to use and the predefined subset of utility data.
- The data integrator integrates all kinds of utility network data and manages the utility cadastre information in a city or country. In the phase of utility cadastre management, the data integrator should provide the required information for the application to users. This role builds a bridge between the data regulatory bodies and users.
- Data users can use utility data for utility cadastre management applications.

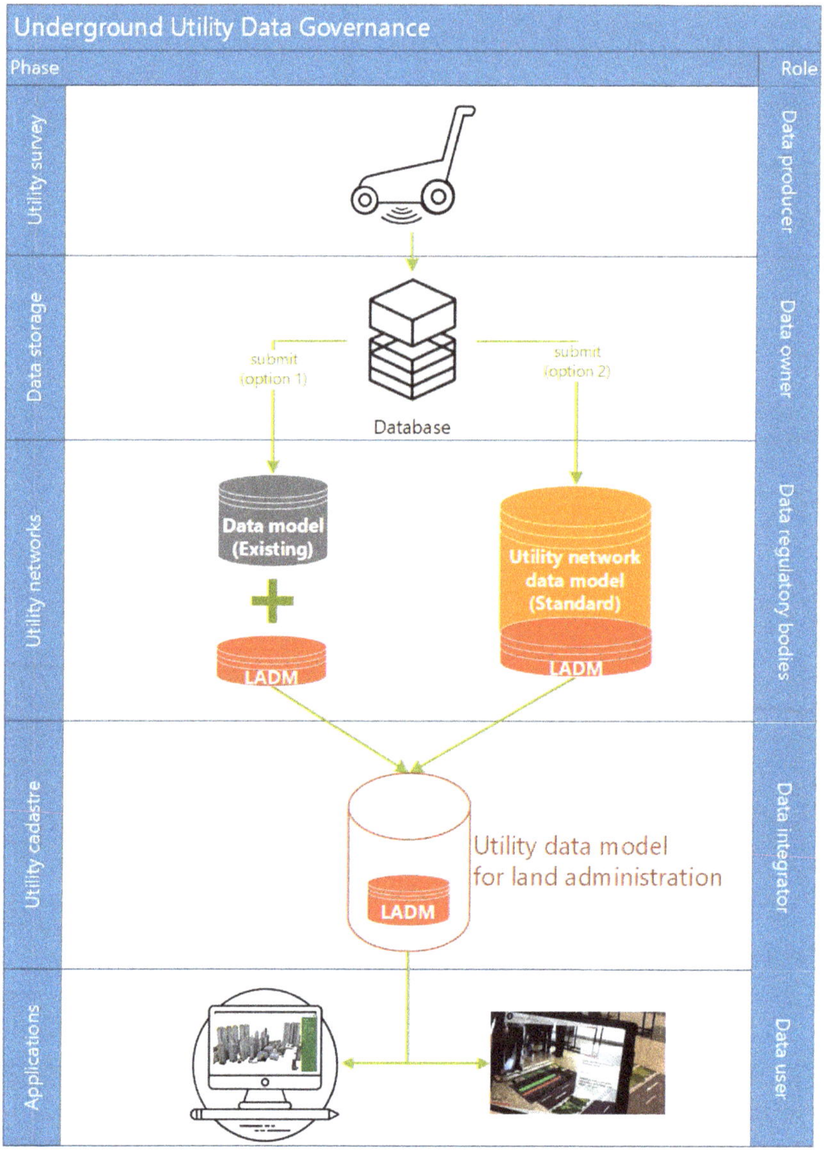

Figure 7. The framework of underground utility network data governance.

In this work process, the surveyor as data producer captures data during the field work. After that, the data will be submitted to data owner (e.g., PUB) who needs to manage their own utility networks data. According to the requirements of government, the utility data will be submitted to data regulatory bodies (e.g., PUB and SLA). There are two options for data submission. A general utility network data model will be designed as a standard to manage underground utility data for data regulatory bodies. If the data regulatory body does not have any utility data model, they can use this standard data model. If they have their utility data model, they can continue to use it or change to use the standard one. A consolidated 3D utility data model will be designed to support utility cadastre management. The data integrator (e.g., SLA) needs to integrate data of different kinds of utility networks. The LADM plays as a connection component to build a relationship between the general utility network data model and utility cadastre data model in the utility cadastre management. Meanwhile, the LADM will connect the underground utility network to the land administration of above ground. Finally, the underground utility data model should support applications in land administrative management.

3.2. 3D Underground Utility Data Model for Land Administration

Current work focuses on the conceptual design of a 3D underground utility data model and connects it to land administration. In order to understand the demands of underground utility data users, a workshop was organized to learn the work process and needs of land administration in Singapore. This studying includes four application domains: land acquisition and purchase, planning and coordination, land transfer and sale, and land leasing. Currently, the existing data sources are the hardcopy of the utility network, 2D CAD and 2D geospatial information. There is an urgent demand of 3D geospatial information of underground utility and space to evaluate underground environment and support reallocation, land sales and the other applications. Therefore, the 3D underground utility data model includes three packages to organize the basic information and structure of utility networks, utility survey information, and the land administration information (Figure 8). In order to connect the 3D underground utility data model to the information of land administration, these three packages inherit from the Singapore cadastral data model and LADM (ISO 19152). Meanwhile, the geometric and spatial definition are inherited from the spatial schema data model ([41]).

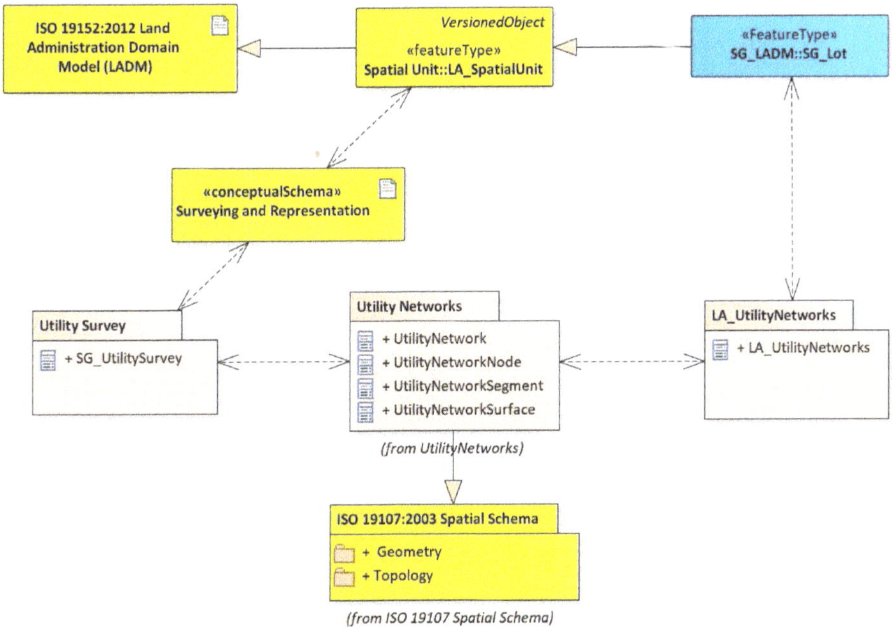

Figure 8. The overview of packages of the 3D underground utility data model.

The *Utility Networks* package describes the basic information of utility networks, which includes geometric, spatial and physical information. Based on the partonomy (part–whole) relationships, this work defines the hierarchy of utility networks in three levels (Figure 9). The macro-level is the whole utility networks, which is described by the *UtilityNetwork* class with the basic information of utility networks, such as the type, and material of utility networks. The meso-level is the surface of the utility networks, which is the part of the utility networks. The surface could be the tunnel, duck, manhole and the other types of space in the utility networks. Hence, the aims of

UtilityNetworkSurface class are to describe the types and 3D geometric information (e.g., diameter) of surface. The micro-level is the basic elements of utility networks, which includes nodes and segments of utilities. The node is a connection point in the network, which is defined by the *UtilityNetworkNode* class. The segment is the line segment of the utility, which is defined by the *UtilityNetworkSegment* class. The relationship between micro and meso level helps to transform 2D to 3D data as well. Figure 10 shows the relationships of different classes in the *Utility Network* package and basic attributes of each class. The values of utility networks type inherit from LA_LegalSpaceUtilityNetwork in the LADM (ISO 19152) [19].

The *LA_UtilityNetworks* class aims to describe the land administration information of utilities. On one side, it connects to the utility network surface in order to identify the land administration information of different parts of utility networks. On the other side, it connects to the cadastral parcel from the Singapore cadastral data model and LADM [19]. The spatial relationship is used to describe the relationship of cadastral parcels and utilities, which includes contain, cross and touch. This class could support ownership management of utilities and land administration management.

The *Utility Survey* class aims to organize utility survey information. It could help to manage survey status and accuracy of data. The *Utility Survey* class inherits attributes of the survey from the Singapore cadastral data model. Furthermore, the ground conditions and survey methods are related to the accuracy of data directly. Hence, the *Utility Survey* class integrates information from Standard and Specification for Utility Survey in Singapore [42]. Meanwhile, the *Utility Survey* class builds the connection between utility networks and *LA_Point*, *LA_BoundaryFace* and *LA_SpatialSource* in the *Surveying and Representation* package. The *Evaluate* attribute describes the method to check the accuracy of surveying data. If the accuracy of the data is unknown, the value of *Evaluate* is null. In future work, the accuracy level should be defined to be based on the depth level, soil condition and survey method.

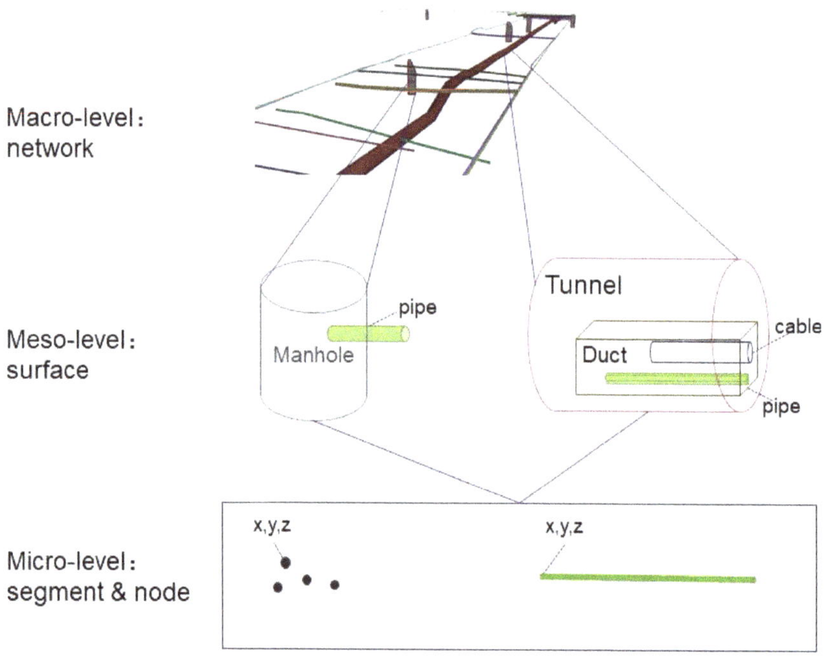

Figure 9. Multilevel structure of utility networks.

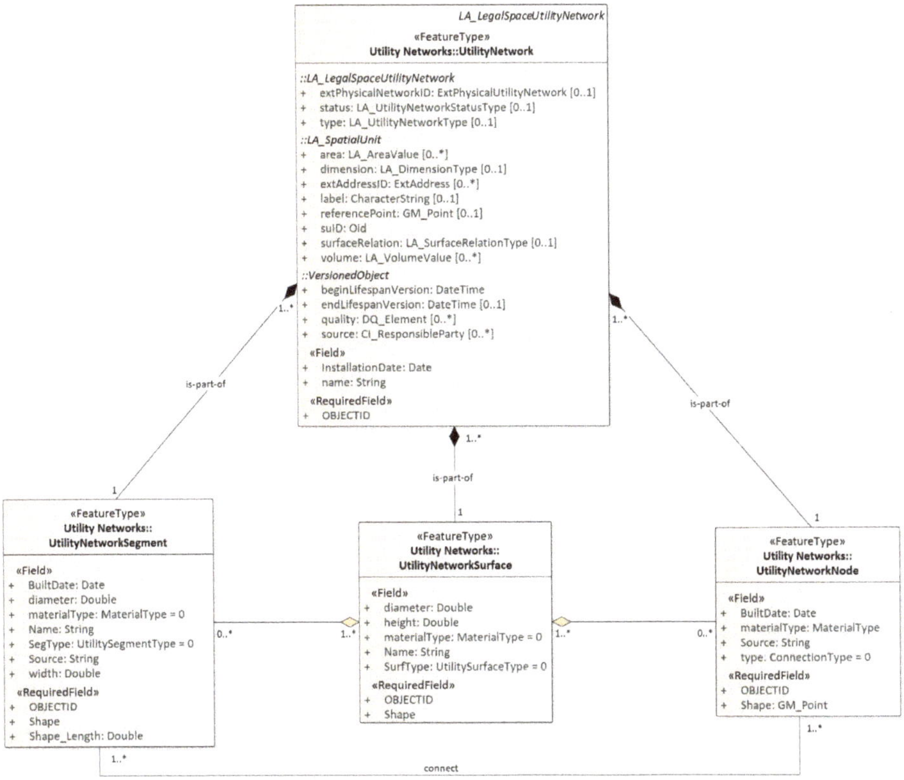

Figure 10. The classes diagram of utility networks.

4. Case Study

This initial study aims to integrate of GPR data and the existing underground utility data and land cadastral data in the form of the geospatial database. It aims to find a reasonable work process to bridge the gap between data capture and application. Moreover, this implementation can help to improve the design of a 3D data model for underground utility.

4.1. Study Area and Datasets

This initial study was conducted at around Lorong 2, 3 and 4 at Toa Payoh, which is located in the northern part of Singapore. This is one of the pilot study sites in our project to deploy a mobile mapping platform, namely Pegasus: Stream (https://idsgeoradar.com/products/ground-penetrating-radar/pegasus-stream) combines a Stream EM GPR (IDS Georadar, part of Hexagon, Switzerland) and Leica Pegasus Two (Leica geosystem AG, part of Hexagon, Switzerland) photo and laser scanner for massive 3D mapping of above and underground features. The data captured by the Pegasus: Stream is geo-referenced using an on-board GNSS receiver and IMU and a distance measurement instrument (DMI). The Stream EM GPR contains a large number of array antennae, with dual frequencies (200 MHz and 600 MHz). The antennae transmit and receive in two distinct polarizations (HH and VV), allowing the reconstruction of a 3D underground utility network with a single pass of the GPR. Table 3 shows the technical specification of the Stream EM GPR.

The scanning site is a 1.8 km long bi-directional 4-lane asphalt road in an inland area of Singapore that has seen development since the 1960s. This study was conducted to investigate the feasibility of GPR for large scale underground utility mapping for the purpose of improving the quality of existing utility map information. The data were collected at a driving speed of about 15 km/h. All of the acquired data were post-processed and interpreted to detect and extract underground utilities using a commercial off-the-shelf processing software along with the GPR system. At the current stage, we do not use point cloud data of the above ground. The identified utilities were then transferred to CAD/GIS format with x, y, z values as points and lines for 3D data modelling and visualization using the same processing software. Figure 11 shows an example of GPR data in CAD (Figure 11a) and GIS (Figure 11b) format.

Table 3. Technical specifications of the Stream EM GPR.

Overall weight	228 kg (500 lbs)
Max. acquisition speed	15 kph (9mph)
Positioning	Survey wheel and/or GPS or Total Station
Scan Rate per Channel (@512 samples/scan)	87 scans/sec
Scan Interval	17 scans/m @ 200 MHz 33 scans/m @ 600 MHz
Antenna Footprint	Width 1.84 m
Number of Channel	38
Antenna Central Frequencies	200 MHz (34 channels) 600 MHz (4 channels)
Antenna Spacing	6 cm
Antenna Polarization	Horizontal (HH) and Vertical (VV)

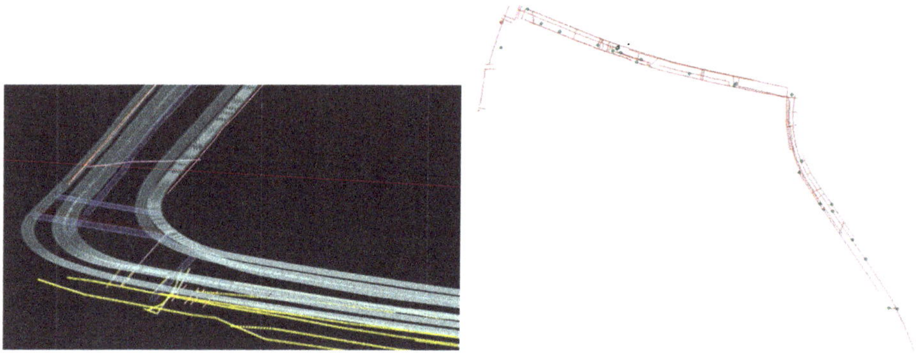

(**a**) CAD (**b**) GIS

Figure 11. GPR data in CAD and GIS format.

The existing datasets from Geospace and cadastral data from Singapore Land Authority were used as secondary data to obtain or improve the attributes of utilities that were extracted from the radargram and to explore the relationship between the above land administration information and underground utilities. These existing utility data are as-build data from utility services (e.g., power, water, gas, telecommunication and sewerage) and cadastral information in 2D form. Of these datasets, it contains only a small portion of the information that has a diameter with updated time and type. It possesses challenges for land planning with such limited information.

4.2. 3D Visualisation

To develop the 3D utility data model for land administration, the underground utilities need to be connected to the land parcels. Figure 12 explains the work process in this case study. The data model is designed in UML and exported to XML format, which can be imported into ArcGIS as a geodatabase schema. Based on the database schema, the GPR data can be loaded as utility network components in polyline and point. According to the information from the existing utility data and GPR data, the utilities can be modelled in 3D (multipath). The 3D modelling is realized manually in the ArcSence and CityEngine.

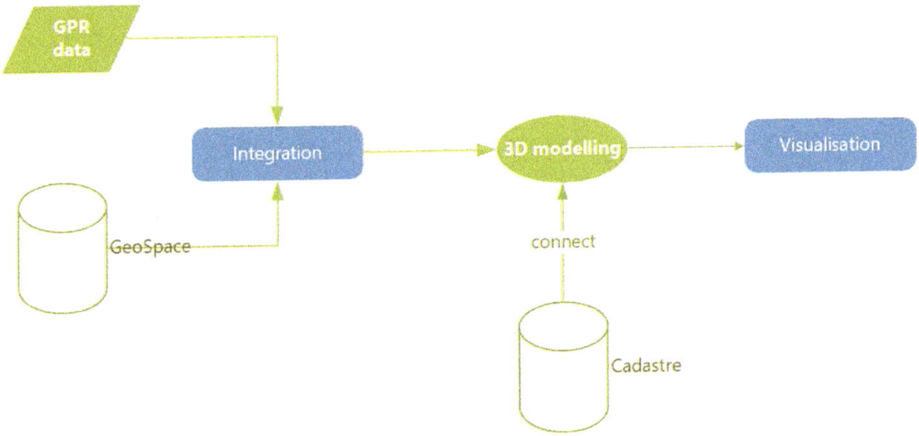

Figure 12. Implementation process.

In order to get the related land administration information, the utility networks data can be integrated with a cadastral parcel through their spatial relationships. Because the existing cadastral data are in 2D, the current work only considers the pipeline within the cadastral parcel in 2D. In order to improve the accuracy of data in 3D, the current cadastral data have to be extended to 3D so as to support more spatial relationships (e.g., cross and touch). Figure 13 shows an example of 3D visualization of utilities with objects above ground. As shown in the figure, the selected pipeline is highlighted in pink. The information shown in the pop window includes spatial data from GPR and other attributes about the underground utility survey and land cadastral information above ground.

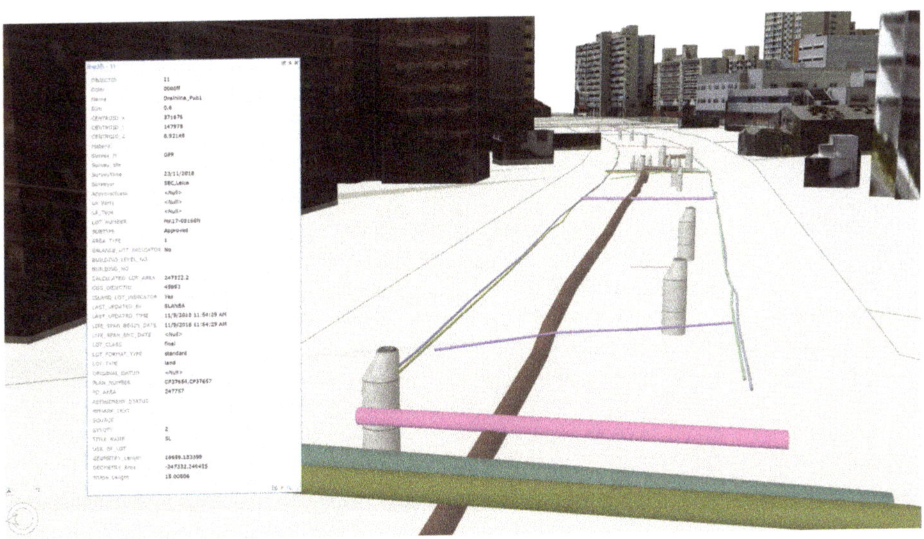

Figure 13. A result example of 3D visualization with land administration information.

4.3. Discussion

This is a simple implementation to explore the work process of 3D modelling of underground utility from the GPR data and existing 2D data. Because GPR cannot capture the diameters, material and some attributes of utilities, it is necessary to extract these information from the GeoSpace database for 3D modelling. Depending on the spatial relationship (e.g., overlap, within) of the GPR data and existing utility data, some of the utilities from GPR data can be connected to the existing utility data. Because of two main limitations, there is a big challenge to improve the accuracy of data during the manual integration of the GPR and existing data. First, the existing utility data are as-build data which may not be reliable enough for updating work. Second, the existing utility data are in 2D data, which is difficult to identify utilities accurately. Hence, the future work needs to find the solution to detect much more attributes of utilities from GPR data. In addition, the tentative integration of underground utility and land cadastral data helps to improve the development of the data model for land administration.

5. Conclusions

This paper proposes to develop a consolidated 3D data model of underground utilities for land administration. The work includes two parts. On the one hand, a framework for data governance is designed to organize the workflow of utility data survey, management and application through five roles. Through the understanding of current workflow in the utility data usage, this work needs to clearly define the operations and rights of each role in the work process of 3D underground utility mapping. On the other hand, a 3D data model of underground utilities is designed with 3D spatial information, i.e., utility survey information, and land administration information of underground utilities. In order to fill the gap between data capture and usage, this data model has the following main tasks:

- Integrating utility networks data from varying non-destructive surveying technologies. Moreover, it proposes an idea to manage the data accuracy based on the parameters, ground condition and other information during the field survey. This is a first step towards bridging the gap between data acquisition and data management for 3D underground utility mapping.
- Integrating the existing data and GPR data. As mentioned earlier, GPR data cannot get the diameters and types of utilities. This way helps to improve the attributes of utilities from GPR data. Moreover, it is also a process to transform utility data from 2D to 3D.
- In the data integration, the key step is to connect the utility network data model with the LADM for 3D cadastral management of underground utility in Singapore. It is useful to support ownership management applications and build the relationship between utilities and land parcels. Such a reliable and consolidated centralized repository of underground utility data will provide a crucial basis for land administration of underground infrastructures.

A case study is implemented based on the GPR data from the large scale mobile underground utility mapping. The initial implementation transform GPR data from CAD to GIS format and 3D visualization of utilities based on the 3D utility data model. In order to get land administration information, the utility networks have been connecting to the cadastral parcel. The accuracy and details of utility networks need to be improved in future work, such as the spatial relationship between utilities and cadastral parcels. To fully support the land administration of underground space, the 3D utility data model should eventually be extended to include other underground objects and infrastructures in the future, such as underground substations, pedestrian links, common services tunnels, road and rail networks, etc.

This is an ongoing work and in the initial stage. Two main aspects of limitations need to be improved in future work. First, for the accuracy of utility data. Obviously, the GPR data are not enough to provide comprehensive 3D underground utility networks. The other kinds of data (e.g., Gyroscope) should be integrated to provide more precise attributes for underground utilities. Moreover, the details of the shapes and structures of utilities need to be improved. Second, the next step of the data model development will improve the definition of land administration for underground utilities. Additionally, in order to develop a comprehensive underground utility database, it is necessary to explore the methods to use the existing data and integrate it with newly collected data. The 3D data model should be extended to be 4D (3D + time) to support data updating. A showcase will be developed to realize land administration of underground utility based on a 3D underground utility data model. This will work with a selected agency as data regulatory body and the preferred data integrator. They will help us to evaluate and improve the framework and definition of the data model. After that, recommendations from this showcase will be used to extend the data model include other underground infrastructures and develop the platform of underground space management to support various applications in Singapore.

Author Contributions: Conceptualization, J.Y. and K.H.S.; methodology, J.Y. and K.H.S.; implementation, J.Y.; Data curation, J.Y. and S.W.J.; writing—original draft preparation, J.Y.; writing—review and editing, S.W.J., K.H.S., G.S. and A.W.; visualization, J.Y.; supervision, G.S. and A.W.

Funding: Underground-related studies and projects fund (USPF), Ministry of National Development and is sponsored by the Singapore Land Authority.

Acknowledgments: This publication has been realized as part of the project "Digital Underground: 3D Mapping of Utility Networks" at the Future Cities Laboratory, established by ETH-Zürich and Singapore's National Research Foundation (NRF), and operating under the auspices of the Singapore-ETH Centre.

Conflicts of Interest: The authors declare no conflict of interest.

References

1. The Urban Redevelopment Authority (URA). *Draft Master Plan 2019—Proposals for an Inclusive, Sustainable and Resilient City*; Urban Redevelopment Authority: Singapore, 2019. Available online: www.ura.gov.sg/Corporate/Media-Room/Media-Releases/pr19-13 (accessed on 20 August 2019).
2. Jaw, S.W.; Van Son, R.; Khoo, V.H.S.; Schrotter, G.; Loo, R.W.K.; Teo, S.S.N.; Yan, J. The Need for a Reliable Map of Utility Networks for Planning Underground Spaces. In Proceedings of the 2018 17th International Conference on Ground Penetrating Radar (GPR), Rapperswil, Switzerland, 18–21 June 2018; pp. 1–6.
3. Yan, J.; Jaw, S.W.; Son, R.V.; Soon, K.H.; Schrotter, G. Three-dimensional data modelling for underground utility network mapping. *ISPRS Int. Arch. Photogramm. Remote Sens. Spat. Inf. Sci.* **2018**, *XLII-4*, 711–715. [CrossRef]
4. Hao, T.; Rogers, C.; Metje, N.; Chapman, D.; Muggleton, J.; Foo, K.; Wang, P.; Pennock, S.; Atkins, P.; Swingler, S.; et al. Condition assessment of the buried utility service infrastructure. *Tunn. Undergr. Space Technol.* **2012**, *28*, 331–344. [CrossRef]
5. Lai, W.W.L.; Dérobert, X.; Annan, P. A review of ground penetrating radar application in civil engineering: A 30-year journey from locating and testing to imaging and diagnosis. *NDT E Int.* **2018**, *96*, 58–78.
6. Ni, S.H.; Huang, Y.H.; Lo, K.F.; Lin, D.C. Buried pipe detection by ground penetrating radar using the discrete wavelet transform. *Comput. Geotech.* **2010**, *37*, 440–448. [CrossRef]
7. Rogers, C.; Hao, T.; Costello, S.; Burrow, M.; Metje, N.; Chapman, D.; Parker, J.; Armitage, R.; Anspach, J.; Muggleton, J.; et al. Condition assessment of the surface and buried infrastructure—A proposal for integration. *Tunn. Undergr. Space Technol.* **2012**, *28*, 202–211. [CrossRef]
8. Lagüela, S.; Solla, M.; Puente, I.; Prego, F.J. Joint use of GPR, IRT and TLS techniques for the integral damage detection in paving. *Constr. Build. Mater.* **2018**, *174*, 749–760. [CrossRef]
9. van Disseldorp, R. Unique Autonomous Pipeline Mapping System. In Proceedings of the 2nd Pipeline Technology Conference, Hannover, Germany, 16–17 April 2007.
10. The British Standards Institution. *PAS 128:2014 Specification for Underground Utility Detection, Verification and Location*; BSI Standards Limited: London, UK, 2014.
11. Sato, M. *GPR and Its Application to Environmental Study*; Center for Northeast Asia Studies (CNEAS), Tohoku University: Miyagi, Japan, 2001.
12. Lunt, I.A.; Hubbard, S.S.; Rubin, Y. Soil moisture content estimation using ground-penetrating radar reflection data. *J. Hydrol.* **2005**, *307*, 254–269. [CrossRef]

13. Lai, W.W.L.; Chang, R.K.W.; Sham, J.F.C. A blind test of nondestructive underground void detection by ground penetrating radar (GPR). *J. Appl. Geophys.* **2018**, *149*, 10–17. [CrossRef]
14. Desai, L. Underground Utility Survey. *Imper. J. Interdiscip. Res. (IJIR)* **2016**, *2*, 4.
15. Yelf, R.J. Application of ground penetrating radar to civil and geotechnical engineering. *Electromagn. Phenom.* **2007**, *7*, 18.
16. van Oosterom, P. Research and development in 3D cadastres. *Comput. Environ. Urban Syst.* **2013**, *40*, 1–6. [CrossRef]
17. Van Oosterom, P.; Lemmen, C. The Land Administration Domain Model (LADM): Motivation, standardisation, application and further development. *Land Use Policy* **2015**, *49*, 527–534. [CrossRef]
18. Pouliot, J.; Girard, P. *3D Cadastre: With or without Subsurface Utility Network?* International Federation of Surveyors (FIG): Athens, Greece, 2016; OCLC: 831214525.
19. International Standards Organization. *ISO 19152: Geographic information—Land Administration Domain Model (LADM)*; ISO Copyright Office: Geneva, Switzerland, 2012.
20. Döner, F.; Thompson, R.; Stoter, J.; Lemmen, C.; Ploeger, H.; van Oosterom, P.; Zlatanova, S. 4D cadastres: First analysis of legal, organizational, and technical impact—With a case study on utility networks. *Land Use Policy* **2010**, *27*, 1068–1081. [CrossRef]
21. Döner, F.; Thompson, R.; Stoter, J.; Lemmen, C.; Ploeger, H.; Oosterom, P.v.; Zlatanova, S. Solutions for 4D cadastre—With a case study on utility networks. *Int. J. Geogr. Inf. Sci.* **2011**, *25*, 1173–1189. [CrossRef]
22. Oosterom, P.V.; Stoter, J.; Ploeger, H. Initial Analysis of the Second FIG 3D Cadastres Questionnaire: Status in 2014 and Expectations for 2018. In Proceedings of the 4th International FIG 3D Cadastre Workshop, Dubai, UAE, 9–11 November 2014.
23. Kantonales Geoinformationsgesetz (KGeoIG). 2018. Available online: www.zhlex.zh.ch/Erlass.html?Open&Ordnr=704.1 (accessed on 26 July 2019).
24. Geoinformation Act, GeoIA. 2007. Available online: www.admin.ch/opc/en/classified-compilation/20050726/index.html#fn1 (accessed on 26 July 2019).
25. Leitungskatasterverordnung (LKV). 2012. Available online: www.zhlex.zh.ch/Erlass.html?Open&Ordnr=704.14 (accessed on 26 July 2019).
26. Kanton Zürich. Leitungskataster-Stadt Zurich. Zürich. 2017. Available online: www.stadt-zuerich.ch/ted/de/index/geoz/geodaten_u_plaene/leitungskataster.html (accessed on 26 July 2019).
27. Geodaten zu Ver-und Entsorgungsleitungen. 2012. Available online: www.webnorm.ch/null/null/sia%20405/d/2012/D/Product (accessed on 26 July 2019).
28. Radulović, A.; Sladić, D.; Govedarica, M.; Ristić, A.; Jovanović, D. LADM Based Utility Network Cadastre in Serbia. *ISPRS Int. J. Geo-Inf.* **2019**, *8*, 206. [CrossRef]
29. Vu, N.; Roi, M.; Markovinovi, D. Towards 3D and 4D Cadastre in Croatia. In Proceedings of the 4th International Workshop on 3D Cadastres, Dubai, UAE, 9–11 November 2014; pp. 261–280.
30. Vučić, N.; Roić, M.; Mađer, M.; Vranić, S.; Van Oosterom, P. Overview of the Croatian Land Administration System and the Possibilities for Its Upgrade to 3D by Existing Data. *ISPRS Int. J. Geo-Inf.* **2017**, *6*, 223. [CrossRef]
31. Tan, L.C.; Looi, K.S. Towards a Malaysian multipurpose 3D cadastre based on the Land Administration Domain Model (LADM)—An empirical study. In Proceedings of the 5th FIG Land Administration Domain Model Workshop, Kuala Lumpur, Malaysia, 24–25 September 2013; pp. 24–25.
32. Becker, T.; Nagel, C.; Kolbe, T.H. Integrated 3D Modeling of Multi-Utility Networks and Their Interdependencies for Critical Infrastructure Analysis. In *Advances in 3D Geo-Information Sciences*; Springer: Berlin/Heidelberg, Germany, 2011; pp. 1–20.
33. Becker, T.; Nagel, C.; Kolbe, T.H. Semantic 3D modeling of multi-utility networks in cities for analysis and 3D visualization. In *Progress and New Trends in 3D Geoinformation Sciences*; Springer: Berlin/Heidelberg, Germany, 2013; pp. 41–62.
34. Hijazi, I.; Kutzner, T.; Kolbe, T.H. *Use Cases and their Requirements on the Semantic Modeling of 3D Supply and Disposal Networks*; Kulturelles Erbe erfassen und bewahren-Von der Dokumentation zum virtuellen Rundgang, 37; Wissenschaftlich-Technische Jahrestagung der DGPF: Würzburg, Germany, 2017; pp. 288–301.
35. Gröger, G.; Kolbe, T.H.; Nagel, C.; Häfele, K.H. *OGC City Geography Markup Language (CityGML) Encoding Standard*; Version 2.0, OGC doc no. 12-019; Open Geospatial Consortium: Wayland, MA, USA, 2012.

36. Scholtenhuis, L.L.o.; Duijn, X.d.; Zlatanova, S. Representing geographical uncertainties of utility location data in 3D. *Autom. Constr.* **2018**, *96*, 483–493. [CrossRef]
37. Liebich, T. IFC 2x Edition 3. In *Model Implementation Guide*, Version 2.0; buildingSMART International: Hertfordshire, UK, 2009.
38. JRC. *D2.8.III.6 INSPIRE Data Specification on Utility and Government Services—Technical Guidelines*; European Commission Joint Research Centre: Brussels, Belgium, 2013.
39. ESRI. What Are Geometric Networks? ESRI, USA, 2017. Available online: https://desktop.arcgis.com/en/arcmap/10.3/manage-data/geometric-networks/what-are-geometric-networks-.html (accessed on 17 August 2019).
40. Lieberman, J.; Ryan, A. *OGC Underground Infrastructure Concept Study Engineering Report*; OGC Engineering Report; Open Geospatial Consortium: Wayland, MA, USA, 2017.
41. International Standards Organization. *ISO 19107:2003(en), Geographic information—Spatial Schema*; ISO Copyright Office: Geneva, Switzerland, 2003.
42. Singapore Land Authority. *Standard and Specifications for Utility Survey in Singapore*; Singapore Land Authority: Singapore, 2017.

remote sensing

MDPI

Article

Deep Fully Convolutional Networks for Cadastral Boundary Detection from UAV Images

Xue Xia, Claudio Persello * and Mila Koeva

Faculty of Geo-Information Science and Earth Observation (ITC), University of Twente, 7522 NB Enschede, The Netherlands

* Correspondence: c.persello@utwente.nl

Received: 12 June 2019; Accepted: 18 July 2019; Published: 20 July 2019

Abstract: There is a growing demand for cheap and fast cadastral mapping methods to face the challenge of 70% global unregistered land rights. As traditional on-site field surveying is time-consuming and labor intensive, imagery-based cadastral mapping has in recent years been advocated by fit-for-purpose (FFP) land administration. However, owing to the semantic gap between the high-level cadastral boundary concept and low-level visual cues in the imagery, improving the accuracy of automatic boundary delineation remains a major challenge. In this research, we use imageries acquired by Unmanned Aerial Vehicles (UAV) to explore the potential of deep Fully Convolutional Networks (FCNs) for cadastral boundary detection in urban and semi-urban areas. We test the performance of FCNs against other state-of-the-art techniques, including Multi-Resolution Segmentation (MRS) and Globalized Probability of Boundary (gPb) in two case study sites in Rwanda. Experimental results show that FCNs outperformed MRS and gPb in both study areas and achieved an average accuracy of 0.79 in precision, 0.37 in recall and 0.50 in F-score. In conclusion, FCNs are able to effectively extract cadastral boundaries, especially when a large proportion of cadastral boundaries are visible. This automated method could minimize manual digitization and reduce field work, thus facilitating the current cadastral mapping and updating practices.

Keywords: deep learning; fully convolutional networks; cadaster boundaries; contour detection; unmanned aerial vehicles

1. Introduction

Cadasters, which record the physical location and ownership of the real properties, are the basis of land administration systems [1]. Nowadays, cadastral mapping has received considerable critical attention. An effective cadastral system formalizes private property rights, which is very important to promote agricultural productivity, secure effective land market, reduce poverty and support national development in the broadest sense [2]. However, it is estimated that over 70% of the world population does not have access to a formal cadastral system [3]. Traditional field surveying approaches to record land parcels are often claimed to be time-consuming, costly and labor intensive. Therefore, there is a clear need for innovative tools to speed up this process.

Since the availability of very high resolution (VHR) satellite or aerial images, remote sensing has been used for mapping cadastral boundaries instead of field surveying, and is advocated by fit-for-purpose (FFP) land administration [4]. In practice, cadastral boundaries are often marked by physical objects, such as roads, building walls, fences, water drainages, ditches, rivers, clusters of stones, strips of uncultivated land, etc. [1]. Such boundaries are visible in remotely sensed images and bear the potential to be automatically extracted through image analysis algorithms, hence avoiding huge fieldwork surveying tasks. According to FFP, boundaries measured through on-site cadastral surveys using total station or the Global Navigation Satellite System (GNSS) with a precise location are considered as fixed, while boundaries delineated from high-resolution imagery are called general [4]. Although less spatially precise, general boundary approaches are much cheaper and faster than conventional cadastral surveys. Typically, high-resolution satellite images (HRSI) have been used for interpreting cadastral boundaries, but there are still obstacles such as high cost, cloudy or dated imagery [5]. Therefore, Unmanned Aerial Vehicles (UAV), renowned for low-cost and high spatial–temporal resolution, as well as being able to fly under clouds, are chosen as the data source for cadastral boundary extraction in this research.

The detection of cadastral boundaries from remotely sensed images is a difficult task. Above all, only visible cadastral boundaries coinciding with physical objects are detectable in the image. Moreover, as visible cadastral boundaries can be marked by different objects, spectral information alone is insufficient for the detection. In other words, there exists a sematic gap between the high-level boundary concept and low-level visual cues in the image. More reliable and informative features should be constructed to bridge the semantic gap, thus more advanced feature extraction techniques are needed.

State-of-the-art methods are mostly based on image segmentation and edge detection [6]. Segmentation refers to partitioning images into disjoint regions, inside which the pixels are similar to each other with regard to spectral characteristics [7]. Researchers claimed that segmentation-based approaches have two general drawbacks: Sensitive to intra-parcel variability and dependent on parameter selection. The latter often requires prior knowledge or trial and error [8]. Multi-Resolution Segmentation (MRS) is one of the most popular segmentation algorithms [9]. Classical edge detection aims to detect sharp changes in image brightness through local measurements, including first-order (e.g., Prewitt or Sobel) and second-order (e.g., Laplacian or Gaussian) derivative-based detection [10]. Derivative-based edge detection is simple but noise sensitive. Amongst others, the Canny detector is justified by many researchers as a predominant one, for its better performance and capacity to reduce noise [6]. More recently, learning-based edge detection stands out as remarkable progress, which combines multiple low-level image cues into statistical learning algorithms for edge response prediction [10]. Globalized Probability of Boundary (gPb) is considered as one of the state-of-the-art methods. It involves brightness, color and texture cues into a globalization framework using spectral clustering [11]. Both MRS and gPb are unsupervised techniques, hence the high-level cadastral boundary is still hard to distinguish from all the detected edges.

Recent studies indicate that deep learning methods such as Convolutional Neural Networks (CNNs) are highly effective for the extraction of higher-level representations needed for detection or classification from raw input [12], which brings in new opportunities in cadastral boundary detection. Traditional CNNs are usually made up of two main components, namely convolutional layers for extracting spatial-contextual features and fully connected feedforward networks for learning the classification rule [13]. CNNs follow a supervised learning algorithm. Large amounts of labeled examples are needed to train the network to minimize the cost function which measures the error between the output scores and the desired scores [14]. Fully Convolutional Networks (FCNs) are a more recent deep learning method. In an FCN architecture, the fully connected layers of traditional CNNs are replaced by transposed convolutions. This is the reason why these networks are called fully convolutional. As compared to CNNs, FCNs are able to perform pixel-wise predictions and accept arbitrary-sized input, thus largely reducing computational cost and processing time [15]. The superiority of FCNs in feature learning and computational efficiency makes them promising for the detection of visible cadastral boundaries, which provides the predominant motivation of this research. To the best of the authors' knowledge, this is the first study investigating FCNs for cadastral boundary extraction.

In the remainder of this article, we apply deep FCNs for cadastral boundary detection based on UAV images acquired over one urban and one semi-urban area in Rwanda. We compare the results of FCNs with two other state-of-the-art image segmentation and edge detection techniques, namely MRS and gPb [9,11]. The performance of these methods is evaluated using the precision-recall framework. Specifically, to provide better insights into the detection results, we provide a separate accuracy assessment for visible and invisible cadastral boundaries, and an overall accuracy assessment for all cadastral boundaries.

2. Study Area

For the purpose of the study two sites within the Musanze district, Amajyaruguru Province of Rwanda, representing an urban and sub-urban setting, respectively, were selected as case-study locations. The selection is based on the availability of UAV images and the morphology of cadastral boundaries. The urban site is located in the Muhoza sector and the sub-urban site is in the Busogo sector. Figure 1 gives an overall view of the study area.

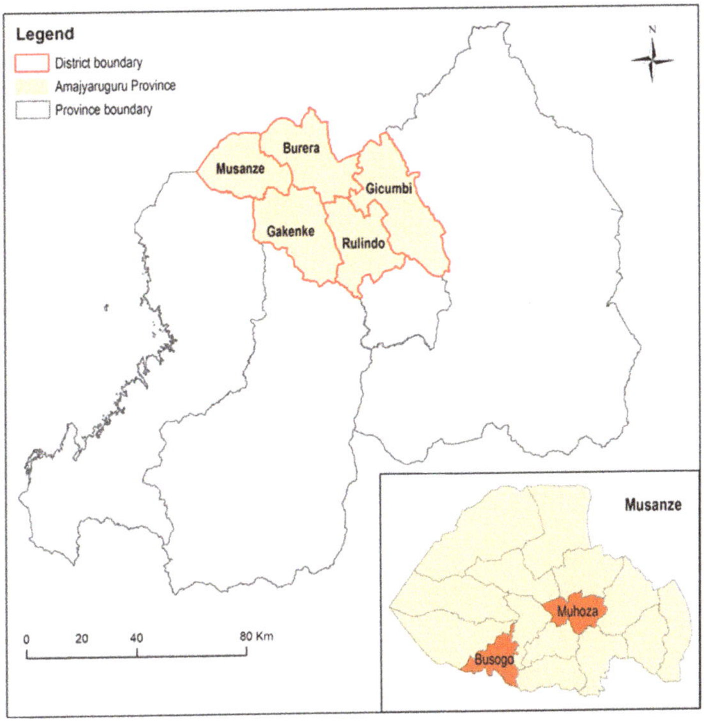

Figure 1. Study areas. Two case study sites in Rwanda were selected, namely Busogo and Muhoza, representing a sub-urban and urban setting, respectively.

In 1962, land ownership in Rwanda had changed from customary law to a system of state ownership. In 2005, a new policy was accepted called Organic Land Law (OLL) with the aim to improve land tenure security. Rwanda is one of the countries which first tested the FFP approach. Since 2008, the country has been fully covered by aerial images acquired and processed by a Dutch company [16]. Even compromising with accuracy, Rwanda generated its national cadastral map based on these aerial images using a participatory mapping approach. However, due to the continuously changing environment, the data is currently outdated. New technologies supporting cheap, efficient and fit-for-purpose accurate cadastral mapping will largely facilitate the data updating practices in Rwanda. Therefore, the selection of the study area has been led by the impending local demand for data update.

3. Materials and Methods

The workflow of this research includes three major parts, namely data preparation, boundary detection and accuracy evaluation. In the first part, we selected three training tiles and one testing tile from each study area and prepared the RGB layer and boundary reference for each tile. In the second part, we applied FCNs, MRS and gPb for cadastral boundary detection and validated their performance based on the testing tiles. For the last part, we employed precision-recall measures for accuracy assessment, with a 0.4 m tolerance from reference data.

3.1. Data Preparation

The UAV images used in this research were acquired for the its4land (https://its4land.com/) project in Rwanda in 2018. All data collection flights were carried out by Charis UAS Ltd. The drone used for data collection in Busogo was a DJI Inspire 2, equipped with Zenmuse X5S sensor. The drone used in Muhoza was a FireFLY6 from BIRDSEYEVIEW, with a SONY A6000 sensor. Both sensors acquire three bands (RGB) and capture nadir images. The flight height above the ground for Busogo was 100 m and for Muhoza 90 m. The final Ground Sampling Distance (GSD) was 2.18 cm in Busogo and 2.15 cm in Muhoza. For more detailed information about flight planning and image acquisition refer to [17].

In this research, the spatial resolution of the UAV images was resampled from 0.02 m to 0.1 m considering the balance between accuracy and computational time. Four tiles of 2000 × 2000 pixels were selected from each study site for the experimental analysis. Among them, three tiles were used for training and one for testing the algorithm. The training and testing tiles in Busogo are named TR1, TR2, TR3 and TS1, and those in Muhoza are named TR4, TR5, TR6 and TS2 (Figure 2).

Figure 2. The Unmanned Aerial Vehicle (UAV) images and boundary reference for selected tiles. TR1, TR2, TR3 and TS1 are selected tiles from Busogo; TR4, TR5, TR6 and TS2 are selected tiles from Muhoza. For each area, the former three were used for training and the last one was used for testing the algorithms. The boundary references in TR1~TR6 are the yellow lines. In TS1 and TS2, we separated the boundary references as visible (green lines) and invisible (red lines).

For each tile, RGB images and the boundary reference were prepared as input for the classification task. The reference data was acquired by merging the 2008 national cadaster and Rwandan experts' digitization. The 2008 national cadaster is currently outdated, hence the experts' digitization is provided as supplements. This acquired reference was presented as polygons in a shapefile format showing the land parcels. However, to feed the FCN, the boundary reference has to be in a raster format with the same spatial resolution as RGB images. Therefore, we first converted the polygons into borderlines and then rasterized the boundary lines with a spatial resolution of 0.1 m. Figure 2 visualizes the RGB images and boundary reference for the selected tiles in Busogo and Muhoza. To have a better understanding of the detection results on the testing tiles, we separated the boundary reference as visible and invisible in TS1 and TS2, which are marked as green and red in the above maps, respectively. We extracted visible cadastral boundaries by following clearly visible features, including strips of stones, fences, edge of the rooftop, road ridges, change in textural pattern and water drainage. Table 1 shows the rules that we followed for extracting visible boundaries in an extraction guide. The rest are considered invisible cadastral boundaries.

Table 1. Extraction guide for visible cadastral boundaries.

Object class	Visible Cadastral Boundary
Input data	0.1 m × 0.1 m UAV image
Reference frame	Coordinate System: WGS 1984 UTM zone 35S Projection: Transverse Mercator False Easting: 500,000 False Northing: 10,000,000 Central Meridian: 27 Scale factor: 0.9996 Latitude of origin: 0.000 Units: Meter
Definition	A visible cadastral boundary is a line of geographical features representing limits of an entity considered to be a single area under homogeneous real property rights and unique ownership.
Identifying visible cadastral boundaries	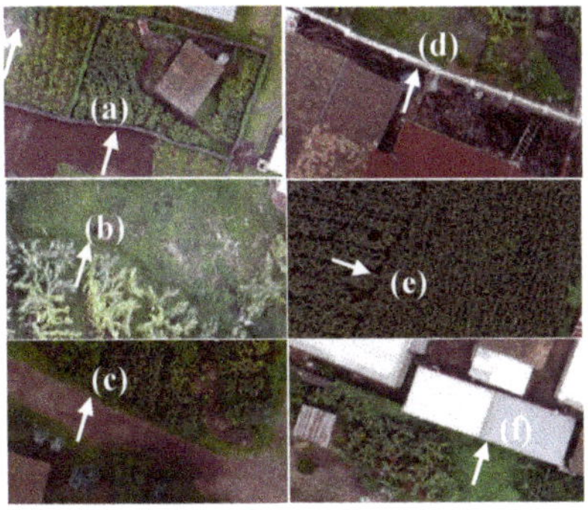 (a) Strip of stone (b) Water drainage (c) Road ridges (d) Fences(e) Textural pattern transition (f) Edge of rooftop
Extraction	• Hypothesize and digitize cadastral boundary by applying expert and contextual knowledge • Digitize the middle of the line features • Use line features for digitization

3.2. *Boundary Detection*

3.2.1. Fully Convolutional Networks

We address boundary detection as a supervised pixel-wise image classification problem to discriminate between boundary and non-boundary pixels. The network used in this research is modified from the FCN with dilated kernel (FCN-DK) as described in [18]. We mainly did three modifications: (1) discarding the max-pooling layers; (2) using smaller-size filters; and (3) constructing deeper networks. A typical max-pooling layer computes the local maximum within the pooling filter, thereby merging the information of nearby pixels and reducing the dimension of the feature map [14]. In most cases, down-sampling is performed using pooling layers to capture large spatial patterns in the image, and then the coarse feature maps extracted through this process are up-sampled back to produce pixel-wise prediction at the resolution of the input image. However, in the structure of FCN-DK, max-pooling layers are designed not to down-sample the input feature map by using a stride of one, therefore avoiding the need for up-sampling in the later stage. Nevertheless, the max-pooling results in a smoothing effect which is reducing the accuracy of boundaries. We therefore discarded it in the proposed network. In [18], Persello and Stein also demonstrated a case study based on FCN-DK using six convolutional layers with a filter size of 5 × 5 pixels. We modified it into 12 convolutional layers with 3 × 3 filters, as two 3 × 3 filters have the same receptive field as one 5 × 5 filter but less learnable parameters. Moreover, compared to single larger-sized filters, multiple small filters are interleaved by activation functions, resulting in better abstraction ability. Therefore, with less learnable parameters and better feature abstraction ability, smaller filters along with deeper networks are preferred.

Figure 3 shows the architecture of the proposed FCN. It consists of 12 convolutional layers interleaved by batch normalization and Leaky Rectified Linear Units (Leaky ReLU). Batch normalization layer is used to normalize each input mini-batch [19], and Leaky ReLU is the activation function of the network [20]. The classification is performed by the final SoftMax layer.

The core components of our network are the convolutional layers. They can extract spatial features hierarchically at different layers corresponding to different levels of abstraction. The 3 × 3 kernels used in the convolutional layers are dilated increasingly from 1 to 12 to capture larger contextual information. As a result, a receptive field of up to 157 × 157 pixels was achieved in the final layer. In each convolutional layer, zero paddings were used to keep the output feature maps at the same spatial dimension as the input. Therefore, the proposed FCN can be used to classify arbitrarily sized images directly and obtain correspondingly sized outputs.

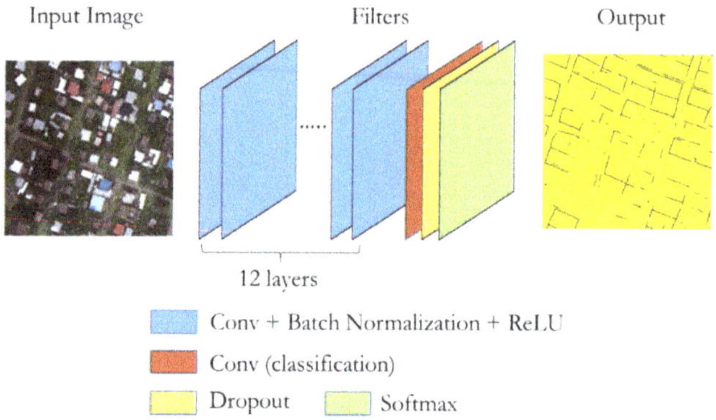

Figure 3. Architecture of the proposed FCN.

To train the FCN, we randomly extracted 500 patches for training and 500 patches for validation from each training tile. All the patches were fully labeled with a patch size of 145 × 145 pixels. Stochastic gradient descent with a momentum of 0.9 was used to optimize the loss function. The training is performed in multiple stages using a different learning rate. We use a learning rate of 10^{-5} for the first 180 epochs and a learning rate of 10^{-6} for another 20 epochs. A sudden decrease can be observed in the learning curves when the learning rate changes (Figure 4). The implementation of the network is based on the MatConvNet (http://www.vlfeat.org/matconvnet/) library. All experiments were performed on a desktop workstation with an Intel Core CPU i7-8750H at 2.2 GHz, 16 GB of RAM, and a Nvidia Quadro P1000 GPU. The training time for the FCN was 6 h for each study area.

3.2.2. Globalized Probability of Boundary (gPb)

Globalized Probability of Boundary (gPb) was proposed by Arbeláez et al. in 2011 [11]. gPb (global Pb) is a linear combination of mPb (multiscale Pb) and sPb (spectral Pb). The former conveys local multiscale Pb signals and the latter introduces global information. Multiscale Pb is an extension of the

Pb detector advanced by Martin, Fowlkes and Malik [21]. The core block of Pb detector is calculating the oriented gradient signal (x,y,θ) from the intensity images. By placing a circular disc at the pixel location $(x,)$ and dividing it into two half-discs at angle θ, we can obtain two histograms of pixel intensity values within each half-disc. (x,y,θ) is defined by the $\chi 2$ distance between the two histograms. For each input image, the Pb detector divides it into four intensity images, including brightness, color a, color b and texture channel. The oriented gradient signals are calculated separately for each channel. Multiscale Pb modifies the Pb detector by considering the gradients at three different scales, which means we give the discs three different diameters. Therefore, we can obtain local cues at different scales, from fine to coarse structures. For each pixel, the final mPb is obtained by combining the gradients of brightness, color a, color b and texture channel on three scales. Spectral Pb combines the multiscale image cues into an affinity matrix which defines the similarity between pixels. The eigenvectors of the affinity matrix which carry contour information are computed. They are treated as an image and convolved with Gaussian directional derivative filters. The sPb is calculated by combing the information from different eigenvectors.

Generally speaking, mPb detects all the edges while sPb extracts only the most salient one from the whole image. gPb combines the two and provides uniformly better performance. After detecting the boundary probability of each pixel using gPb, we also applied a grouping algorithm using the Oriented Watershed Transform and Ultrametric Contour Map (gPb–owt–ucm) to extract connected contours [11].

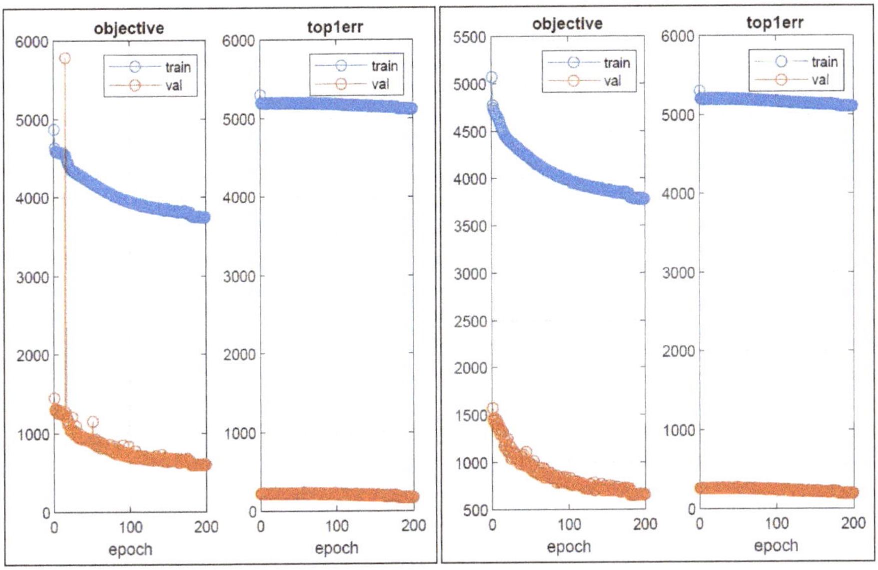

Figure 4. Learning curves of the FCNs in Busogo (**left**) and Muhoza (**right**).

3.2.3. Multi-Resolution Segmentation (MRS)

We conducted MRS in eCognition software (version 9.4). MRS is a region-merging technique starting from each pixel forming one image object or region [22,23]. The merging criteria is local homogeneity, which describes the similarity between adjacent image objects. The merging procedure stops when all the possible merges exceed the homogeneity criteria.

MRS relies on several parameters, which are image layer weights, scale parameter (SP), shape and compactness. Image layer weights define the importance of each image layer to the segmentation process. In this research, we had three layers (RGB) in the input image. We gave them equal weights. Scale parameter is the most important parameter, which controls the average image object size [9].

A larger scale parameter allows higher spectral heterogeneity within the image objects, hence allowing more pixels within one object. Defining the proper SP is critical for MRS. In our research, we selected the SP resorting to the automatic Estimation of Scale Parameters 2 (ESP2) tool, which was advanced in [9]. Shape parameter ranges from 0 to 1. It indicates a weighting between the object's shape and its spectral color. A high value in shape parameter means less importance is put on spectral information. We set a shape parameter of 0.3 in our research. Compactness defines how compact the segmented objects are. The higher the value, the more compact the image objects may be. It was set to 0.5 in this research.

3.3. Accuracy Assessment

The results in this research were evaluated considering the detection accuracy on the testing tiles. While validating the predicted results with reference data, a certain tolerance is often used in cadastral mapping. According to the International Association of Assessing Officers (IAAO), the horizontal spatial accuracy for cadastral maps in urban environments is usually 0.3 m or less, and in rural areas an accuracy of 2.4 m is sufficient [24]. Besides, FFP approaches advocates the flexibility in terms of accuracy to best accommodate social needs [3]. Complying with FFP land administration, we chose a 0.4 m tolerance for urban and peri-urban environments in this research. When adopted for other applications, this number can be adjusted correspondingly and according to the demands.

The accuracy assessment resorted to precision-recall measures, which are a standard evaluation technique especially for boundary detection in computer vision fields [21]. Precision (P), also called correctness, measures the ratio of correctly detected boundary pixels to the total detected boundary pixels. Recall (R), also called completeness, indicates the percentage of correctly detected boundaries to the total boundaries in the reference. The F measure (F) represents the harmonic mean between precision and recall [25]. As F combines both precision and recall, it can be regarded as an overall quality measure. The range of these three measures is between 0 and 1. Larger values represent higher accuracy.

Specifically, the accuracy assessment was done by overlapping the detected boundary with buffered reference data (0.4 m buffer). With the following table and formulas, we indicated how to calculate precision, recall and the F-score. Pixels labelled as boundary class in both detection and the buffered reference are called True Positive (TP), while pixels labelled as boundary in detection but non-boundary in the buffered reference are called False Positive (FP). The term False Negative (FN) and True Negative (TN) are defined similarly (Table 2). By overlaying the detection result with the buffered reference, we can obtain the value of TP, FP, TN and FN, respectively. TP stands for the number of correctly detected boundary pixels, and the sum of TP and FP indicates the number of total detected boundary pixels. Hence, we can calculate the value of precision through Formula 1. However, the sum of TP and FN stands for the number of total boundaries in the buffered reference, rather than the original reference. Therefore, we need to divide the sum of TP and FN by 8, which is the width of the buffered reference, to get the number of total boundaries in the original reference. This is because the buffered reference has a uniform width of 8 pixels, while the original reference is only single-pixel wide. Equations (2) and (3) show how to calculate recall and the F-score.

Table 2. Confusion matrix for binary classification.

	Positive Prediction	**Negative Prediction**
Positive Class	True Positive (TP)	False Negative (FN)
Negative Class	False Positive (FP)	True Negative (TN)

In order to know the capability of different methods in detecting visible and invisible cadastral boundaries, we calculated the classification accuracy for the visible cadastral boundary, invisible cadastral boundary and all cadastral boundaries, separately. By overlapping the detected boundary with the buffered reference of only visible, invisible or all cadastral boundaries, we obtain three sets of accuracy assessment results for each algorithm on every testing tile.

$$P = TP/(TP + FP), \tag{1}$$

$$R = 8 \cdot TP/(TP + FN), \tag{2}$$

$$F = 2 \cdot P \cdot R/(P + R). \tag{3}$$

4. Results

The proposed method along with the competing methods were implemented on both study sites, Busogo and Muhoza, to test their generalization ability. The results are evaluated considering the classification accuracy on the testing tiles using the precision-recall framework. The visual and numerical results of the testing tiles are demonstrated in the following table and figures.

Table 3 presents the separate accuracy for visible and invisible boundaries, as well as the overall accuracy for all cadastral boundaries of each algorithm on TS1 and TS2. Taking the classification accuracy of FCN on visible cadastral boundaries in TS1 as an example, FCN achieves 0.75 in precision, which means the ratio of truly detected visible boundaries to the total detected boundaries is 75%. The value of recall is 0.65, indicating 65% of visible cadastral boundaries among all the visible boundaries in the reference are detected. The final F-score of FCN is 0.70, which can be regarded as an overall measure of quality performance. Other results from Table 3 could be interpreted in the same way. Interestingly, according to the mathematical implications of precision, the sum of the P value on visible and invisible boundaries should be equal to the P value on all cadastral boundaries. We can easily verify this through the six sets of data in Table 3, with a small tolerance of ±0.03.

Table 3. Classification accuracies of the Fully Convolutional Network (FCN), Globalized Probability of Boundary–Oriented Watershed Transform–Ultrametric Contour Map (gPb–owt–ucm) and Multi-Resolution Segmentation (MRS) on TS1 and TS2. Three kinds of accuracies are calculated by comparing the detected boundary to the reference of visible, invisible and all cadastral boundaries.

Algorithm	Reference	TS1			TS2		
		P	R	F	P	R	F
FCN	visible	0.75	0.65	0.70	0.74	0.45	0.56
	invisible	0.06	0.07	0.06	0.06	0.09	0.07
	all	0.78	0.39	0.52	0.79	0.35	0.48
gPb-owt-ucm	visible	0.21	0.87	0.34	0.23	0.93	0.37
	invisible	0.03	0.19	0.06	0.04	0.39	0.07
	all	0.24	0.57	0.33	0.26	0.78	0.39
MRS	visible	0.19	0.82	0.31	0.18	0.90	0.30
	invisible	0.05	0.27	0.08	0.04	0.56	0.08
	all	0.23	0.57	0.33	0.22	0.80	0.35

According to Table 3, FCN achieves an F-score of 0.70 on visible boundaries and 0.06 on invisible boundaries in TS1. The score of the former is much higher than the latter. A similar situation can also be witnessed in TS2, which indicates that FCN detects mainly visible cadastral boundaries. The F-score of FCN on all boundaries in TS1 is 0.52, larger than that in TS2 (0.48). We can interpret this result considering the proportion of visible cadastral boundaries (The proportion of visible cadastral boundaries in each tile are calculated by computing the ratio of the total length of visible cadastral boundaries to that of all cadastral boundaries) in each tile, which is 57% in TS1 and 72% in TS2. Surprisingly, with more visible cadastral boundaries, TS2 get poorer detection results. According to the R value of visible boundaries, 65% of visible cadastral boundaries are detected in TS1 and the number is only 45% in TS2. It means that although TS2 has more visible cadastral boundaries, less of them are detected. The main difference in detection ability of the FCN in TS1 and TS2 can be understood by considering the various types of visible cadastral boundaries. TS1 is located in a sub-urban area, with fences and strips of stones being the most predominant visible boundaries, whereas TS2 is in an urban area, where building walls and fences play the leading role. The better performance of FCN in TS1 indicates that FCN is good in detecting visible boundaries like fences and strips of stones, while cadastral boundaries that coincide with building walls are more difficult for FCN to recognize. Based on the above analysis we can conclude that FCN detects mainly visible cadastral boundaries, especially those demarcated by fences or strips of stones.

Comparing FCN to gPb–owt–ucm and MRS, the most salient finding is that under the same situation, like the detection accuracy for visible boundary in TS1 or all boundaries in TS2, the P value of FCN is always larger than that of gPb–owt–ucm and MRS, while the R value of FCN is always smaller. FCN always achieves the highest F-score. These results show that gPb–owt–ucm and MRS can detect large proportion of cadastral boundaries, but also many false boundaries. FCN has a very high precision rate, leading to the best overall performance.

Figure 5 shows the visible and invisible boundary references and the detected results of the investigated algorithms. According to Figure 5, the missing boundary fragments in the FCN classification output are mainly invisible boundaries. Besides, FCN has a more regular and cleaner output than gPb–owt–ucm and MRS. Although the outlines of buildings and trees correspond to strong edges, they are not confused by FCN with cadastral boundaries.

Figure 5. Reference and classification maps obtained by the investigated techniques. The visible boundary references are the green lines; the invisible are the red lines; and the detected boundaries are the yellow lines.

Figure 6 presents the error map of the detection results. By overlapping the detection map with the boundary reference, the correctly detected boundaries are marked as yellow; the false detection rea marked as red; and the missing boundaries are marked as green. Figure 6 supplies a better intuition of the detection results. Fewer red lines can be observed in the FCN output as compared to the other two algorithms, once again proving that FCN has higher precision.

The difference in computational cost between these methods is also worth highlighting. As mentioned earlier, it takes 6 h to train the FCN. However, once trained well, one tile is classified in 1 min with the proposed FCN, whereas it takes 38 min for MRS and 1.5 h for gPb–owt–ucm, respectively.

Figure 6. The error map of the investigated techniques. Yellow lines are TP; red lines are FP; and green lines are FN.

5. Discussion

Considering the constraints of the invisible cadastral boundaries, land administration professionals perceived that a 40 to 50 percent automatic delineation would be very significant in reducing time and labor involved in cadastral mapping practices [26]. This goal is reached by FCN with a 0.4 m tolerance in both Busogo and Muhoza, indicating good generalization and transferability of the proposed approach in cadastral boundary mapping. From Section 4, the numerical results and visual results proves and supplements each other, suggesting two main findings: (1) The true positives detected by the FCN are mainly visible boundaries like fences and strips of stones; and (2) gPb–owt–ucm and MRS have high recall, while FCN has high precision and better overall performance.

Compared with alternative edge detection and image segmentation approaches, FCN achieved better overall performance. The reason lies in the strong feature learning and abstraction ability of FCN. Lacking abstraction ability, standard edge detection and image segmentation cannot fill the semantic gap between the high-level cadastral boundary concept and low-level image features. As a result, gPb–owt–ucm and MRS achieved high recall but low precision. In other words, gPb–owt–ucm cannot determine cadastral boundaries from all the detected contours, while MRS cannot eliminate over-segmentation caused by the spectral differences within one cadastral parcel. It is also worth noticing that FCN performs supervised boundary detection, while MRS and gPb–owt–ucm are both unsupervised techniques. This may explain their differences in the abstraction ability. Being supervised, the proposed FCN-based detector is trained to detect cadastral boundaries and to disregard other irrelevant edges like building outlines.

FCN can supply high precision, while gPb–owt–ucm and MRS can supply high recall. Therefore, for further study, we can consider a combination of these methods. We can combine them in two ways. The first one is to involve the output of gPb or MRS along with UAV images as input for FCN. FCN has a strong feature learning ability. It is possible that FCN can determine cadastral boundaries from the outputs of gPb or MRS, hence increasing both precision and recall. The second way is to apply an approach similar to [27], where the boundary map of FCN and the probability map of gPb are linearly combined and followed by the owt–ucm procedure to extract connected boundaries.

6. Conclusions

The deep FCN proposed in this research is capable of extracting visible cadastral boundaries from raw UAV images. Experiments carried out on both study sites achieved an F-score around 0.5. Very clean and clear boundaries were extracted by the proposed method, avoiding the effect of messy building contours. In both study sites, the proposed method performed better than contending algorithms. The knowledge of the local experts is needed to correct the extracted boundaries and include them in a final cadastral system. We conclude that the proposed automated method followed by experts' final correction and verification can reduce the processing time and labor force of the current cadastral mapping and data updating practices.

So far, the proposed technique is mainly suitable when a large proportion of boundaries are visible. Detecting invisible boundaries, i.e., not demarcated by physical objects, from remotely sensed images is obviously extremely challenging. In other researches, invisible boundaries are identified by manual digitization via post-processing steps. The its4land project proposed a QGIS plugin (https://its4land.com/automate-it-wp5/) which supports an interactive, semi-automatic delineation to expedite the process. In the beginning of this research, the authors also attempted to fill this research gap by considering the fact that a cadastral boundary often lies in between two buildings. We tried to introduce building information as an additional input to train the FCNs for identifying the invisible boundaries. However, our experimental results obtained so far showed that there was no obvious improvement by adding building information in cadastral boundary detection. Further research is required in this direction. In future research, we will consider to improve the capability to detect cadastral boundaries (visible and invisible) using Generative Adversarial Networks (GANs) [28]. Within this framework, the generative model and discriminative model form an adversarial training, which is sharpening the training process by focusing on the most critical samples to learn. This approach is expected to improve the accuracy in boundary detection tasks which are often characterized by scarce training data.

Author Contributions: M.K., C.P., and X.X. conceptualized the aim of this research; X.X. wrote the majority of the paper; C.P. and M.K revised and edited the paper over several rounds; and X.X. set up and performed the experimental analysis using parts of the software codes developed by C.P. and under the supervision of both C.P. and M.K.

Funding: This research received no external funding.

Acknowledgments: The authors would like to thank the team of "its4land" research project number 687828 which is part of the Horizon 2020 program of the European Union, for providing UAV data for this research.

Conflicts of Interest: The authors declare no conflict of interest.

References

1. Luo, X.; Bennett, R.M.; Koeva, M.; Lemmen, C. Investigating Semi-Automated Cadastral Boundaries Extraction from Airborne Laser Scanned Data. *Land* **2017**, *6*, 60. [CrossRef]
2. Williamson, I. The justification of cadastral systems in developing countries. *Geomatica* **1997**, *51*, 21–36.
3. Enemark, S.; Mclaren, R.; Lemmen, C.; Antonio, D.; Gitau, J.; De Zeeuw, K.; Dijkstra, P.; Quinlan, V.; Freccia, S. *Fit-For-Purpoes Land Administration: Guiding Principles for Country Implementation*; GLTN: Nairobi, Kenya, 2016.
4. Enemark, S.; Bell, K.C.; Lemmen, C.; McLaren, R. *Fit-For-Purpose Land Administration*; International Federation of Surveyors: Copenhagen, Denmark, 2014; p. 44.
5. Ramadhani, S.A.; Bennett, R.M.; Nex, F.C. Exploring UAV in Indonesian cadastral boundary data acquisition. *Earth Sci. Inform.* **2018**, *11*, 129–146. [CrossRef]

6. Crommelinck, S.; Bennett, R.; Gerke, M.; Nex, F.; Yang, M.; Vosselman, G. Review of Automatic Feature Extraction from High-Resolution Optical Sensor Data for UAV-Based Cadastral Mapping. *Remote Sens.* **2016**, *8*, 689. [CrossRef]
7. Pal, N.R.; Pal, S.K. A review on image segmentation techniques. *Pattern Recognit.* **1993**, *26*, 1277–1294. [CrossRef]
8. García-Pedrero, A.; Gonzalo-Martín, C.; Lillo-Saavedra, M. A machine learning approach for agricultural parcel delineation through agglomerative segmentation. *Int. J. Remote Sens.* **2017**, *38*, 1809–1819. [CrossRef]
9. Drăguţ, L.; Csillik, O.; Eisank, C.; Tiede, D. Automated parameterisation for multi-scale image segmentation on multiple layers. *ISPRS J. Photogramm. Remote Sens.* **2014**, *88*, 119–127. [CrossRef] [PubMed]
10. Li, Y.; Wang, S.; Tian, Q.; Ding, X. A survey of recent advances in visual feature detection. *Neurocomputing* **2015**, *149*, 736–751. [CrossRef]
11. Arbeláez, P.; Maire, M.; Fowlkes, C.; Malik, J. Contour Detection and Hierarchical Image Segmentation. *IEEE Trans. Pattern Anal. Mach. Intell.* **2011**, *33*, 898–916. [CrossRef] [PubMed]
12. Zhu, X.X.; Tuia, D.; Mou, L.; Xia, G.-S.; Zhang, L.; Xu, F.; Fraundorfer, F. Deep Learning in Remote Sensing: A Comprehensive Review and List of Resources. *IEEE Geosci. Remote Sens. Mag.* **2017**, *5*, 8–36. [CrossRef]
13. Bergado, J.R.; Persello, C.; Gevaert, C. A Deep Learning Approach to the Classification of sub-decimetre Resolution Aerial Images. In Proceedings of the IEEE International Geoscience and Remote Sensing Symposium, Beijing, China, 10–15 July 2016; pp. 1516–1519.
14. LeCun, Y.; Bengio, Y.; Hinton, G. Deep learning. *Nature* **2015**, *521*, 436–444. [CrossRef] [PubMed]
15. Shelhamer, E.; Long, J.; Darrell, T. Fully Convolutional Networks for Semantic Segmentation. *IEEE Trans. Pattern Anal. Mach. Intell.* **2017**, *39*, 640–651. [CrossRef] [PubMed]
16. Maurice, M.J.; Koeva, M.N.; Gerke, M.; Nex, F.; Gevaert, C. A photogrammetric approach for map updating using UAV in Rwanda. In Proceedings of the GeoTechRwanda 2015, Kigali, Rwanda, 18–20 November 2015; pp. 1–8.
17. Stöcker, C.; Ho, S.; Nkerabigwi, P.; Schmidt, C.; Koeva, M.; Bennett, R.; Zevenbergen, J. Unmanned Aerial System Imagery, Land Data and User Needs: A Socio-Technical Assessment in Rwanda. *Remote Sens.* **2019**, *11*, 1035. [CrossRef]
18. Persello, C.; Stein, A. Deep Fully Convolutional Networks for the Detection of Informal Settlements in VHR Images. *IEEE Geosci. Remote Sens. Lett.* **2017**, *14*, 2325–2329. [CrossRef]
19. Ioffe, S.; Szegedy, C. Batch Normalization: Accelerating Deep Network Training by Reducing Internal Covariate Shift. In Proceedings of the International Conference on Machine Learning, Lille, France, 6–11 July 2015.
20. He, K.; Zhang, X.; Ren, S.; Sun, J. Delving Deep into Rectifiers: Surpassing Human-Level Performance on ImageNet Classification. In Proceedings of the 2015 IEEE International Conference on Computer Vision (ICCV), Santiago, Chile, 7–13 December 2015; pp. 1–11. [CrossRef]
21. Martin, D.R.; Fowlkes, C.C.; Malik, J. Learning to detect natural image boundaries using local brightness, color, and texture cues. *IEEE Trans. Pattern Anal. Mach. Intell.* **2004**, *26*, 530–549. [CrossRef] [PubMed]
22. Baatz, M.; Schäpe, A. *Multiresolution Segmentation: An Optimization Approach for High Quality Multi-Scale Image Segmentation*; Wichmann Verlag: Karlsruhe, Germany, 2000.
23. Nyandwi, E.; Koeva, M.; Kohli, D.; Bennett, R.; Nyandwi, E.; Koeva, M.; Kohli, D.; Bennett, R. Comparing Human Versus Machine-Driven Cadastral Boundary Feature Extraction. *Remote Sens.* **2019**, *11*, 1662. [CrossRef]
24. IAAO. *Standard on Digital Cadastral Maps and Parcel Identifiers*; International Association of Assessing Officers: Kansas City, MO, USA, 2015.
25. Hossin, M.; Sulaiman, M.N. Review on Evaluation Metrics for Data Classification Evaluations. *Int. J. Data Min. Knowl. Manag. Process* **2015**, *5*, 1–11. [CrossRef]
26. Wassie, Y.A.; Koeva, M.N.; Bennett, R.M.; Lemmen, C.H.J. A procedure for semi-automated cadastral boundary feature extraction from high-resolution satellite imagery. *J. Spat. Sci.* **2018**, *63*, 75–92. [CrossRef]

27. Persello, C.; Tolpekin, V.A.; Bergado, J.R.; de By, R.A. Delineation of agricultural fields in smallholder farms from satellite images using fully convolutional networks and combinatorial grouping. *Remote Sens. Environ.* **2019**, *231*, 111253. [CrossRef]
28. Goodfellow, I.J.; Pouget-Abadie, J.; Mirza, M.; Xu, B.; Warde-Farley, D.; Ozair, S.; Courville, A.; Bengio, Y. Generative Adversarial Networks. In Proceedings of the Neural Information Processing Systems, Montreal, QC, Canada, 8–13 December 2014; pp. 2672–2680.

remote sensing

MDPI

Article

Extraction of Visible Boundaries for Cadastral Mapping Based on UAV Imagery

Bujar Fetai *, Krištof Oštir, Mojca Kosmatin Fras and Anka Lisec

Faculty of Civil and Geodetic Engineering, University of Ljubljana, Jamova cesta 2, 1000 Ljubljana, Slovenia
* Correspondence: bujar.fetai@fgg.uni-lj.si; Tel.: +386-1-476-8560

Received: 8 May 2019; Accepted: 24 June 2019; Published: 26 June 2019

Abstract: In order to transcend the challenge of accelerating the establishment of cadastres and to efficiently maintain them once established, innovative, and automated cadastral mapping techniques are needed. The focus of the research is on the use of high-resolution optical sensors on unmanned aerial vehicle (UAV) platforms. More specifically, this study investigates the potential of UAV-based cadastral mapping, where the ENVI feature extraction (FX) module has been used for data processing. The paper describes the workflow, which encompasses image pre-processing, automatic extraction of visible boundaries on the UAV imagery, and data post-processing. It shows that this approach should be applied when the UAV orthoimage is resampled to a larger ground sample distance (GSD). In addition, the findings show that it is important to filter the extracted boundary maps to improve the results. The results of the accuracy assessment showed that almost 80% of the extracted visible boundaries were correct. Based on the automatic extraction method, the proposed workflow has the potential to accelerate and facilitate the creation of cadastral maps, especially for developing countries. In developed countries, the extracted visible boundaries might be used for the revision of existing cadastral maps. However, in both cases, the extracted visible boundaries must be validated by landowners and other beneficiaries.

Keywords: land plot; land cadastre; cadastral boundaries; cadastral maps; UAV; image processing; image segmentation; feature extraction

1. Introduction

Establishing a complete land cadastre and keeping it up-to-date is a contemporary challenge for many developing and developed countries, respectively [1,2]. In this research, the distinction between 'developing' and 'developed' countries is considered from a land administration perspective. A developing country refers to a country with low cadastral coverage. A developed country refers to full coverage of a country's territory with defined cadastral land plot boundaries and associated land rights. According to the International Federation of Surveyors (FIG) and the World Bank, only one-quarter of people's land rights across the world are formally recognized by cadastral or other land recording systems [1]. Thus, in developing countries, initial efforts are directed to accelerating cadastral mapping as a basis for defining and recording land rights boundaries and formalizing land-related rights aiming to guarantee land tenure security [3,4]. In developed countries, beyond the initial adjudication stage or establishment of a cadastre, another challenge is the maintenance of person-right-land relation attributes and keeping the cadastral systems up-to-date [5,6]. In countries with a tradition and long history of developing a cadastral system, conventional ground-based cadastral surveying techniques and high positional accuracy of boundary surveying were required. Decades were needed to complete the process of cadastral surveying/mapping and registration [1,6]. Although land cadastres were established, some of the cadastral systems could not be maintained, which led to outdated cadastral maps. Person-right-land relationship is complex and dynamic. Keeping the cadastral system up-to-date (continuous recording of person-right-land relation attributes, in any land related event, as close as possible to real-time) also requires a flexible and dynamic cadastral system [2,7]. Proposed cadastral surveying techniques are mostly indirect ones rather than ground-based. Ground-based techniques are often argued as being time-consuming and labor intensive [1,5,8].

Emerging tools are mapping techniques based on remote sensing data, in particular, data acquired with sensors on Unmanned Aerial Vehicles (UAVs) [9–18]. Cadastral maps are usually defined as a spatial representation of recorded land plot boundaries or other spatial units that the land rights concern [19]. In general, sensors on UAVs provide low-cost, efficient, and flexible high-resolution spatial data acquisition systems enabling the production of point clouds, Digital Surface Models (DSM) and orthoimages [20,21]. In cadastral applications, UAVs have gained increasing popularity due to the high cadastral mapping potential in a different setting, in rural and urban areas, for developing and developed countries [22]. In addition, UAVs are used for both the creation and updating of cadastral maps [22]. In developing countries, UAV-based cadastral mapping usually serves as a tool for the creation of a formal cadastral system [11–13]. In developed countries, the case studies focus on the assessment of UAVs' data positional accuracy estimation and its conformity with local positional accuracy requirements aiming to use the UAV data for updating existing cadastral maps [14–18]. Here, updating in most cases refers to the comparison of two cadastral maps—one representing the database state, the other recently acquired data. The term updating can be used as a synonym for a "revision" of existing cadastral maps [23]. However, in all case studies reported in [22], cadastral boundaries are manually delineated.

It is argued that a large number of cadastral boundaries are visible and coincide with natural or manmade physical object boundaries [2,24,25]. In the land administration domain, automatic extractions of visible cadastral boundaries have been a recent topic of investigation. The latest studies, though limited in number, assert that visible boundaries, such as hedges, land cover boundaries, etc., which might indicate cadastral boundaries, could be automatically extracted using methods such as algorithms that detect object boundaries in images [22,26–29]. In fact, not all visible cadastral boundaries can be automatically detected—certain boundaries would require a semi-automatic approach, especially in urban areas where the morphology of cadastral boundaries is complex [7]. Nevertheless, the potential of computer vision methods for automatic detection and extraction of visible objects in the images is promising for cadastral applications, especially due to the urgent global need for accelerating and facilitating cadastral mapping as a basis for registration of land rights and following the dynamics of land tenure and land use.

1.1. Visible Boundary Detection and Extraction for Cadastral Mapping

Automatic feature extraction methods from images acquired with high-resolution optical sensors have already proved to be useful for the extraction of boundaries of linear features such as roads and rivers [30–34], and to a much lesser degree, they have also been explored for the purpose of cadastral boundary delineation. A recent study from Crommelinck et al. [22] provides an overview of computer vision methods that might be applicable in the land administration domain for automatic detection and extraction of object boundaries from images acquired with high-resolution optical sensors. Additionally, the general workflow for automatic detection and extraction of visible object boundaries for UAV-based cadastral mapping is provided [22]. The general workflow consists of (i) image pre-processing, (ii) image segmentation, (iii) line extraction, (iv) contour or boundary generation, and (v) image and/or boundary post-processing. Image pre-processing usually includes image conversions, such as resampling or tiling, in order to fit the requirements of a chosen computer vision method. Image segmentation refers to the process of dividing a digital image into non-overlapping objects, which represent homogeneous areas [35]. The third workflow step is the extraction of lines or edges from the segmented images [36]. The next step, contour generation, refers to the extraction of a closed object outlines in the image. In computer vision, they are usually defined as object boundaries, which are derived from connecting edges or lines. An 'object boundary' should encompass an 'object' in an image, and due to this, both terms are used synonymously in this study. In cadastral applications, objects are usually defined as polygon-based spatial units. The final step, post-processing, includes interventions on the image such as vectorization and/or simplification of automated extraction of objects [26,37]. However, only a limited number of studies have investigated the automatic extraction of objects from images acquired with high-resolution optical sensors for cadastral boundary delineation.

The work by Babawuro and Zou [38] tested Canny and Sobel edge-detection algorithms for the extraction of visible cadastral boundaries from high-resolution satellite imagery (HRSI). In addition, the Hough Transform feature extraction method was used to connect edges and to identify straight lines. The visual presentation of the results showed that the proposed approach can detect agricultural land boundaries, but there were no quantity measures on quality assessment. Kohli et al. [28,29] investigated the use of an object-based approach, namely the multi-resolution segmentation (MRS) and estimation of scale parameter (ESP) to extract visible cadastral boundaries from HRSI. An object-based approach refers to the extraction of object outlines based on a grouping of pixels with similar characteristics and is applied to high-level features which represent shapes in an image [22]. The accuracy assessment in Kohli et al. [28] was pixel-based, and the detection quality in terms of error of commission and omission for MRS were 75% and 38%, respectively. For ESP, the error of commission was 66% and the error of omission 58%. The localization quality for MRS was 71%, whereas it was 73% for ESP, within a 41–200 cm distance from the reference boundaries. Another case of the automatic extraction of visible boundaries based on HRSI is described in Wassie et al. [27]. The study explored the potential of mean-shift segmentation for the extraction of visible cadastral boundaries. The mean-shift segmentation algorithm is a QGIS open source plugin [27]. The object-based measures were applied for the accuracy assessment. Within a buffer distance of 2 m, the percentage indicated the correctness was 34%, while for the completeness it was 83% [27]. The extractions with mean-shift segmentation were closed object boundaries (polygon-based) in vector format and topologically correct. The mean-shift segmentation was applied to a full extent of satellite images. Accordingly, some of the automatic object extraction methods were applied also using UAV images.

The study from Crommelinck et al. [26] outlines the potential of the Global Probability of Boundary (gPb) contour detection method for an automatic boundary delineation based on UAV imagery. gPb is open-source and available as pre-compiled Matlab package. The method was found to be applicable only for processing images of fewer than 1000 x 1000 pixels due to the demanding computing process [26]. The contour map or detected objects were in raster format and required vectorization. Furthermore, Crommelinck et al. [37] discuss the interactive method of visible boundary extractions. The interactive method combines the gPb contour detection, simple linear iterative clustering (SLIC) super pixels and random forest classifier, which allow a semi-automatic approach for the delineation of visible boundaries. The interactive method was tested on visible road outlines based on UAV datasets. The results show that the approach is much more efficient than manual boundary delineation, and all road boundaries were delineated comprehensively.

All the case studies reviewed, both automatic boundary extractions from HRSI and UAV images, have been tested in rural areas since it is argued that most of the cadastral boundaries are visible in such areas [26]. However, not all computer vision automatic feature extraction methods suitable for visible cadastral boundary delineation have already been tested.

Another tool that is also referred to as the 'state-of-the-art' for automatic detection and extraction of features from images is the ENVI feature extraction (FX) module [39,40]. ENVI FX is an object-based module for detecting and extracting multiple object outlines from high-resolution multispectral or panchromatic digital images. The extraction is based on spectral (brightness and color), texture, and spatial characteristics [41]. To the best of the authors' knowledge, there have been no previous publications, nor evidence, that the ENVI FX module has been applied for detecting and extracting visible cadastral boundaries on UAV images.

The justification for using this method is based on Crommelinck et al. [22], in which general workflow and feature extraction methods appropriate for cadastral mapping are provided. The main aim of this study is not to compare automatic feature extraction methods already used for cadastral mapping. Instead, the study focuses on the potential of a feature extraction method which has not been tested yet in cadastral applications. The study can be seen as an important contribution to land administration discussions focusing on cadastral mapping, as there have been a limited number of studies for automatic visible cadastral boundary delineation from imageries acquired using high-resolution optical sensors.

1.2. Objective of the Study

The study is based on the assumption that many cadastral boundaries are visible [2]. The study's main objective is to outline the potential of the ENVI FX module as well as its limitations for the automatic delineation of visible object boundaries for UAV-based cadastral mapping. It investigates which processing steps (scale level and merge level) using the ENVI FX module need to be applied for UAV-based cadastral mapping. The automatic delineated visible boundaries on UAV images, similarly as manual delineations, can be used for both the creation and updating/revision of cadastral maps.

Overall, the study addresses the whole of the UAV-based cadastral mapping workflow steps, which include image pre-processing, automatic detection and extraction of visible object boundaries on the UAV image, and post-processing of extracted boundaries to more closely approximate cadastral boundaries.

2. Materials and Methods

2.1. UAV Data

To achieve the objective of the study, a rural area in Slovenia was selected as the number of visible (cadastral) boundaries in such areas is higher compared to dense urban ones. In addition, the selected rural area includes roads, agricultural field outlines, fences, hedges, and tree groups, which are assumed to indicate cadastral boundaries [22]. The UAV images of the case study area were indirectly geo-referenced, using an even distribution of ground control points (GCP) within the field as criteria. The GCPs were surveyed with real-time kinematic (RTK) by using Global Navigation Satellite System (GNSS) receiver, Leica Viva, connected in the Slovenian GNSS network, SIGNAL. The signals were received from satellite constellations of GPS and GLONASS. The total number of GCPs was 12. The Position dilution of precision (PDOP) values ranged from 1.2 to 1.7. The flight altitude was 80 m and 354 images were taken to cover the study area. The images were captured on October 19th, 2018 in the noon time (good weather conditions, clear sky) at solar zenith angle of approximately 35 degrees. The study site had a coverage area of 25 ha. The planimetric accuracy assessment of the UAV orthoimage was based on comparison between GCPs coordinates surveyed with the GNSS receiver and the coordinates of GCPs on the UAV orthoimage. The estimated root-mean-square-error (RMSE) was 2.5 cm. Table 1 shows the specifications of data capture and Figure 1 shows the UAV orthoimage of the study area.

Table 1. Specification of unmanned aerial vehicle (UAV) dataset for the selected study area in Slovenia.

Location	UAV Model	Camera/Focal Length [mm]	Overlap Forward/Sideward [%]	Flight altitude [m]	GSD [cm]	Pixels
Ponova vas, Slovenia	DJI Phantom 4 Pro	1″ CMOS 20mp/24	80/70	80	2.0	35,551 × 31,098

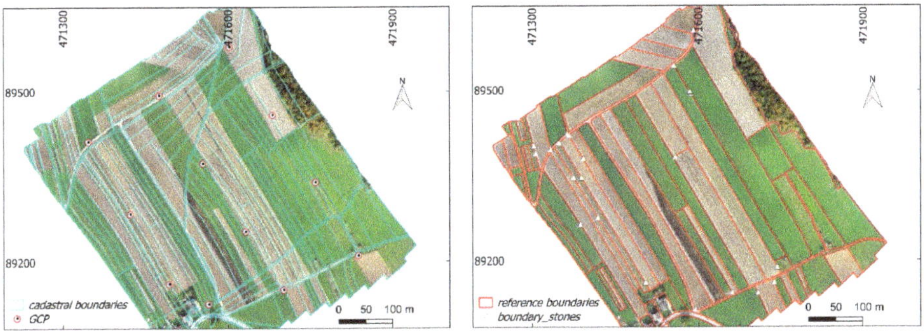

(a) (b)

Figure 1. (**a**) Cadastral map and ground control points (GCPs). (**b**) Manually delineated object visible boundaries used as reference data to determine the detection/extraction quality. (**a**,**b**) Overlaid on UAV orthoimage of Ponova vas, Slovenia (EPSG 3794).

2.2. Reference data

The current cadastral map for the selected area was retrieved from the e-portal of the Slovenian Surveying and Mapping Authority, which is an online platform for requesting official cadastral data [42]. The cadastral map was overlaid on the UAV orthoimage (Figure 1a). The visual interpretation of the combined dataset showed immediately that the cadastral map does not correspond with the visible objects that indicate land possession (land cover) boundaries (roads, agricultural field outlines) on the UAV orthoimage. From the initial analyses, it appeared that only 8% of cadastral boundaries correspond with the manually digitized visible boundaries (at 25 cm tolerance). This is because the current official cadastral map was created by digitizing previous analog cadastral maps whose origin goes back in the first half of 19th century. Due to the underestimated need for cadastral map updating as well as due to the land reforms in the 20th century (i.e., land nationalization and denationalization) the current possession land boundaries do not correspond with cadastral boundaries. Considering this, as reference data, manually digitized boundaries were used instead of the official cadastral data, as the aim of this research is to automatically delineate visible object boundaries from a UAV orthoimage and, at the same time, study the potential of the ENVI FX solution for the automatic detection of visible boundaries. Moreover, during the manual digitization of reference boundaries, some white stones considered as possession boundary signs were used as a guide for proper digitization (Figure 1b). The placement of white stones is a common practice in the selected study area, and for this reason, they were considered as reliable information during the manual digitization. In addition, the confidence in white stones as boundary signs is based on the authors' experiences in professional cadastral surveying.

2.3. Visible Boundary Delineation Method and Workflow

2.3.1. ENVI Feature Extraction (FX)

The investigated tool, ENVI FX, is a combined process of image segmentation and classification. The focus of this study is only at image segmentation and calculating spatial attributes for each segmented object [41]. In addition to spatial attributes, spectral and textural attributes are often used by users for further image classification analysis.

The first step, image segmentation, is based on the technique developed by Jin [43] and involves calculating a gradient map, calculating cumulative distribution function, modification of the gradient map by defining a scale level, and segmentation of a modified gradient map by using the Watershed Transform [44]. A gradient is calculated for each band of the image. The ENVI FX module uses two approaches: edge method and intensity. The edge method calculates a gradient map using the Sobel edge detection algorithm [44]. The Intensity method converts each pixel to a spectral intensity value by averaging it across the selected image bands [44]. The edge method is used for detecting features with distinct boundaries and is considered in this study. In contrast, the Intensity method is suitable for digital elevation models, images of gravitational potential and images of electromagnetic fields [44]. After a gradient map is calculated, a density function of gradients over the whole map is calculated in the form of a cumulative relative histogram [43]. Once the cumulative distribution function has been calculated, it can be used along with the gradient map to calculate the gradient scale space [43]. The gradient map can be modified by changing the scale level. The scale level is the relative threshold on the cumulative relative histogram from which the corresponding gradient magnitude can be determined [43]. For example, at a scale level of 50, the lowest 50 percent of gradient magnitude values are discarded from the gradient image [44]. Increasing the scale level results in fewer segments and keeps objects with the most distinct boundaries [41]. Once the scale level is selected the Watershed Transform algorithm is applied to the modified gradient map. The Watershed Transform is based on the concept of hydrologic watersheds [22,35]. In digital imagery, the same process can be similarly explained as the darker a pixel, the lower its "elevation" (minimum pixel). The algorithm categorizes a pixel by increasing the greyscale value, then begins with the minimum pixels and "floods" the image, dividing the image into objects with similar pixel intensities. The result is a segmented image and each segmented object is assigned with a mean spectral value of all the pixels that belong to that object [44].

The second step is merging. This step aggregates over-segmented areas by using the ENVI FX default full Lambda schedule algorithm. The algorithm is meant to aggregate object outlines within larger, textured areas, such as trees and, fields, based on a combination of spectral and spatial information [41,45]. The merge level represents the threshold Lambda value. Merging occurs when the algorithm finds a pair of adjacent objects such that the merging cost is less than a defined threshold Lambda value—if the merge level is set to 20, it will merge adjacent objects with the lowest 20 percent of Lambda values [45]. When a merge level of 0 is selected no merging will be performed. In this step, the selection of Texture Kernel Size is optional, i.e., the size of a moving box centered over each pixel value. The ENVI FX default Texture Kernel Size is 3, and the maximum is 19 [45].

The final step is the export of object boundaries in a vector format and a segmented image in a raster format. Moreover, each extracted object consists of spatial, spectral, and texture information in the attribute table [41].

2.3.2. Visible Boundary Delineation Workflow

The visible boundary delineation workflow (Figure 2) consists of four main steps. In the following, each workflow step is described in detail with additional comments based on our own preliminary studies aiming to understand and justify the selection of the parameters and algorithms used. The first and second steps were implemented in ENVI 5.5 image analysis software [46] by using the ENVI FX [47] tool. The other steps were implemented using QGIS [48] and GRASS [49] functions.

1. *Image pre-processing:* The first step is resampling the UAV orthoimage. The UAV orthoimage was resampled from 2 cm to lower spatial resolutions—25 cm, 50 cm and 100 cm ground sample distances (GSD). The selected GSDs allowed the identification of the impact of different GSDs on the results of automatic boundary extractions. The pixel average method was used for resampling the UAV orthoimage as it provides a smoother image. In addition, further resampling methods (nearest neighbor and bilinear) were tested and did not provide significant differences in the number of automatic object boundary extractions—at higher scale and merge levels of the ENVI FX algorithm. The resampling step was also applied in [26], to make transferable the investigated method to a UAV orthoimage for cadastral mapping purposes. In addition, extracting objects from a UAV orthoimage of lower spatial resolution is computationally less expensive.
2. *Boundary detection and extraction:* The ENVI FX module was applied to each down-sampled UAV orthoimage. The detection and extraction of visible boundaries from the UAV orthoimage was based on the ENVI FX scale and merge level values. The texture kernel size was set to default,

i.e., 3. In addition, further object extractions were tested at the highest texture kernel size and no differences in the number and locations of extracted objects were identified. Scale level values ranged from 50 to 80 and merge level values from 50 to 99. The initial incremental value for both scale and merge levels was 10. In cases where a jump in the total number of extracted objects was detected the incremental value was dropped for both scale and merge levels. In order to identify the optimal scale and merge values for the detection and extraction of visible objects for cadastral mapping, all possible range values of scale and merge combinations were tested. For each extraction information about the total number of extracted objects and processing time was stored. This resulted in 50 boundary maps for each resampled UAV orthoimage. The boundary map consisted of extracted objects (polygon-based), which were in digital vector format.

3. *Data post-processing:* The process included two steps: (i) the filtering of extracted objects, and (ii) the simplification of extracted objects. (i) The minimum object area and the total number of objects identified in the reference data (Figure 1b) were used to determine optimal scale and merge levels. The minimum reference object area was 204 m2, and the total number of objects was 68. All extracted objects that were smaller than the minimum object area from reference data were filtered out (removed). The total number of remaining objects was compared with the total number of objects from the reference data and the tolerance of +/- 10 objects was set—those parameters that produced numbers of objects that were closest to those found in the reference data, i.e., within defined tolerance, were deemed optimal. The boundary maps from which smaller objects were removed were labeled as filtered objects. The output of filtered objects consisted of holes, i.e., due to polygon-based geometry of objects, which were mostly present either in the forest or individual trees and are of less relevance for cadastral applications—a boundary between adjacent objects belongs to both. (ii) Extracted and filtered object boundaries were smoothed and simplified to be used for the interpretation of possession boundaries aiming to support a cadastral mapping (i.e., land plot restructuring in this case, as the situation requires a new cadastral survey or land consolidation). The smoothing of extracted/filtered object boundaries was done by using the Snakes algorithm [49]. The Douglas–Peucker algorithm was applied to the smoothed object boundaries in order to further simplify the object boundaries [22,49]. These objects both smoothed and simplified were labeled as simplified extracted/filtered objects.
4. *Accuracy assessment:* The accuracy assessment was object-based since the results were in vector format. The buffer overlay method was used for accuracy assessment. The method is described in detail in Heipke et al. [50]. The accuracy assessment was based on computing the percentages of extracted (or reference) boundary lengths which overlapped within a buffer polygon area generated around the reference (or extracted) boundaries (Figure 3) [50]. To determine the completeness, correctness, and quality of extracted boundaries, calculated boundary lengths of true positives (TP), false positives (FP), and false negatives (FN) were used. The completeness refers to the percentage of reference boundaries which lie within the buffer around the extracted boundaries (matched reference). The correctness refers to the percentage of extracted boundaries, which lie within the buffer around the reference boundaries (matched extraction). The accuracy assessment was performed on buffer widths of 25 cm, 50 cm, 100 cm, and 200 cm. The selection of buffer widths is in line with other studies and was based on the most common tolerances regarding boundary positions in land administration, especially for rural areas [26,27]. The percentage indicating the overall quality was generated from the previous two by dividing the length of the matched extractions with the sum of the length of extracted data and the length of unmatched reference [50]. The accuracy assessment was applied to automatic extracted objects, simplified extracted objects, filtered objects, and simplified filtered objects (Figure 2).

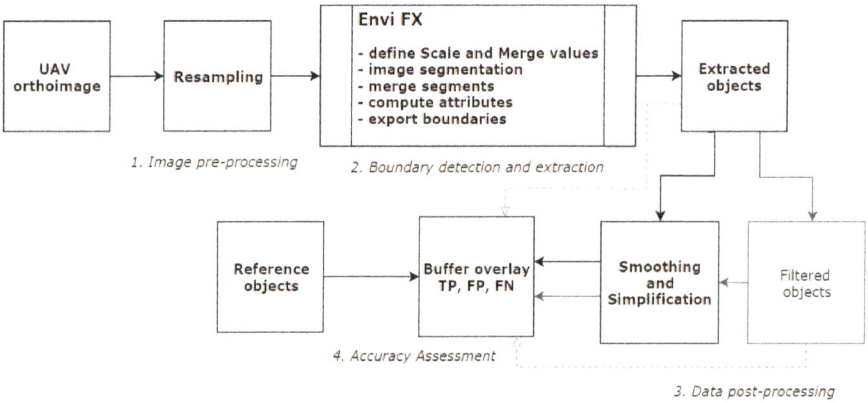

Figure 2. Cadastral mapping workflow based on the automatic detection and extraction of visible boundaries from UAV imagery.

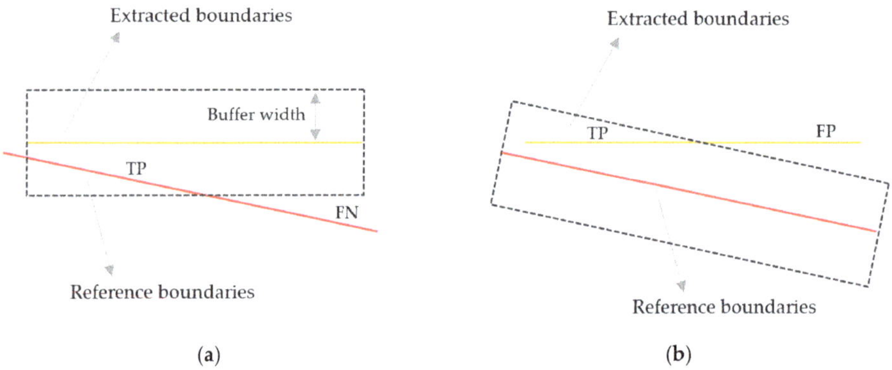

Figure 3. Object-based accuracy assessment method—buffer overlaying method. (**a**) Matched reference. (**b**) Matched extraction. (**a**,**b**) Calculation of boundary lengths of true positives (TP), false positives (FP) and false negatives (FN) (Adapted from [50]).

3. Results

Resampling the UAV orthoimage to a lower spatial resolution, i.e., a larger value of GSD, resulted in fewer and faster extractions of object boundaries compared to the number of extracted object boundaries generated at the original size of the UAV orthoimage. The processing time for one boundary map was 1–2 min. A larger GSD, at the same scale and merge values, resulted in fewer boundary extractions (Table 2, Figures 4 and 5).

Table 2. Ground sample distance (GSD) and number of pixels after image pre-processing.

GSD [cm]	Pixels	Resampling Method
25	2856 × 2498	Pixel average
50	1428 × 1249	Pixel average
100	714 × 625	Pixel average

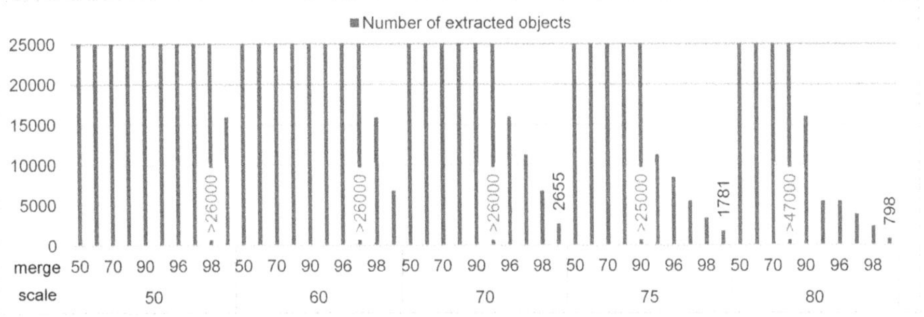

(a)

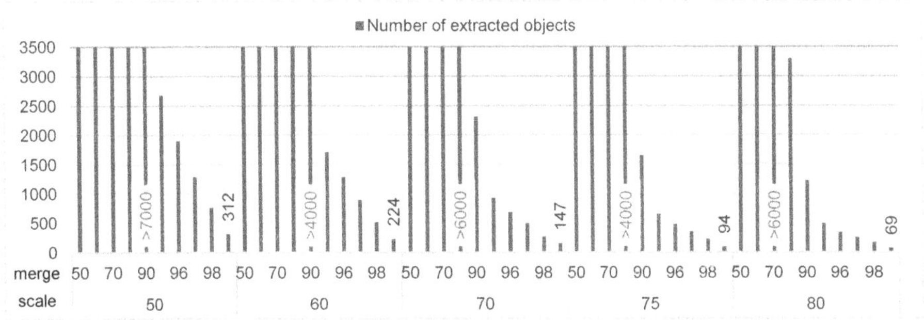

(b)

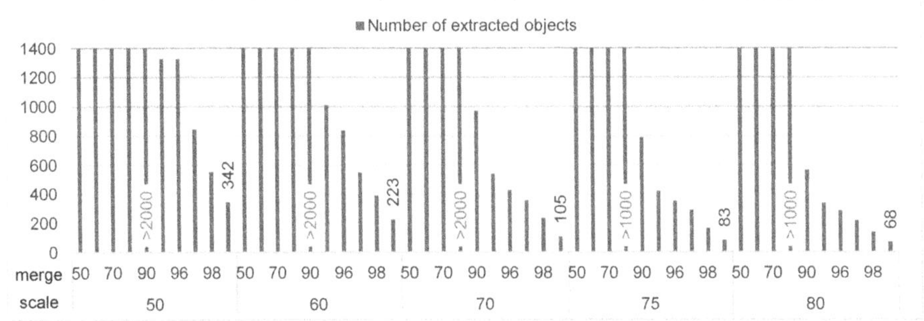

(c)

Figure 4. Scale/merge level and number of extracted objects from the resampled UAV orthoimages (**a**) ground sample distance (GSD) 25 cm, (**b**) GSD 50 c, and (**c**) GSD 100 cm. (**a**–**c**) Grey labels—number of extracted objects outside the range, black labels—the lowest number of extracted objects per scale and merge parameter value.

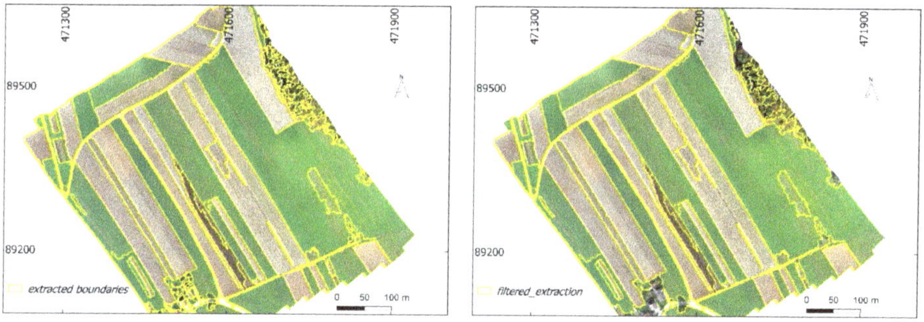

(e) (f)

Figure 7. (**a**,**c**,**e**) Extracted objects at scale level 80 and merge level 95 for (**a**) GSD 25 cm, (**c**) GSD 50 cm, and (**e**) GSD 100 cm. (**b**,**d**,**f**) Filtered objects of scale level 80 and merge level 95 based on minimum object area from the reference data.

A simplification algorithm was applied to both extracted objects and filtered objects. The results showed that if extracted objects are smoothed and smoothed objects are later simplified, the localization of simplified objects is almost equal to that of the extracted ones (Figure 8). The initial tests show that possible shifts in location are possible when a direct implementation of the simplification algorithm to extracted visible objects is used.

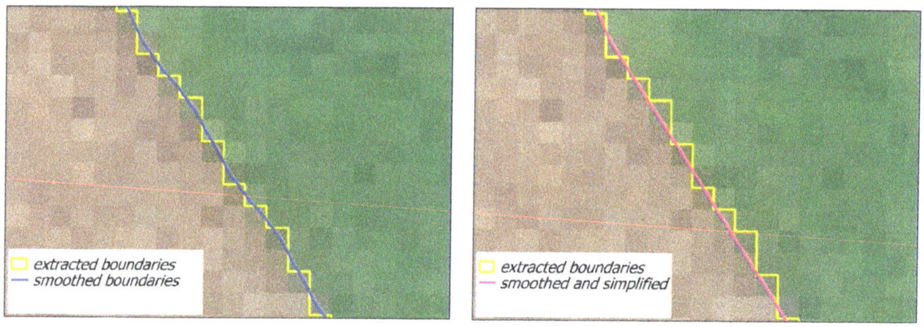

(**a**) (**b**)

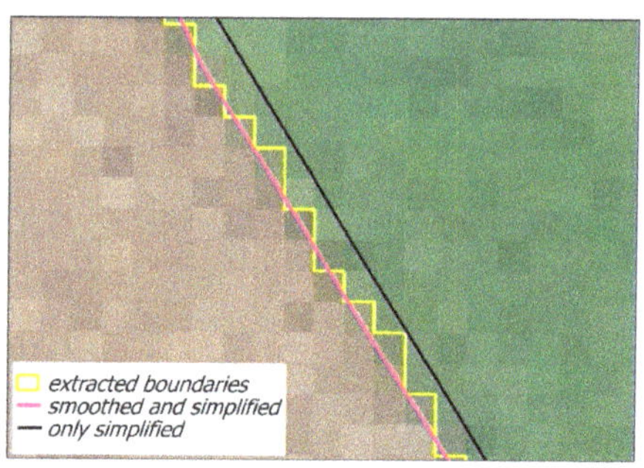

(**c**)

Figure 8. (**a**) Extracted objects smoothed by making use of the Snakes algorithm. (**b**) Simplification of extracted objects by making use of the Snakes smoothing algorithm and Douglas–Peucker simplification algorithm. (**c**) Extracted objects simplified with Douglas–Peucker algorithm (in black) and compared to object simplifications on (**b**).

The buffer overlay method was used for the accuracy assessment. The accuracy assessment method was applied to the extracted objects, simplified extracted objects, filtered objects and simplified filtered objects. The results show that there is no significant difference in accuracy assessment results when comparing extracted and simplified objects (Table 3, Table 4, and Table 5). At a buffer width of 2 m, a GSD of 50 cm and a GSD of 100 cm provide a higher percentage of correctly extracted objects compared to a GSD of 25 cm. The percentage of correctly extracted objects was 66% for both a GSD of 50 cm and a GSD of 100 cm (Tables 4 and 5). However, the filtering approach contributed to the increased correctness (decreased completeness) and overall quality, for all GSDs. From the filtered objects, the best results for correctness were recorded at a GSD of 50 cm (Figure 9). The percentage indicated the correctness was 77%, while for the completeness it was 67%.

Table 3. Accuracy assessment of boundary extractions for a GSD of 25 cm, scale 80, merge 95.

Buffer width [cm]	Completeness [%]		Correctness [%]		Quality [%]	
	Extracted	Filtered	Extracted	Filtered	Extracted	Filtered
25	58	37	18	26	16	20
50	73	48	28	39	26	31
100	78	56	38	50	36	41
200	**81 (81)** [1]	61 (62) [1]	**48 (49)** [1]	59 (61) [1]	**46 (46)** [1]	50 (48) [1]

[1] Percentages of simplified boundaries.

Table 4. Accuracy assessment of boundary extractions for a GSD of 50 cm, scale 80, merge 95.

Buffer width [cm]	Completeness [%]		Correctness [%]		Quality [%]	
	Extracted	Filtered	Extracted	Filtered	Extracted	Filtered
25	45	40	28	35	21	23
50	64	55	46	56	38	41
100	71	61	57	68	48	52
200	**75 (74)** [1]	65 (67) [1]	**65 (66)** [1]	76 (77) [1]	**56 (53)** [1]	59 (56) [1]

[1] Percentages of simplified boundaries.

Table 5. Accuracy assessment of boundary extractions for a GSD of 100 cm, scale 80, merge 95.

Buffer Width [cm]	Completeness [%]		Correctness [%]		Quality [%]	
	Extracted	Filtered	Extracted	Filtered	Extracted	Filtered
25	31	27	21	24	14	15
50	53	47	39	43	29	30
100	67	59	58	64	47	47
200	**73 (71)** [1]	63 (67) [1]	**66 (66)** [1]	72 (73) [1]	**55 (52)** [1]	55 (52) [1]

[1] Percentages of simplified boundaries.

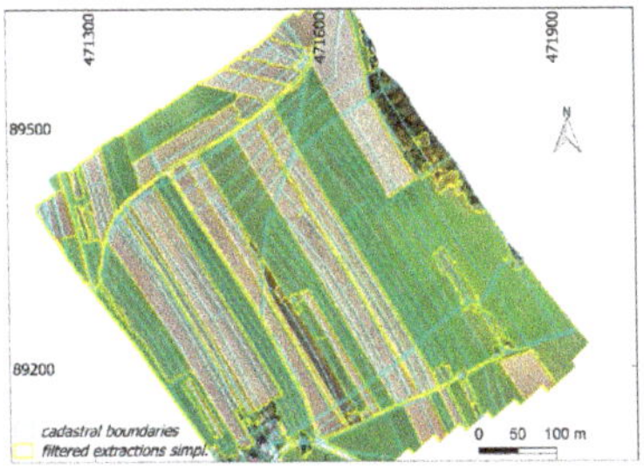

(a)

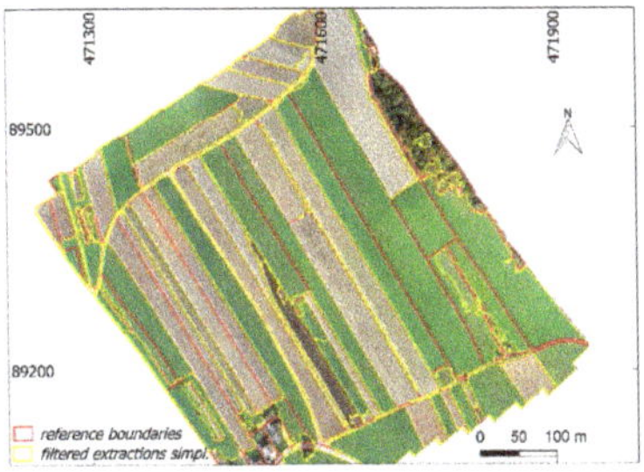

(b)

Figure 9. (**a**,**b**) Filtered objects of scale level 80 and merge level 95—simplified, compared with (**a**) cadastral map and (**b**) manually delineated visual object boundaries used as reference data.

4. Discussion

4.1. The Developed Workflow

The developed workflow aimed to provide a solution for UAV-based cadastral mapping using automatic visible boundary extraction with the ENVI FX module (Figure 2). The developed workflow consisted of four steps: (i) image pre-processing, (ii) boundary detection and extraction, (iii) data post-processing, and (iv) accuracy assessment.

The first workflow step includes resampling of a UAV orthoimage. Here, the results of our case study showed that larger GSDs provided faster and fewer extractions of visible object boundaries compared to the original GSD of a UAV orthoimage. For higher spatial resolutions, i.e., smaller GSDs, considering the selected Scale level and Merge level values, the total number of extracted objects was higher.

The second step, which includes object boundary detection and extraction, is dependent on the scale and merge level. The results, presented in Figure 4, showed that lower values of scale and merge levels resulted in a higher number of extracted objects, which led to over-segmentation by reaching thousands of extracted objects. Considering the total number of the reference objects, it is important to note that a scale and a merge level that provide object extractions close to the total number of objects from reference data are important for automatic delineation of visible cadastral boundaries.

The following step, data post-processing, aimed to investigate optimal scale and merge levels and to simplify the extracted objects. The optimal values based on a filtering approach showed that for all tested GSDs in this study, most suitable scale and merge level values for automatic delineation of visible cadastral boundaries were 70–80 and 95–99, respectively. These values can be considered as optimal scale and merge levels for rural areas in general or areas with characteristics similar to the study area of this research. However, to validate the proposed workflow and optimal Scale and Merge levels in areas with different characteristics, such as areas with a larger number of buildings or areas with trees covering parts of boundaries, further experiments are needed. Hence, the scale and merging levels appropriate for cadastral mapping have been determined and this step can be skipped from the workflow step of data post-processing (Figure 2). The use of the Snakes algorithm for smoothing and the Douglas–Peucker algorithm for simplifying has been shown to be very effective (Figure 8a,b). This approach, when combining both smoothing and simplification algorithms, gives better results in terms of a simplified boundary position compared to directly implementing the Douglas–Peucker simplification algorithm, where undesired shifting in boundary position was observed (Figure 8c). In [22], it was reported that the direct implementation of the Douglas–Peucker algorithm was used as a post-processing method in many papers to improve the output by optimizing the shape of objects. However, the simplification approach applied in this study was not examined in the previous papers.

The final step of the workflow was accuracy assessment (see also Section 4.2). The accuracy assessment was based on the buffer overlay method. By increasing the width of the buffer, more extracted boundaries appear to be within the buffer area, which impacts the completeness, correctness, and the overall quality—larger the buffer, the better the results. To have a uniform assessment for all tested GSDs the results were compared at a buffer distance of 2 m. From the reviewed publications presented in Section 1.2, a buffer width of 2 m was also applied in [26–28] as most suitable for the presentation of accuracy assessment results and to avoid uncertainties from resampling effects. However, for comparison to cadastral data, buffer widths should be based on local accuracy requirements [26].

The workflow developed, overall, is in accordance with the general workflow for the cadastral mapping based on suitable computer vision methods for automatic visible boundary extraction provided in [22]. In addition, it provides an additional step and method in data post-processing, such as filtering out irrelevant and small objects from the boundary map, which improves overall quality assessment. Furthermore, it suggests a combined approach for the simplification of extracted object boundaries.

4.2. Quality Assessment

Bringing the scale and merge levels to the maximum resulted in some unextracted and fewer visible objects for the whole extent of the image. Although some of the visible objects were left unextracted, the maximum scale and merge level enabled the detection of a group of objects such as a group of tree boundaries, especially at GSDs of 50 cm and 100 cm (Figure 5f,i). In both cases, the balance between completeness and correctness was hard to achieve. This issue was also reported in [26,28]. For this reason, the filtering approach was applied. It was based on the minimum object size as well as on the total number of the objects, both defined based on the reference data. This allowed us to reduce the risk that some of the visible object boundaries remained unextracted as well as over-segmented.

The optimal scale level of 80 and a merge level of 95, were chosen for all three GSDs, to investigate the impact of the same scale and merge level in different resampling. The selection was based on common scale and merge levels per GSD (Figure 6). However, this does not mean that the chosen scale and merge level provided the best object boundary extraction for each of the GSDs. For instance, for small GSDs, the correctness of extracted boundaries is higher at the maximum scale and merge levels (e.g., Figure 5c). For the same scale and merge level, the correctness grows significantly from a GSD of 25 cm to a GSD of 50 cm. The correctness for a GSD of 100 cm was almost equal to the one for a GSD of 50 cm. Considering that more optimal scale and merge levels were applicable for a GSD of 50 cm (Figure 6) and the difference insignificant when compared to the results obtained for a GSD of 100 cm, a GSD of 50 cm appeared to be better in detecting visible boundaries compared to the other two GSDs.

The quantitative method applied for accuracy assessment to automatically extracted objects, filtered objects and to their simplifications, showed that there was no significant difference between extracted objects and simplified objects. This result indicates that the method applied for simplification can be considered appropriate, i.e., the original location of extracted objects was maximally maintained. Although there was no difference in accuracy assessment, the simplification of extracted (or filtered) objects is significant for proper cadastral mapping. Cadastral boundaries usually are defined by straight lines with fewer vertices.

The percentage of suitable extracted boundaries (compared to reference data), for a scale level of 80 and a merge level of 95, resulted in 74% for the assessment of the completeness and 66% for the assessment of the correctness for the extracted object boundaries having a GSD of 50 cm. However, the filtering approach strongly influenced the accuracy assessment. For filtered extractions, the level of completeness was 67%, and the level of correctness was 77%. These results show that the filtering approach increased the correctness of automatically extracted boundaries, and it reaches almost 80% (Table 4). This was due to filtering out small object boundaries from the boundary map. The excluded small objects were mostly present in tree and built-up areas on the UAV orthoimage, i.e., only outlines of group objects were retained (Figure 7c,d). In road extractions, the achieved values for extractions are around 85% for correctness and around 70% for completeness to be of real practical importance [26,34]. Such percentages can hardly be achieved by the workflow developed for automatic delineation of all visible boundaries since the morphology of cadastral boundaries is usually more complex and not all cadastral boundaries are visible, unlike road boundaries.

The accuracy assessment was based on the manually delineated boundaries, which were defined as reference data (Figure 1b). The visible boundaries were manually delineated on the ground truth UAV orthoimage. It is argued that manually delineated boundaries influence the overall results of the accuracy assessment since different human operators might digitize differently [26]. However, in the selected case study, most of the object boundaries were sharp and the presence of white stones at outlines of the agricultural field contributed to the objectivity of manual digitalization. In addition, the real cadastral data could not be used since they did not correspond with the object boundaries on the image (Figure 1a) and it would not have been possible to outline the potential of the ENVI FX. However, the approach of automatic extraction of visible boundaries is case dependent. To reliably avoid the influence of manually digitized reference data, the following studies should consider a case study where the cadastral map is up to date.

4.3. Strengths and Limitations of the Automatic Extraction Method Used

The ENVI FX module handled the full extent of the resampled UAV orthoimages, and no additional image tiling or image conversions were required. ENVI FX provided closed object boundaries directly in vector format, topologically correct polygons. Therefore, no additional image post-processing step, such as vectorization of detected object boundaries, was needed (Figure 5). Thus, the visible object boundaries generated can be directly used for further processing and analysis within geographic information systems (GIS). Additionally, the final output consists of spatial, spectral and textural attributes which are assigned automatically to each extracted object and saved in the attribute table. The vectorized and geo-referenced visible object boundaries, as interpreted in this research, are crucial in cadastral applications especially for the purposes of land plot boundary delineations. Overall, ENVI FX has the potential to automatically delineate visible cadastral boundaries, especially in rural areas.

A comparison of the results regarding the accuracy assessment obtained in this study and the accuracies obtained in the studies [26–28] cannot be done at this time for a number of reasons. First, not all the reviewed feature extraction methods have been applied to UAV imagery. Second, different UAVs may provide different quality of orthoimages. Third, the nature, size, location, and the characteristics of the study objects are far too different. In order to make a reliable comparison on accuracy assessments of different feature extraction methods, first of all, each method has to be studied individually and later tested at the same study area(s). However, the image processing approach of different feature extractions methods may be comparable.

From the reviewed feature extraction methods that have already been applied for detection of visible cadastral boundaries, it can be seen that the MRS method, ESP method, and mean-segmentation method also do not require further image tiling and the final output of the boundary map was in vector format [27,28]. In contrast, vectorization of detected object boundaries was needed for the gPb contour detection method. In addition, it was reported that the method is inapplicable when processing UAV images of more than 1000 pixels in width and height [26]. Similar issues regarding the vectorization of detected object boundaries were reported in [38], where Canny and Sobel edge detection algorithms were used. In order to obtain topologically correct polygons, an additional feature extraction method was used aiming to connect the edges.

ENVI FX allowed some shadow areas in the UAV orthoimage to be extracted as boundaries; however, these do not represent real boundaries in the field. In order to minimize the influence of shadows on feature extraction, it is recommended to capture images in the local time where the solar zenith angle has the smallest possible value. However, the solar zenith angle depends on the geographic location of the study area. Additionally, some other factors such as weather conditions also influence the quality of captured images. To avoid such issues, it is preferable to capture images on a cloudy day without wind. Although ENVI FX has proved to be efficient, one of its limitations is that it is not an open-source tool like mean-shift segmentation, gPb contour detection, Canny, and Sobel, which might be a reason why it is not often used in the land surveyor community. In addition, the extracted objects from the resampled UAV orthoimages were following the pixel borders and further shape simplification was required to make them comparable to spatial units in cadastral applications.

Considering that the morphology of cadastral boundaries is complex [7], compared to physical boundaries, such as boundaries of roads or rivers, delineation of cadastral boundaries cannot be fully automated at this time, and additionally, the verification of the results has to be done with the participation of landowners and other land rights holders. The limitations on extracting only visible object boundaries lie in the fact that not all visible boundaries (land cover boundaries) represent cadastral boundaries (land right boundaries). For instance, when two agricultural cadastral units leased to the same farmer are farmed as one unit, and vice versa. However, visible object boundaries which coincide with the land right boundaries can be automatically detected and used in cadastral applications. In addition, the UAV-based spatial data acquisition is usually affected by special operational regulations that restrict the use of this technology, in particular in urban areas [18].

4.4. Applicability of the Developed Workflow

The developed workflow provided geo-referenced boundary maps in a format compliant with the formats that are used in GIS environments. This shows that the extracted objects can be easily transferable, and applicable in GIS for cadastral purposes. In cases where cadastral maps are rarely present and the concept of fit-for-purpose cadastre is in place, the workflow, with the selected method for automatic extraction of visible boundaries, shows the potential for the automation of the visible cadastral boundary delineation procedure [1]. Thus, the approach developed generally contributes to the acceleration and facilitation of the creation of cadastral maps (Figure 9b), especially in developing countries, where general boundaries are accepted, and positional accuracy is of lesser importance [25]. However, the approach is suitable for the areas where the boundaries of physical objects are visibly detectable on a UAV orthoimage, for instance, in rural areas. The workflow might be applicable for both the creation and updating/revision of cadastral maps, similar to the manual delineations of cadastral boundaries on a UAV orthoimage. In addition, the workflow developed might lower the costs and time compared to the manual delineation of cadastral boundaries, especially in rural areas [26].

Furthermore, in developed countries, the approach based on automatic extraction of visible boundaries might be used for a revision of current cadastral maps (Figure 9a). In this case, the extracted visible boundaries can be used as a basis for a new cadastral survey or land rearrangements, depending on the discrepancy between cadastral maps and land possession (as shown in the case study). Although the beneficiaries agree with the visible boundaries, if higher accuracy is required, the revised objects (spatial units) can later be manually delineated from a UAV orthoimage or re-surveyed with ground-based surveying techniques. It must be emphasized that the extracted visible boundaries, both for the creation of cadastral maps and updating, should be inspected by the local community and all beneficiaries (landowners, other land rights holders) in order to be legally validated.

5. Conclusions

The overall aim of this study was to provide an UAV-based cadastral mapping workflow based on the ENVI FX module for automatic detection of visible boundaries. The study first investigated, which processing steps are required for a cadastral mapping workflow following the potential and limitations of the ENVI FX for automatic visible boundary detection and extraction.

The results showed that more correct visible object boundaries, suitable for the interpretation of land cover (cadastral) boundaries, were extracted at larger values of GSD. In addition, the identified optimal scale and merge levels for detection and extraction of visible cadastral boundaries were between 70 and 80 and 95 and 99, respectively. The identification of the optimal parameters for cadastral mapping was based on the defined minimum object area and the total number of objects from the reference data using the so-called filtering approach. The filtering approach contributed to the increased correctness of automatically extracted boundaries. The best results were recorded at the resampled UAV orthoimage with a GSD of 50 cm, and the percentage of correctness indicated was 77%, while for the completeness it was 67%. It must be emphasized that the workflow developed is applicable mostly for rural areas where the number of visible boundaries is higher compared to complex urban areas.

The workflow can be used in developing countries to accelerate and facilitate the creation of cadastral maps aiming to formalize a land tenure system and guarantee legal security to land rights holders. In developed countries, the extracted visible boundaries based on this workflow might be used for efficient revision of existing cadastral maps. However, in both cases, the extracted visible boundaries have to be validated by landowners and other beneficiaries. The extraction of visible objects can be considered as only one step in the facilitation of cadastral mapping, as extracting these is not enough for complete and correct cadastral mapping. It is worth highlighting that cadastral boundaries may, in fact, be completely inside the property and that some boundaries between properties may not be clearly visible. In order to use the proposed workflow in the cadastral domain, the approach can be expanded. Additional steps should focus on methods for the possible involvement of current landowners in the process of cadastral mapping. The extension of the current workflow is one of the aims of the authors' further research.

Author Contributions: Conceptualization, B.F. and A.L.; methodology, B.F., K.O. and A.L.; software, B.F.; validation, B.F., K.O. and A.L.; formal analysis and data processing, B.F.; investigation, B.F.; resources, B.F., A.L., K.O. and M.K.F.; writing—original draft preparation, B.F.; writing—review and editing, B.F., A.L., K.O. and M.K.F.; visualization, B.F.; supervision, A.L.

Funding: The authors acknowledge the financial support of the Slovenian Research Agency (research core funding No. P2-0406 Earth observation and geoinformatics).

Acknowledgments: We acknowledge Klemen Kozmus Trajkovski and Albin Mencin for capturing the UAV data and for the technical support during the fieldwork.

Conflicts of Interest: The authors declare no conflict of interest.

Abbreviations

The following abbreviations are used in this article:

DSM	Digital Surface Model
ESP	Estimation of Scale Parameter
FIG	International Federation of Surveyors
FN	False Negative
FP	False Positive
FX	ENVI Feature Extraction
GCP	Ground Control Point
GIS	Geographic Information System
GNSS	Global Navigation Satellite System
gPb	Global Probability of Boundary
GSD	Ground Sample Distance
HRSI	High-Resolution Satellite Imagery
MRS	Multi-Resolution Segmentation
PDOP	Position Dilution of Precision
RMSE	Root-Mean-Square-Error
RTK	Real-Time Kinematic
SLIC	Simple Linear Iterative Clustering
TP	True Positive
UAV	Unmanned Aerial Vehicle

References

1. Enemark, S. International Federation of Surveyors. In *Fit-For-Purpose Land Administration: Joint FIG/World Bank Publication*; FIG: Copenhagen, Denmark, 2014; ISBN 978-87-92853-10-3.
2. Luo, X.; Bennett, R.; Koeva, M.; Lemmen, C.; Quadros, N. Quantifying the Overlap between Cadastral and Visual Boundaries: A Case Study from Vanuatu. *Urban Sci.* **2017**, *1*, 32. [CrossRef]
3. Zevenbergen, J. A Systems Approach to Land Registration and Cadastre. *Nord. J. Surv. Real Estate Res.* **2004**, *1*, 11–24.
4. Simbizi, M.C.D.; Bennett, R.M.; Zevenbergen, J. Land tenure security: Revisiting and refining the concept for Sub-Saharan Africa's rural poor. *Land Use Policy* **2014**, *36*, 231–238. [CrossRef]
5. *Land Administration for Sustainable Development*, 1st ed.; Williamson, I.P. (Ed.) ESRI Press Academic: Redlands, CA, USA, 2010; ISBN 978-1-58948-041-4.
6. Zevenbergen, J.A. *Land Administration: To See the Change from Day to Day*; ITC: Enschede, The Netherlands, 2009; ISBN 978-90-6164-274-9.
7. Luo, X.; Bennett, R.M.; Koeva, M.; Lemmen, C. Investigating Semi-Automated Cadastral Boundaries Extraction from Airborne Laser Scanned Data. *Land* **2017**, *6*, 60. [CrossRef]
8. Enemark, S. Land Administration and Cadastral Systems in Support of Sustainable Land Governance—A Global Approach. In Proceedings of the Re-Engineering the Cadastre to Support E-Government, Tehran, Iran, 4–26 May 2009.

9. Maurice, M.J.; Koeva, M.N.; Gerke, M.; Nex, F.; Gevaert, C. A Photogrammetric Approach for Map Updating Using UAV in Rwanda. Available online: https://bit.ly/2FyhbEi (accessed on 25 June 2019).
10. Wayumba, R.; Mwangi, P.; Chege, P. Application of Unmanned Aerial Vehicles in Improving Land Registration in Kenya. *Int. J. Res. Eng. Sci.* **2017**, *5*, 5–11.
11. Ramadhani, S.A.; Bennett, R.M.; Nex, F.C. Exploring UAV in Indonesian cadastral boundary data acquisition. *Earth Sci. Inform.* **2018**, *11*, 129–146. [CrossRef]
12. Mumbone, M.; Bennet, R.; Gerke, M. Innovations in Boundary Mapping: Namibia, Customary Lands and UAVs. In Proceedings of the Linking Land Tenure and Use for Shared Prosperity, Washington, DC, USA, 23–27 March 2015; p. 22.
13. Volkmann, W.; Barnes, G. Virtual Surveying: Mapping and Modeling Cadastral Boundaries Using Unmanned Aerial Systems (UAS). In Proceedings of the FIG Congress 2014, Kuala Lumpur, Malaysia, 16–21 June 2014; p. 13.
14. Rijsdijk, M.; van Hinsbergh, W.H.M.; Witteveen, W.; ten Buuren, G.H.M.; Schakelaar, G.A.; Poppinga, G.; van Persie, M.; Ladiges, R. Unmanned Aerial Systems in the process of Juridical verification of Cadastral borde. *ISPRS Int. Arch. Photogramm. Remote Sens. Spat. Inf. Sci.* **2013**, *XL-1/W2*, 325–331. [CrossRef]
15. Mesas-Carrascosa, F.J.; Notario-García, M.D.; Meroño de Larriva, J.E.; Sánchez de la Orden, M.; García-Ferrer Porras, A. Validation of measurements of land plot area using UAV imagery. *Int. J. Appl. Earth Obs. Geoinf.* **2014**, *33*, 270–279. [CrossRef]
16. Manyoky, M.; Theiler, P.; Steudler, D.; Eisenbeiss, H. Unmanned Aerial Vehicle in Cadastral Applications. *ISPRS Int. Arch. Photogramm. Remote Sens. Spat. Inf. Sci.* **2012**, *XXXVIII-1/C22*, 57–62. [CrossRef]
17. Kurczynski, Z.; Bakuła, K.; Karabin, M.; Kowalczyk, M.; Markiewicz, J.S.; Ostrowski, W.; Podlasiak, P.; Zawieska, D. The posibility of using images obtained from the UAS in cadastral works. *ISPRS Int. Arch. Photogramm. Remote Sens. Spat. Inf. Sci.* **2016**, *XLI-B1*, 909–915. [CrossRef]
18. Cramer, M.; Bovet, S.; Gültlinger, M.; Honkavaara, E.; McGill, A.; Rijsdijk, M.; Tabor, M.; Tournadre, V. On the use of RPAS in National Mapping—The EuroSDR point of view. *ISPRS Int. Arch. Photogramm. Remote Sens. Spat. Inf. Sci.* **2013**, *XL-1/W2*, 93–99. [CrossRef]
19. Binns, B.O.; Dale, P.F. Cadastral Surveys and Records of Rights in Land. Available online: http://www.fao.org/3/v4860e/v4860e03.htm (accessed on 20 March 2019).
20. Watts, A.C.; Ambrosia, V.G.; Hinkley, E.A. Unmanned Aircraft Systems in Remote Sensing and Scientific Research: Classification and Considerations of Use. *Remote Sens.* **2012**, *4*, 1671–1692. [CrossRef]
21. Colomina, I.; Molina, P. Unmanned aerial systems for photogrammetry and remote sensing: A review. *ISPRS J. Photogramm. Remote Sens.* **2014**, *92*, 79–97. [CrossRef]
22. Crommelinck, S.; Bennett, R.; Gerke, M.; Nex, F.; Yang, M.; Vosselman, G. Review of Automatic Feature Extraction from High-Resolution Optical Sensor Data for UAV-Based Cadastral Mapping. *Remote Sens.* **2016**, *8*, 689. [CrossRef]
23. Heipke, C.; Woodsford, P.A.; Gerke, M. Updating geospatial databases from images. In *Advances in Photogrammetry, Remote Sensing and Spatial Information Sciences: 2008 ISPRS Congress Book*; Baltsavias, E., Li, Z., Chen, J., Eds.; CRC Press: London, UK, 2008.
24. Bennett, R.; Kitchingman, A.; Leach, J. On the nature and utility of natural boundaries for land and marine administration. *Land Use Policy* **2010**, *27*, 772–779. [CrossRef]
25. Zevenbergen, J.; Bennett, R. The visible boundary: More than just a line between coordinates. In Proceedings of the GeoTech Rwanda, Kigali, Rwanda, 18–20 November 2015; pp. 1–4.
26. Crommelinck, S.; Bennett, R.; Gerke, M.; Yang, M.; Vosselman, G. Contour Detection for UAV-Based Cadastral Mapping. *Remote Sens.* **2017**, *9*, 171. [CrossRef]
27. Wassie, Y.A.; Koeva, M.N.; Bennett, R.M.; Lemmen, C.H.J. A procedure for semi-automated cadastral boundary feature extraction from high-resolution satellite imagery. *J. Spat. Sci.* **2018**, *63*, 75–92. [CrossRef]
28. Kohli, D.; Crommelinck, S.; Bennett, R. Object-Based Image Analysis for Cadastral Mapping Using Satellite Images. In *Proceedings of the International Society for Optics and Photonics, Image Signal Processing Remote Sensing XXIII*; SPIE. The International Society for Optical Engineering: Warsaw, Poland, 2017.
29. Kohli, D.; Unger, E.-M.; Lemmen, C.; Koeva, M.; Bhandari, B. Validation of a cadastral map created using satellite imagery and automated feature extraction techniques: A case of Nepal. In Proceedings of the FIG Congress 2018, Istanbul, Turkey, 6–11 May 2018; p. 17.

30. Singh, P.P.; Garg, R.D. Road Detection from Remote Sensing Images using Impervious Surface Characteristics: Review and Implication. *ISPRS Int. Arch. Photogramm. Remote Sens. Spat. Inf. Sci.* **2014**, *XL-8*, 955–959. [CrossRef]
31. Kumar, M.; Singh, R.K.; Raju, P.L.N.; Krishnamurthy, Y.V.N. Road Network Extraction from High Resolution Multispectral Satellite Imagery Based on Object Oriented Techniques. *ISPRS Ann. Photogramm. Remote Sens. Spat. Inf. Sci.* **2014**, *II-8*, 107–110. [CrossRef]
32. Wang, J.; Qin, Q.; Gao, Z.; Zhao, J.; Ye, X. A New Approach to Urban Road Extraction Using high-resolution aerial image. *ISPRS Int. J. Geo-Inf.* **2016**, *5*, 114. [CrossRef]
33. Paravolidakis, V.; Ragia, L.; Moirogiorgou, K.; Zervakis, M. Automatic Coastline Extraction Using Edge Detection and Optimization Procedures. *Geosciences* **2018**, *8*, 407. [CrossRef]
34. Mayer, H.; Hinz, S.; Bacher, U.; Baltsavias, E. A test of Automatic Road Extraction approaches. *ISPRS Int. Arch. Photogramm. Remote Sens. Spat. Inf. Sci.* **2006**, *36*, 209–214.
35. Dey, V.; Zhang, Y.; Zhong, M. A review of image segmentation techniques with remote sensing perspective. In Proceedings of the ISPRS TC VII Symposium, Vienna, Austria, 5–7 July 2010; Volume XXXVIII. Part 7a.
36. Mueller, M.; Segl, K.; Kaufmann, H. Edge- and region-based segmentation technique for the extraction of large, man-made objects in high-resolution satellite imagery. *Pattern Recognit.* **2004**, *37*, 1619–1628. [CrossRef]
37. Crommelinck, S.; Höfle, B.; Koeva, M.N.; Yang, M.Y.; Vosselman, G. Interactive Cadastral Boundary Deliniation from UAV data. *ISPRS Ann. Photogramm. Remote Sens. Spat. Inf. Sci.* **2018**, *IV-2*, 81–88. [CrossRef]
38. Babawuro, U.; Zou, B. Satellite Imagery Cadastral Features Extractions using Image Processing Algorithms: A Viable Option for Cadastral Science. *IJCSI Int. J. Comput. Sci. Issues* **2012**, *9*, 30–38.
39. Wang, J.; Song, J.; Chen, M.; Yang, Z. Road network extraction: A neural-dynamic framework based on deep learning and a finite state machine. *Int. J. Remote Sens.* **2015**, *36*, 3144–3169. [CrossRef]
40. Poursanidis, D.; Chrysoulakis, N.; Mitraka, Z. Landsat 8 vs. Landsat 5: A comparison based on urban and peri-urban land cover mapping. *Int. J. Appl. Earth Obs. Geoinf.* **2015**, *35*, 259–269. [CrossRef]
41. *ITT Visual Information Solutions ENVI Feature Extraction User's Guide*; Harris Geospatial Solutions: Broomfield, CO, USA, 2008.
42. The Surveying and Mapping Authority of the Republic of Slovenia e-Surveying Data. Available online: https://egp.gu.gov.si/egp/?lang=en (accessed on 29 May 2019).
43. Jin, X. Segmentation-Based Image Processing System. U.S. Patent 8,260,048, 4 September 2012.
44. ENVI Segmentation Algorithms Background. Available online: https://www.harrisgeospatial.com/docs/backgroundsegmentationalgorithm.html#Referenc (accessed on 21 March 2019).
45. ENVI Merge Algorithms Background. Available online: https://www.harrisgeospatial.com/docs/backgroundmergealgorithms.html (accessed on 21 March 2019).
46. ENVI Development Team. *ENVI The Leading Geospatial Analytics Software*; Harris Geospatial Solutions: Broomfield, CO, USA, 2018.
47. Extract Segments Only. Available online: https://www.harrisgeospatial.com/docs/segmentonly.html (accessed on 25 March 2019).
48. QGIS Development Team. *QGIS a Free and Open Source Geographic Information System, Version 2.18*; Las Palmas de, G.C., Ed.; Open Source Geospatial Foundation: Beaverton, OR, USA, 2018.
49. GRASS GIS Development Team. *GRASS GIS Bringing Advanced Geospatial Technologies to the World, Version 7.4.2*; Open Source Geospatial Foundation: Beaverton, OR, USA, 2018.
50. Heipke, C.; Mayer, H.; Wiedemann, C. Evaluation of Automatic Road Extraction. *Int. Arch. Photogramm. Remote Sens.* **1997**, *32*, 151–160.

remote sensing

MDPI

Article

Unmanned Aerial System Imagery, Land Data and User Needs: A Socio-Technical Assessment in Rwanda

Claudia Stöcker [1,*], Serene Ho [2,3], Placide Nkerabigwi [4], Cornelia Schmidt [5], Mila Koeva [1], Rohan Bennett [6,7] and Jaap Zevenbergen [1]

1 University of Twente, Faculty of Geo-Information Science and Earth Observation (ITC), 7522 NB Enschede, The Netherlands; m.n.koeva@utwente.nl (M.K.); j.a.zevenbergen@utwente.nl (J.Z.)
2 RMIT University, School of Science, Melbourne VIC 3000, Australia; serene.ho2@rmit.edu.au or serene.ho@kuleuven.be
3 KU Leuven, Public Governance Institute, 3000 Leuven, Belgium
4 INES-Ruhengeri, Institute of Applied Sciences, Musanze, Rwanda; nkerplac@yahoo.fr
5 Esri Rwanda Ltd., Kigali, Rwanda; c.schmidt@esri.rw
6 Swinburne Business School, University of Technology Swinburne, Hawthorn VIC 3122, Australia; rohanbennett@swin.edu.au
7 Kadaster International, 7311 KZ Apeldoorn, The Netherlands
* Correspondence: e.c.stocker@utwente.nl; Tel.: +31534894099

Received: 14 March 2019; Accepted: 24 April 2019; Published: 1 May 2019

Abstract: Unmanned Aerial Systems (UAS) are emerging as a tool for alternative land tenure data acquisition. Even though UAS appear to represent a promising technology, it remains unclear to what extent they match the needs of communities and governments in the land sector. This paper responds to this question by undertaking a socio-technical study in Rwanda, aiming to determine the match between stakeholders' needs and the characteristics of the UAS data acquisition workflow and its final products as valuable spatial data for land administration and spatial planning. A needs assessment enabled the expression of a range of land information needs across multiple levels and stakeholder sectors. Next to the social study, three different UAS were flown to test not only the quality of data but the possibilities of the use of this technology within the current institutional environment. A priority list of needs for cadastral and non-cadastral information as well as insights into operational challenges and data quality measures of UAS-based data products are presented. It can be concluded that UAS can have a significant contribution to match most of the prioritized needs in Rwanda. However, the results also reveal that structural and capacity conditions currently undermine this potential.

Keywords: UAS; UAV; needs assessment; cadastre; aerial photogrammetry; land administration; fit-for-purpose

1. Introduction

Since the early 2000s, Unmanned Aerial Systems (UAS) have become increasingly significant for both scientific as well as commercial applications [1,2]. The advent of this low cost, reliable, and user-friendly platform, as well as recent developments in digital photogrammetry and structure from motion (SfM) image processing software solutions, create new opportunities for collecting timely, tailored, detailed, and high-quality geospatial information. Studies on the various surveying technologies provide evidence that UAS can fill the data acquisition gap between time-consuming but highly accurate ground surveys, and faster yet relatively expensive classical aerial surveys [3]. Evidence of numerous UAS-based data acquisition missions across the globe prove the capabilities of this innovative technique. The platform has been applied to various domains such as agriculture [4,5], geosciences [6–10], and disaster risk management [11,12]. [13] provides a detailed review of remote sensing applications based on UAS. General advantages of UAS as remote sensing platform are the flexibility regarding application and usage, the high resolution of images, the ease-of-use and the immediateness of the results. Common drawbacks are regulatory uncertainties [14] and time-consuming ground control measurements if real-time kinematic (RTK) or post-processing kinematic (PPK) based workflows are not an option. Resulting data products include true orthoimages, digital elevation models and 3D point clouds, which are increasingly harnessed as a spatial framework to accomplish land administration processes. The applicability of UAS for various cadastral purposes has been tested in various pilots, e.g., meeting juridical boundary requirements in western Europe [15–17], mapping customary land rights in Namibia [18], and boundary mapping in Indonesia [19]. All pilot studies remained at a small-scale, reaching from several households to entire neighbourhoods.

Compared to other remote sensing techniques such as satellite images or classical aerial images, UAS data has clear advantages in the resolution, which is often below 10 cm and provides a high level of detail. To reach low ground sampling distances, flight height is usually set to less than 100 m–a limitation which is mostly also given by the national UAS regulations. Thus, the scale of one UAS missions is very low, reaching from a few hectares up to hundreds of hectares, depending on the platform, the field of view of the sensor, image overlap and flight pattern. Thus, aerial/satellite images are more suitable for large-scale mapping. With a particular focus on land rights recording, [20] concluded additional advantages of UAS data collection workflows: reliability of the data, open and transparent data collection procedure and the ease of implementation. The latter parameters are of particular importance to the implementation of fit-for-purpose land administration tool with a strong focus on developing countries.

While UAS appear to be a promising technology, there has been little discussion in the literature as to what extent this technology can match the needs of communities and governments especially when land administration is absent, incomplete, or in a state of decay. A flexible and pragmatic approach to meet the needs of people and their relationship to land refers to the key principle of recent land administration approaches [21–23]. Unlike leveraging technical standards, these approaches advocate that the data acquisition method of the underlying spatial framework should have a strong focus on managing current land issues in a specific context. Little has been done to study how different innovative geospatial technologies fit different needs.

Therefore, this paper aims to critically examine the match between stakeholders' needs and the characteristics of the UAS data acquisition workflow and its final products as valuable spatial information for land administration. This was achieved through undertaking a case study in Rwanda where a mixed methods approach was applied. First, the needs of potential end-users were investigated; second, the UAS technology was trialled in Rwanda and third, the performance of the entire UAS-based data acquisition workflow and its ability to match end-user requirements was assessed. A combined analysis of qualitative, as well as quantitative results, provides the empirical basis for discussing the degree of fitness of UAS technology for matching users' needs. The integration of the results in a socio-technical discussion [24] makes this paper a significant contribution as it reveals the opportunities and limitations of UAS technology in the context of current discourses in land administration.

The remainder of this article is organized as follows. After a short overview of the study area in Rwanda, the third section describes the research methodology. The fourth section presents the results focusing on the needs assessment, the UAS test flight missions, and a synthesis, which ultimately debates the fitness of use of UAS technology to attain land administration and spatial planning processes. The discussion relates the results of this study to existing scientific investigations and further reflects on the significance of the work. The conclusion with opportunities and remaining challenges as well as a future outlook complete the article.

2. Study Area

Rwanda, with an area of over 26,000 km^2 and a population of almost 12 million people, is the most densely populated country in Africa (467 per km^2) [25]. The population of Rwanda is still mostly rural, with 83% living in rural areas [25] with the majority depending on subsistence farming although less than half the population own less than 0,5 ha of land or none at all [26]. Despite its land scarcity and prevalence of hilly landscapes, the country continues to be highly reliant on agriculture as a form of employment and subsistence, and an increasing population exerts a growing demand for housing and infrastructure. After independence in 1962, land ownership in the country has evolved from customary law to a system of state ownership. This shift was formalized with the implementation of a new land policy in 2004 and the Organic Land Law (OLL) in 2005, which aimed to improve tenure security through land registration, facilitate the development of an equitable land market in Rwanda and promote the sustainable use of land. In approximately 2013, a country-wide land tenure regularization program (LTRP) was completed where more than 11 million parcels were demarcated and almost 9 million parcels titled to offer Rwandan citizens a range of perceived social, legal and economic benefits. The LTRP used 96% aerial images captured in 2008 and 2009 and 4% satellite imagery as base data to demarcate and adjudicate parcel boundaries in a community-mapping exercise [27]. Geo-information derived from the LTRP has also enabled the development of a national cadastral map (title-based land administration system), which now underpins a range of purposes [28]. However, base data has not been updated since and geo-information is still based on orthoimages from 2008/2009.

When it comes to the organized use of UAS, Rwanda can be considered as progressive in comparison to other East African countries. At the 2017 World Economic Forum in Davos, high-level delegates from the Government of Rwanda promoted Rwanda as the first country to adopt performance-based UAS regulations. They further outline that development of infrastructure and policy frameworks will spur business growth and social impact. In October of 2016, Zipline and the Government of Rwanda launched the world's first national drone delivery service to make on-demand emergency blood deliveries to transfusion clinics across the country. After initial difficulties to receive the permission to operate beyond visual line of sight, the business experienced constant growth. In addition to introducing new products, Zipline plans to build a second distribution centre in the country's east [29]. Besides foreign businesses, local UAS companies such as Charis UAS Ltd. provide professional services in various UAS industries including mapping, crop monitoring, surveying and aerial photography.

3. Material and Methods

This paper employs a mixed methods approach including qualitative and quantitative data to assess the potential of UAS-technology to meet land administration requirements in developing countries. The research framework addresses both the social/institutional as well as the spatial/technical perspective (Figure 1). On the one hand, land information needs of various stakeholder groups are identified through a needs assessment process. On the other hand, case studies of multiple test flights provide input to evaluate the institutional environment and data quality of UAS-based orthoimages. Results are synthesized and jointly discussed to give a better understanding of UAS-technology as a fit-for-purpose tool in the context of land administration [21] and how policies can build on this.

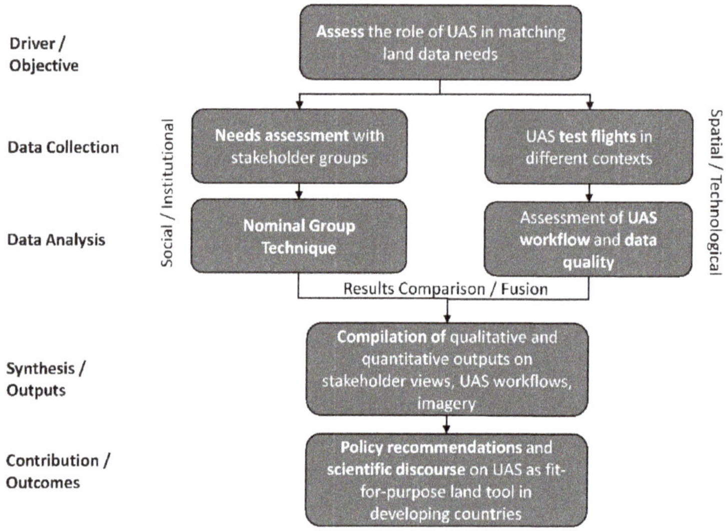

Figure 1. Conceptual framework of the multi-disciplinary approach.

3.1. *Needs Assessment*

Land information needs assessment for Rwanda was conducted using a form of group interview known as the Nominal Group Technique (NGT). NGT was selected as it facilitates a balanced input from all participants, taking advantage of individuals' knowledge and experience to provide deep and meaningful results ranked by importance to the topic of interest [30]. NGT is an effective approach when an identified problem requires a group's ideas and evaluation and therefore well-suited for conducting a needs assessment [31–33]. During the session, only one to two questions are posed to the group as each question takes around two hours to complete. A response to the question in terms of ideas are generated individually then gathered and combined as a group. Group consensus is reached through two rounds of individual voting, a process which prioritizes ideas and provides insight into the extent individual participants agree or disagree with the consensus vote. This structured process has been proven to be effective in addressing power imbalances or dominant behaviour in group data collection like some participants being more vocal than others [34–36].

Validity in the method is accounted for by recruiting participants who are considered experts on the topic [37]. Hence participants were identified by local land administration experts using purposive and snowball sampling. Thirty-eight *organizations* were contacted; of these, 22 participated (58% response rate). Three workshops were held at local and national levels. Invited organisations included national and local (district, sector and cell levels) public sector organisations associated with land (e.g., planning, housing, registration, infrastructure, development), non-statutory organisations, private sector organisations (e.g., leading geospatial consultancies), and several universities (Table 1). Invitations were sent to senior executives within organizations and it was left to the organization to send the most appropriate representative to the workshops. For national workshops, attending participants tended to be middle- or senior-level managers; at the local level, attending participants tended to be frontline operational staff.

At each workshop, one nominal question was posed (due to time limitations): "What land tenure and land-related information are still needed for sustainable *urbanization*?". This was followed by a discussion on how UAS might meet these needs. Cell (smallest administrative entity in Rwanda) officials who could not attend the workshops were interviewed individually using an adapted version of the NGT. Data collection ceased after six interviews when no significantly different insights were gained after four interviews.

Table 1. Types of stakeholders participating in data collection workshops.

Stakeholder Class	Organisations Contacted	Organisations Participated
Public sector organizations specific to land administration (national, province, district, sector, cell levels)	12	12
Public sector organizations (adjacent domains to land)	12	3
Non-statutory organizations	1	1
Private sector organizations	3	3
NGOs, Not-for-profit/Donors and Development partners;	6	0
Research & Development (R&D)	4	3

3.2. UAS Data Collection

In general, the UAS-based data acquisition workflow includes both technical and non-technical aspects. As shown in Figure 2, the UAS itself, the UAS pilots as well as the legal permission to conduct UAS flights refer to the main requirements to proceed with the data acquisition. In this research investigation, the UAS data collection aimed to provide an accurate orthophoto of the study area. Flight planning, data acquisition, and data processing were executed accordingly.

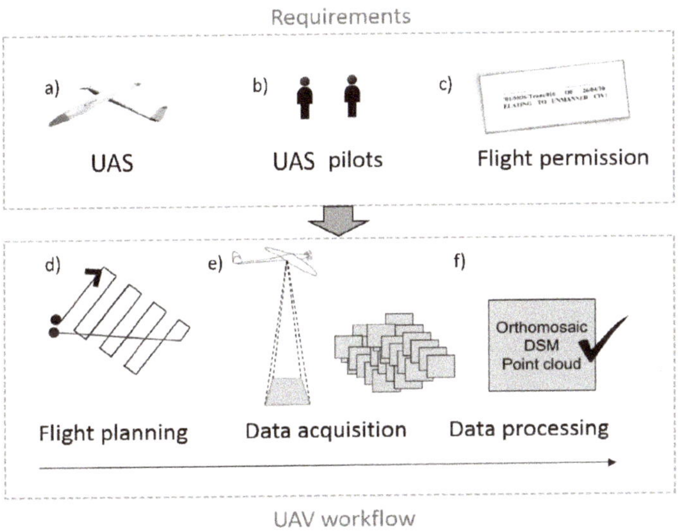

Figure 2. Unmanned Aerial Systems (UAS) data collection–requirements and data acquisition workflow. (**a**) UAS equipment, payload and the ground control station; (**b**) trained staff to pilot and operate the UAS; (**c**) legal permission to conduct the UAS flight mission which can set its own requirements according to the national jurisdiction; (**d**) flight planning with an appropriate software and definition of flight characteristics; (**e**) acquisition of UAS images in the area of interest; (**f**) data cleansing and photogrammetric processing including quality assessment.

3.2.1. UAS Regulations in Rwanda

UAS related regulations are a vital requirement in the safe and successful use of UAS technology. In May 2016, the Ministerial Regulations N°01/MOS/Trans/016 relating to the use of UAS in Rwanda were officially gazetted [38]. Respective regulations are very prescriptive and contain subparts dealing with UAS registration and marking, privacy and safety, airworthiness certification, operating rules and pilot licensing [14]. Before any commencement of activities, the UAS needs to be registered and marked with a unique identifier. Furthermore, pilots, as well as operating agencies, need to hold specific licenses issued by Rwanda Civil Aviation Authority. These requirements demand a high standard of UAS professionality and make it a challenge for external companies and institutions to obtain legal flight permissions. At the time of writing, the authors were yet to complete the administrative procedure required (despite commencing the process in 2017) to operate UAS in Rwanda. Therefore all data collection flights were carried out by Charis UAS Ltd., a Rwandan company specialized in UAS services and the first UAS certified company in Rwanda. The experiences of the authors with the UAS regulations and respective governmental institutions point at very high institutional barriers for market entry. There is only one company which is a certified UAS operator (for land-related mapping) and arguably has a monopoly position. For the specific case related to the work at hand, processes were not transparent and slow with limited access and availability of authoritative, unambiguous and assured information. Although UAS regulations are in place, gaps and lack of capacity can be seen when it comes to both enforcement and implementation.

Besides requirements towards pilot certification, UAS registration, and operator certification, Rwandan UAS regulations outline several operational limitations that have to be taken into account during all UAS flight missions (Table 2). In general, most specifications reflect common restrictions [39] except for the lateral distance between the pilot and the UAS. Even though the visual line of sight remains undefined, the maximum lateral distance of the pilot to the UAS in operation was set to 300 m in 2016. This imposed a substantial constraint to UAS mapping projects. However, in the course of 2018, UAS regulations were revised, and the maximum lateral distance disappeared from the restrictions and the flight height was lifted to 120 m [40]. Specifications of restricted areas and requirements towards distances to structures and people are comparable to standard practice.

Table 2. Operational limitations for UAS flight missions in Rwanda according to Rwandan regulations [38,40].

Operational Limitation	Specification
Maximum take-off weight	25 kg
Time for UAS operation	Only daylight operation
Minimum distance to aerodrome	10 km
Maximum flight height	100 m (increased to 120 m in 2018)
Visual Line Of Sight	Required but undefined
Maximum lateral distance pilot to UAS	300 m (abolished in 2018)
Minimum lateral distance to people, vessels, animals, building and structures	30 m (increased to 100 m in 2018)
Restricted areas	Congested areas of cities, towns or settlements
Ethics and privacy	Respect privacy of oth ers, surveillance of peo ple and property with out their consent is pro hibited

3.2.2. UAS Equipment

Three different types of UAS were tested in this study to assess the variety of UAS as a platform: one rotary-wing UAS (Inspire 2), one hybrid UAS (FireFLY 6) and one fixed-wing UAS (DT18). The consciously chosen platforms have different specifications in terms of operability, coverage, price, and necessity of ground control measurements. This study set-up reflected the broad spectrum of commonly used UAS and allowed to acknowledge these varieties within the assessment of the fitness of use. All platforms were equipped with an RGB sensor to capture nadir images (Table 3). The Inspire2 from DJI refers to semi-professional UAS with a focus for filmmaking. Both, the FireFLY6 and the DT18 PPK are survey-grade UAS of which the FireFLY6 presents a lower cost solution, and the DT18 PPK refers to a professional UAS with high-end components. The DT18 PPK is equipped with a combined Inertial Measurement Unit/ GNSS solution from Applanix (APX15) which allows direct georeferencing and minimizes the need for ground control measurements.

Table 3. Specifications of UAS used in this study.

Name	Inspire 2 (DJI)	FireFLY6 (BIRDSEYEVIEW)	DT18 PPK (Delair Tech)

Type	Rotary wing UAS	Hybrid UAS	Fixed-wing UAS
Sensor	Zenmuse X5S	SONY A6000	DT18 3Bands PPK
Sensor size	13 x 17.3 mm	23.5 x 15.6 mm	8.45 x 7.07 mm
Pixel pitch	2.48 μm	3.92 μm	3.45 μm
Sensor resolution	5280 x 3956 (20.1MP)	6000 x 4000 (24 MP)	2448 x 2048 (5MP)
Area	Busogo (50 ha)–3 flights	Muhoza (94 ha)–2 flights	Gahanga (14 ha)–1 flight
Data collection	497 nadir images (total flight time: 45 min)	991 nadir images (total flight time: 60 min)	372 nadir images (total flight time: 20 min)
Main features	Versality Requires only small space for landing	Flight stability Requires only small space for landing Long endurance	Long flight endurance PPK-capable Automatic flight and landing mode

During flight planning, the first step for the UAS data collection, areas for take-off and landing, the UAS trajectory and the flying height are specified. A typical procedure to create the flight trajectory with strips is 80% forward overlap and 40–80% side overlap [1] since redundancy can compensate for aircraft instabilities. The flight planning configurations in this study were constraint by the regulations (operational limitations), requirements for accurate data in an urban environment and external flight conditions such as wind and illumination. Therefore, taking all these factors into account, the flight missions were carried out with 80% forward and 70% side overlap. According to the regulations, flight height was set to 100m above the surface. All UAS were equipped with a RGB sensor and resulting ground sampling distances varied between 2–3cm depending on the specification of the camera.

To emphasize the diversity of possible flight configurations, data collection included three different contexts–one urban study area, one peri-urban and one rural study area. The location of the study sites is visualized in Figure 3.

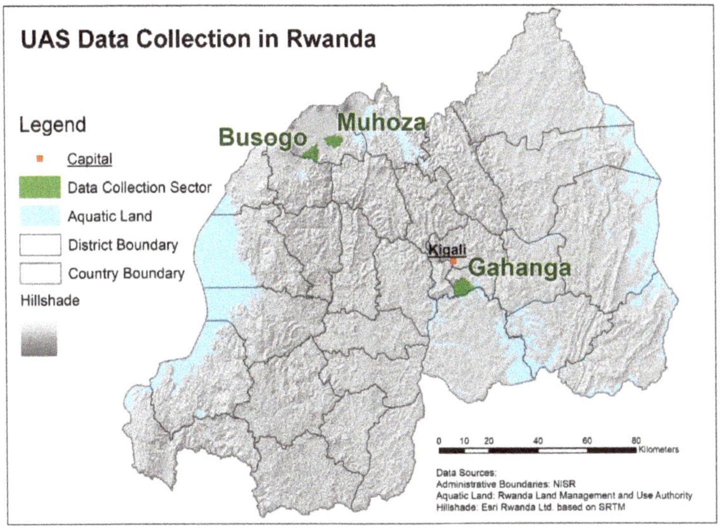

Figure 3. Overview of UAS data collection sites in Rwanda.

Due to the limited availability of open spaces, a hybrid UAS for the data collection of the densely populated urban environment in Muhoza Sector was chosen. In contrast, the rotary wing UAS was used to collect the images of the peri-urban area in Busogo Sector, as it is located on volcano slopes and highly affected by local wind systems which strongly influences the stability of fixed-wing UAS. Due to the regulatory restrictions and the difficulty to find sufficient large landing sites, the fixed-wing flight was conducted from a cricket stadium 20 km south of the City of Kigali. At the time of data acquisition, the maximum lateral distance of the pilot to the UAS was limited to 300m. Since only one area provided sufficient space for landing, the DT18 could only capture images over the cricket stadium which is embedded in a rural area in Gahanga Sector. Both, the Inspire 2 and the FireFLY6 were equipped with a consumer-grade GNSS antenna allowing geotagging of all images. However, the measurements of additional Ground Control Points (GCPs) indispensable. In contrast, the DT18 stood out for its high-quality navigation sensor that records precise attitude logs including both, angular observations (< 10 arcmins) as well as camera positions (<2.3 cm) [41]. However, former test flights showed, that the DT18 requires additional GCPs to correct for (minor) systematic errors [41], particularly when no GNSS corrections are applied.

3.2.3. Ground Control Measurements

Reference points were deployed and measured in all three case locations to include known point coordinates for georeferencing as well as a means for quality control. Clearly visible ground marker had a quadratic shape with an edge length of 30 cm showing a black and white chess pattern (cf. Figure 4) which were evenly distributed in the area of interest. As specified in Table 4, ground truth measurements were carried out with two different GNSS devices. The first was a pair of Leica CS10 stations used in a base and rover set with a final RTK measurement accuracy of 2 cm. The second device used was a handheld Trimble GeoXH receiving RTK corrections via the Rwandan Continuously Operating Reference Station (CORS) GeoNet with a final measurement accuracy of 10 cm. Whereas GCPs were included as a weighted observation during the photogrammetric image processing [42], checkpoints were not taken into account during image processing and present as a classical way to evaluate the geometric accuracy. The georeferenced orthomosaic has been generated following two different block orientation methods. The Gahanga dataset was processed by means of an integrated sensor orientation method [43] that uses the information of camera positions and attitude as well as object coordinates of GCPs for the Bundle Block Adjustment. Since no attitude measurements were available for the FireFLY6 and for the Inspire 2, the block orientation of the Muhoza and Busogo dataset followed the GNSS-supported Aerial Triangulation method based on information on camera positions and object coordinates of GCPs [cf. 43].

Figure 4. Measurement of reference points.

Table 4. Block orientation method and ground truthing information of all datasets.

Dataset	Block Orientation Method	GNSS Device for Ground Truthing Measurements	Count GCPs	Count Checkpoints
Muhoza	GNSS-supported Aerial Triangulation (GNSS-AT)	Leica CS10 and Trimble GeoXH	9	20
Busogo	GNSS-supported Aerial Triangulation (GNSS-AT)	Leica CS10	9	9
Gahanga	Integrated Sensor Orientation	Trimble GeoXH	5	8

Although less than ten reference points are sufficient to achieve high geometric accuracies, redundancy in deployed points has proven to be the preferable option as ground marker might get vanished or destroyed. Due to unforeseen administrative problems, the time between the deployment of the ground marker and the UAS flight itself was almost 5h. This can explain the fact that nearly 25% of all deployed points in the area of Muhoza were taken away. As summarised in Table 5, in the peri-urban and rural areas of Busogo and Gahanga the authors experienced less time delay as well as fewer losses of ground marker.

Table 5. Number of deployed reference points - count before and after the UAS flight.

Area	Teams Deployed	Reference Points Measured Pre-Flight	Reference Points Remained Post-Flight	Time between Measurement and Final Collection of Ground Marker
Muhoza	2	39	30	5 h
Busogo	1	22	18	3 h
Gahanga	1	13	13	2 h

3.2.4. Software and Hardware Requirements

UAS data has been processed with the commercially available software Pix4D (www.pix4d.com) which refers to a well-established professional photogrammetric software. Next to commercially available software, freely available software for UAS image processing, such as Open Drone Map (www.opendronemap.org), offer viable alternatives. Open Drone Map follows a structure-from-motion pipeline which is based on Open SFM. Whereas previous open source software most often had the deficiency to not provide an intuitive, user-friendly user interface, Open Drone Map can be used as a command line tool, with a live USB or via a user-friendly Web-based graphical user interface. Recommended system requirements are similar for Pix4D and Open Drone Map, and refer to 16GB RAM for small projects over with 100–500 images, 32 GB RAM for projects over 500 images and 64 GB RAM for very large projects with over 2000 images. The photogrammetric processing in this paper was completed with Pix4D and took several hours for the Gahanga and Busogo dataset and more than a day for the Muhoza dataset.

4. Results

4.1. What Land Information do Rwandan Stakeholders Need?

The results of the needs assessment revealed that land information needs were not merely about data, but also other enabling requirements such as access, functionality and tool types. Tables 6 and 7 show the outcomes of workshops with government and non-government stakeholders, and how final decisions around land information needs were prioritised. The column, 'relative importance' reflects the proportion of votes awarded to a specific need, while 'popularity' reflects the proportion of participants who voted for that need.

Table 6. Land information needs as identified by government stakeholders.

National Level Government	Relative Importance	Popularity
High accuracy satellite/aerial imagery	18.7	0.8
To know who owns what spatial data	14.7	0.8
Current land use information	9.3	0.4
3D cadastral data	8.0	0.4
Utility supply data	6.7	0.6
Convert existing web-based system to opensource	6.7	0.2
Match land parcel to admin boundary	6.7	0.2
Monitor operation of utilities and projects	6.7	0.4
Integration of utility supply data	6.7	0.4
Existing development at parcel level	4.0	0.2
Sub-national level Government (District)		
Highly accurate spatial data (incl. imagery)	29.63	1.00
More mobile tools	11.85	0.56
Physical characteristics of land	11.85	0.44
Access to information	10.37	0.56
Geological data	8.15	0.33
Land use	5.93	0.56
Implementation of masterplan and DLUP in an efficient way	4.44	0.22
Parcel boundaries	2.96	0.22
Location of underground infrastructure	2.96	0.22
All transactions made on parcel	1.48	0.11
Information to stakeholders	0.74	0.11
Wireless infrastructure	0.74	0.11
Local level Government (Cell)		
Spatial dataset of master plan and land parcels		0.67
GIS software		0.67
Soft copy of master plan		0.5
Soft copy of the DLUP		0.33
Integration of land use map with land information database		0.33
Information about planned infrastructure		0.17

Table 7. Land information needs identified by non-government stakeholders.

Non-Government	Relative Importance	Popularity
Value of land (valuation process)	22.67	0.8
Accessible open data	18.67	0.6
Consultative process around land use planning	12.00	0.4
More detailed (sub-use) land use planning in Master Plan	10.67	0.6
Actual land use information	9.33	0.6
History of land Information to resolve conflict between infrastructure development and arable land	9.33	0.6
Integrated demographic information	6.67	0.6
Sub-meter accuracy of parcel boundaries (urban/peri-urban)	6.67	0.4
Information about proposed infrastructure development and potential risks	4.00	0.4
Maintained web-based Master Plans	2.67	0.2

4.1.1. Government Stakeholders' Needs

Spatial data with a high accuracy (although this was not quantified by participants) related to land tenure and other land information was a priority for both national and sub-national government stakeholders, attracting almost 20 and 30 percent of the total votes respectively. At both national and sub-national levels, the emphasis was less on land tenure information and more on other types of land information such as utilities, existing developments and land use, climatic and topographic data.

Additionally, there was an emphasis on capacity needs (usability and accessibility), indicated in needs such as integration of land parcel other types of land information (e.g., utilities and administrative boundaries), the desire to transition to open source systems and have greater transparency around data custodianship and access rights, and implementation of the District Land Use Plan (DLUP) and masterplan.

4.1.2. Local Government and Communities' Needs

Similar needs were identified by local government, who prioritized needs around digital data and supporting software, and a related desire for land tenure information to be more readily integrated with existing or planned land use and infrastructure. In general, land use tended to accord with the use defined on the title, but inconsistency is starting to be an issue in areas transitioning to more urban land use types. Here, it is common for the community to be found not using their land as zoned, or in some instances, attributed to the information on the title not being updated. For example, in one Cell, despite being rezoned for urban land use (residential), some land titles still reflected the previously planned agricultural land use. In these instances, it appeared that land records were only updated if the landowner formally seeks a building permit or other land-related services; otherwise, the land title remains unharmonised with the Master Plan. Also, although most of the land in Rwanda has been demarcated and titled during the LTRP, some plots (or owners who occupy the plots) remain untitled due to information gaps at the time of the LTRP, e.g.,, lack of identification, family disputes, etc. The use of general boundaries during the LTRP for demarcation for titling has also not been updated accurately due to the resolution of GPS receivers (3 meters) or lack of GPS receivers, leading to the on-ground practices like pacing by foot to resolve conflicts.

Lack of information, or lack of access to information, about the Master Plan (i.e., information about proposed new development) was identified. This inhibits the ability of local government to play a role in plan implementation. Additionally, given that the Master Plan plays a crucial role in setting out future development, it appears that local community consultation is ad hoc. For example, in some villages, local communities do not participate in the establishment of the master plan/LUP: in some others, only Cell officers are contacted, whereby it then falls onto them to inform the community; in yet others, local consultation has been undertaken.

In summary, it appears that at the Cell level, land information needs can be generalised as lack of access to development information (which affects land use practices and enforcement of intended land use types) and lack of up-to-date spatial and administrative information about individual parcels or persistent gaps in information.

4.1.3. Non-Government Stakeholders' Needs

In contrast, non-government stakeholders' needs were less focused on land tenure information and more on other information needs. Information needs like land value, land use, history of land, and proposed infrastructure were identified; the only tenure-related information was sub-meter accuracy of parcel boundaries in urban and peri-urban areas. It is no surprise that capacity needs around data accessibility, stakeholder engagement (e.g., consultation) and up-to-data web-based masterplans were all identified and prioritized.

4.2. What Data Quality can be Achieved with UAS-Technology?

The data quality can be derived from checkpoint residuals and a visual evaluation of the final orthomosaic. Final checkpoint residuals are outlined in Table 8. Almost pixel-level geometric accuracy was achieved with the Busogo dataset. Both Gahanga and Busogo show more than 10cm RMS error of horizontal residuals. Differences regarding final geometric accuracy can be attributed to the UAS equipment and sensor as well as to the device and conditions for the measurements of reference points. Nonetheless, obtained orthomosaics are of high geometric accuracy and comparable to results in other scientific contributions [44,45].

Table 8. Specifications of final results.

Area	Ground Sampling Distance	RMS Error of CP Residuals (X/Y/Z)
Muhoza	2.16 cm	0.122m/0.086m/0.467m
Busogo	2.18 cm	0.033m/0.031m/0.349m
Gahanga	2.63 cm	0.127m/0.170m/0.244m

The visual evaluation revealed commonalities but also some differences in the datasets. Figure 5 presents the final orthomosaics of all three datasets. It is evident that sufficient overlap was considered during the flight missions as no gaps were present and the area of interest was entirely covered by the reconstructed scene. Differences of the visual quality are evident in the close-up views. Here, the image quality was best for the Muhoza dataset, as most features including rooftops, as well as vegetation, were well exposed in the orthomosaic. In contrast, the Busogo dataset showed a lower image quality, visible in over-exposed roofs and problems to fulfil a proper histogram matching during image processing. This can be attributed to the adverse lighting conditions during data capture. Even though meteorological conditions were perfect for flying during data capture of the Gahanga dataset, the sensor showed substantial difficulties to deal with bright and dark image features. Especially a large part of the parking area is very overexposed, even though the surface was covered with reddish gravel.

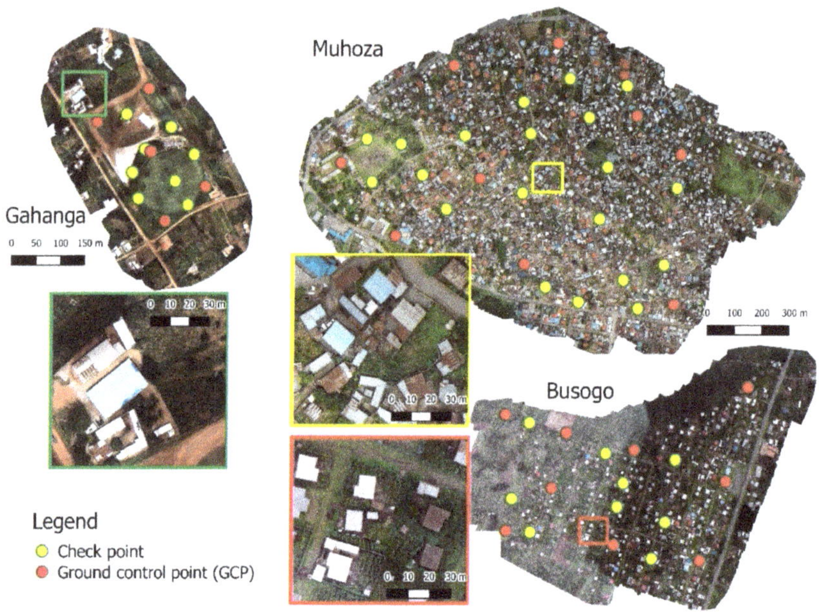

Figure 5. Overview of the generated orthomosaics and GCP/checkpoint distribution.

4.3. Can UAS Respond to the Needs Expressed by Different Stakeholders?

To draw conclusions on the ability of UAS data and UAS-based workflows to satisfy prioritized needs, the results of Tables 6 and 7 were categorised and integrated in a matrix which distinguishes between (a) needs where UAS data has no significant contribution toward the achievement and (b) needs that can be matched with UAS data (Figure 6). The latter category was further associated to one of the four key characteristics of UAS data: high geometric accuracy, provision of up-to-date data, high spatial and/or temporal resolution, and high level of interpretability.

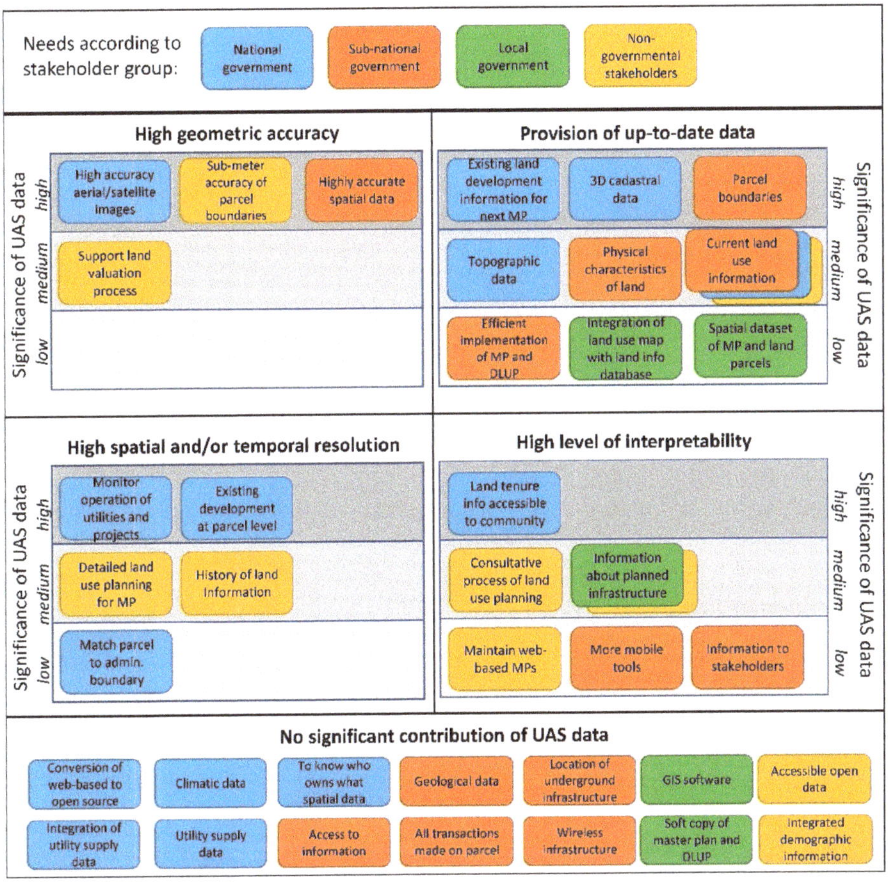

Figure 6. Prioritized needs classified by the ability of UAS data to match stated needs with further association to key characteristics of UAS data and ranking according to the significance of the contribution.

Figure 6 reveals that UAS data can have a significant contribution to match 27 out of 41 prioritised needs. The remaining 14 needs mainly refer to access to data, information, and software. A high and medium significance of UAS data was found mostly among national-level stakeholders, both governmental (eight needs) and non-governmental (seven needs) organisations. The needs of the local government could be met with medium (one need) or low (two needs) significant contributions of UAS data. Most of the prioritised needs of the sub-national government can only partially be fulfilled by UAS data (i.e., medium or low significance). A comparison of the four different characteristics shows that the provision of up to date data and the high level of interpretability are key in contributing to matching the stated needs. However, both aspects are highly interrelated to high geometric accuracy as well as high spatial resolution–otherwise, the data would not show such high level of detail which itself leads to high interpretability and its significant contribution to derive land use and topographic information. Although the quadrants in Figure 6 feature unique characteristics, all are interrelated and are therefore considered overlapping as well.

UAS regulations were found to have considerable impact on the scale of the utilisation of UAS in the context of land administration. Especially flight height and line of sight restrictions limit one data collection to several tens of hectares. Mapping larger areas would thus require constant moving of the ground control station with an adverse impact on the mapping efficiency. Geometric accuracy, was found to be less affected by UAS regulations. In contrast, the high level of interpretability and high spatial resolutions could be an issue when it comes to privacy and ethical constraints. Even though not the case in Rwanda, some countries demand public consent for the data collection of private property. A condition that requires sound data collection preparations and might put large restrictions on the UAS missions, particularly in urban and peri-urban areas.

4.3.1. High Geometric Accuracy

More specifically, the expressed need by government stakeholders for highly accurate spatial data can entirely be met by UAS imagery as shown by the low RMSE of checkpoint residuals in this study. Even though the national CORS in Rwanda cannot be considered as a source of GNSS corrections for PPK workflows, different means of georeferencing have proven to hit similar accuracies. This data characteristic facilitates the manual or digital delineation of parcel boundaries and support valuation and taxation processes–two needs which were prioritised by non-government stakeholders. The current cadastral map is based on the LTR programme which followed a general boundary approach which sometimes shows several meters offset to the correct location of the boundary. Most disputes arose during land transactions in densely populated areas, where plot sizes are small, and people depend on their land for subsistence farming. In those cases, UAS data ultimately facilitates a reliable and geometrically correct database to correct existing cadastral data.

4.3.2. Provision of up-to-Date Data

A comparison of the obtained UAS-based orthomosaic of Muhoza and the corresponding orthomosaic which is based on classical aerial photos from 2009 shows a high number of clearly visible changes (Figure 7), where 13 large buildings and 28 small buildings/annexes were newly constructed, 5 buildings were demolished, and 28 large buildings and 10 small buildings/annexes remained. Especially urban and peri-urban areas face numerous changes with regard to development and urbanization. The high level of detail and the immediate availability of aerial photos provide the geospatial basis to extract up-to-date land use, land development, and topographic information of small scale areas which is crucial to implement current urban development plans efficiently. Similarly, the timely provision of UAS data could support the delineation of parcel boundaries based on orthomosaics or regular base map updating activities. Especially up-to-date 3D point cloud data obtained from UAS images was identified as input for 3D cadastre. A data type, which can neither be provided by satellite images nor by aerial images.

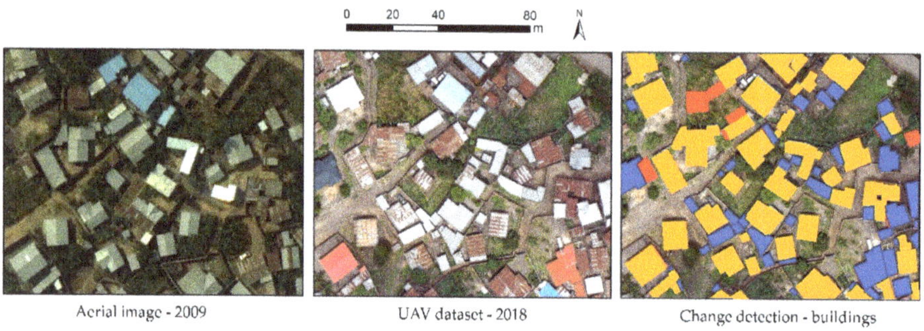

Figure 7. Left: Orthomosaic based on aerial images from 2009; centre: Orthomosaic based on UAS images from 2018; right: Change detection of buildings (orange: buildings remained the same, red: buildings got demolished, blue: new building constructions).

A low significant contribution of UAS data can be seen in contributing to a multi-purpose spatial data infrastructure which enables the integration of different data, which can further support the implementation of spatial development plans. In general, UAS-based data acquisition workflows allow stakeholders to gradually upgrade existing base-maps at a small scale, without the need for significant financial outlay upfront–two fundamental aspects of fit-for-purpose approaches.

4.3.3. High Spatial and/or Temporal Resolution

The proven flexibility of UAS data acquisition supports the collection of a multi-temporal base data for on-going and current tasks such as the revision of the National Development Plans (i.e., Master Plans) for Secondary Cities or development plans for towns and villages. Frequent changes in land use projects can be tracked and monitored using repetitive UAS data collections, especially since plot sizes and administrative areas in Rwanda tend to be small and lend themselves well to UAS data capture. Additionally, disputes about former land ownership and land use can be solved more efficiently with evidence from a multi-temporal database (i.e., history of land information). Next to the temporal aspect, the high spatial resolution of UAS data allows users to extract information about developments at parcel level or a more detailed land use planning which also includes sub-uses in urban and peri-urban areas.

4.3.4. High Level of Interpretability

This fundamental characteristic is attributed to the high level of detail in the generated UAS products which itself allows for a straightforward interpretation of the aerial dataset. People are more likely to correctly interpret the orthophoto as they recognise specific textures of the surface or physical features such as bushes, hedges or particular buildings. This allows UAS data to have a significant contribution in providing the database for visualising land tenure data or planned infrastructure–an asset which supports participatory mapping activities for land administration or urban planning. The authors observed, that in many cases, de-jure land rights do not represent de-facto land rights as the cadastral maps show little details on the physical extent (except for the parcel boundary). The integration of an orthophoto in the cadastral map would support the alignment of de-facto and de-jure land rights as it would spatially outline adjudicated land rights that are easy to interpret even for laymen. Furthermore, UAS data could aid consultative processes of land use planning with clear and understandable background data. A profound significance of UAS data was found in support of maintaining a web-based spatial plan, promoting more mobile tools, and sharing information with stakeholders.

5. Discussion

This study was designed to determine the match between stakeholders' needs and the characteristics of the UAS data acquisition workflow and its final products as useful spatial base data for land administration and spatial planning, particularly within the discourse of a more fit-for-purpose land administration.

5.1. Opportunities of UAS Data Collection to Match Land Information Needs

The socio-technical assessment revealed that the technical capabilities of UAS-based data are well-placed to match most of the prioritised needs in Rwanda. These needs did not only reflect the type of data (e.g., land use data, geological data, utility supply data, etc.) but also on characteristics of data and processes (e.g., geometric accuracy, spatial resolution, custodian of data, data integration, accessibility, etc.). This enabled the matching of the characteristics of UAS data to a particular type of data as well as the specific requirements of the data such as temporal resolution or geometric accuracy. The synthesis as shown in Figure 6 demonstrates that there is a high number of needs where UAS data could potentially have a significant contribution. The results suggest that UAS as a data acquisition device could most likely be adopted by national-level stakeholders or sub-national government stakeholders, which can be attributed to the system in Rwanda where the national government is the main provider of geospatial data. However, with UAS as a low-cost and flexible data acquisition platform, sub-national or local government stakeholders could increase their share of data provision, especially with regard to small scale mapping or multi-temporal flight missions in a local context. This would facilitate the co-production of land information in a decentralized way, a finding that is also reflected by [19]. The opportunity of using UAS-based images to delineate or enhance the accuracy of parcel boundaries is in line with the guiding principles for building fit-for-purpose land administration systems in developing countries [46]. Here, UAS were specifically outlined to provide the large scale image maps to map spatial units in densely populated areas (urban central, informal settlements and small towns). Results further suggest that UAS data can fulfil multiple needs across different domains such as planning and surveying. This is contrary to conventional ground surveying with GNSS or total station, where acquired data only serves a single purpose. Although not explicitly prioritised as a need, the UAS test flights showed that the (nearly) immediate availability of orthophotos could promote citizen participation in the adjudication process, a critical result which was also outlined by [19,47]. Even though the Rwandan land administration information system is very advanced in comparison to other African countries, it was found that the digital nature of a generated UAS-based orthomosaic can easily be integrated in existing spatial data infrastructures to be used by numerous GIS applications, or if absent, support the modernization of current paper-based land registration systems.

5.2. Challenges of UAS Data Collection to Match Land Information Needs

Aside from those advantages, the UAS test flights in Rwanda also reveal four main challenges with regard to the implementation of UAS as a data acquisition tool to match land information needs.

Firstly, it needs to be noted that the terrain in Rwanda–the country of the thousand hills–is a very challenging testbed. Fixed wing drones have only limited climbing rates, and flight planning must be aligned with the physical environment. The availability of sufficient open space for appropriate landing strips is an essential precondition which was found to be challenging to fulfil. Hybrid UAS and rotary wing UAS are likely a more suitable instrument for small scale mapping activities. Current limitations with regard to battery capacity and flight time make hybrid UAS more effective for mapping tasks as they have aFRTK better flight endurance. In contrast, rotary wing UAS should be preferred to monitor the operation of utilities.

The second hurdle refers to the UAS regulations in Rwanda. With an operational limitation to fly only in visual line of sight, scaled application of UAS-mapping activities remain aspirational. Acknowledging the plans of the Government of Rwanda, legislation with a more performance-based orientation may soon be drafted and implemented more effectively. This development could pave the way for broader use of UAS-based data acquisition that supports land tenure recording, as well as extensive land information collection for development purposes, as envisaged in [48].

The third hurdle includes the topic of ground truthing. It has been shown that especially in an urban environment, the collection and measurement of reference points are challenging and means of ground marking should be context-specific. PPK and RTK capable UAS can provide an answer to this challenge as they minimise or even eliminate the need for ground control measurements. However, the availability of professional GNSS equipment or a national network of existing GNSS reference stations is an essential precondition for RTK or PPK-based workflows. If the national CORS is not reliable or not existing, other means of accurate GNSS measurements such as Precise Point Positioning should be taken into consideration.

The fourth challenge refers to soft- and hardware requirements for data processing. Experiences of the authors in Rwanda revealed that the majority of employees of the Rwanda Land Management and Use Authority have a machine which could be able to process smaller datasets up to 500 images. However, to facilitate the processing of the data of an entire township, machines with more RAM and disk space would be needed–ideally a server environment. Cloud-based processing is seen very critical, as internet connections are very often subject to outages. Financial barriers to purchase the required hardware and a commercial software such as Pix4D or similar were perceived as very high–costs that are likely to exceed the procurement costs of the UAS equipment. At the same time, current open source software cannot reproduce the same data quality as commercially available software. However, given the rapid development of Open Drone Map, and the increasing number of users, the software algorithm is likely to mature in the close future.

5.3. Limitations of this Research and Future Work

The scope of this study was limited regarding the validation of the UAS technology in a relevant environment but not in an operational environment as the UAS flights were only trialled without the direct implementation of the data. Thus, this study did not address gaps and challenges on how the expressed stakeholder needs were actually met in the context of Rwanda. Future work could address this with a strong focus on implementation to evaluate the fitness-of-use of UAS workflows with due consideration of the entire innovation chain including GIS applications. This could be coupled with the design and evaluation of appropriate governance and capacity building models to allow the prototype demonstration of UAS-based workflows in an operational environment. Additionally, further research will be needed to explore the role of UAS compared to other geospatial technologies such as satellite data and classical aerial photographs in providing base data that serves as a spatial framework for the various land administration functions [49].

6. Conclusions

The presented work highlights the capabilities of UAS technology to match the needs of land professionals in Rwanda. Results of a sound needs assessment across different stakeholder groups demonstrate prioritised demands of respective respondents. Although ranked differently, the need for high-resolution, up-to-date land information is consistently identified in the final lists of all group discussions. Across the globe, UAS have become an attractive technology and only the upcoming years will show whether multiple governmental and non-governmental stakeholders can capitalise on the numerous benefits of this data acquisition method. The flight missions in Rwanda showed that UAS as a platform to remotely capture images have clear advantages in terms of fit-for-purpose data provision by facilitating accurate, up-to-date data with a potentially high spatial as well as temporal resolution. However, the integration of the needs assessment and the UAS test flights indicates that structural and capacity conditions currently undermine the vast potential of the UAS data acquisition method. Therefore, a key policy priority should be to implement country-specific capacity development and governance strategies; otherwise, scaled implementation and increasing technology uptake might remain wishful thinking. Notwithstanding the outlined challenges, the results of this study show that UAS technology has the potential to be an appropriate land tool with a significant contribution in catering the base data for most of the prioritized land information needs in Rwanda.

Author Contributions: Conceptualization, C.S. (Claudia Stöcker), M.K., R.B. and J.Z.; Data curation, C.S. (Claudia Stöcker); Formal analysis, C.S. (Claudia Stöcker) and S.H.; Funding acquisition, R.B.; Investigation, C.S. (Claudia Stöcker), S.H., P.N. and C.S. (Cornelia Schmidt); Project administration, M.K.; Supervision, M.K. and J.Z.; Visualization, C.S. (Claudia Stöcker); Writing–original draft, C.S. (Claudia Stöcker) and S.H.; Writing–review & editing, C.S. (Claudia Stöcker), S.H., P.N., C.S. (Cornelia Schmidt), M.K., R.B. and J.Z.

Funding: The research described in this paper was funded by the research project "its4land," which is part of the Horizon 2020 program of the European Union, project number 687828.

Acknowledgments: The authors thank Charis UAS Ltd. in Rwanda for the technical realization and support of all UAS test flights for this research investigation.

Conflicts of Interest: The authors declare no conflict of interest.

References

1. Colomina, I.; Molina, P. Unmanned aerial systems for photogrammetry and remote sensing: A review. *ISPRS J. Photogramm. Remote Sens.* **2014**, *92*, 79–97. [CrossRef]
2. Nex, F.; Remondino, F. UAV for 3D mapping applications: A review. *Appl. Geomat.* **2014**, *6*, 1–15. [CrossRef]

3. Mesas-Carrascosa, F.J.; Notario-García, M.D.; de Larriva, M.D.N.G.; de la Orden, M.S.; Porras, A.G.F. Validation of measurements of land plot area using UAV imagery. *Int. J. Appl. Earth Obs. Geoinf.* **2014**, *33*, 270–279. [CrossRef]
4. Honkavaara, E.; Kaivosoja, J.; Mäkynen, J.; Pellikka, I.; Pesonen, L.; Saari, H.; Salo, H.; Hakala, T.; Marklelin, L.; Rosnell, T. Hyperspectral Reflectance Signatures and Point Clouds for Precision Agriculture By Light Weight Uav Imaging System. *Ann. Photogramm. Remote Sens. Spat. Inf. Sci.* **2012**, *I-7*, 353–358. [CrossRef]
5. De Castro, A.I.; Torres-Sánchez, J.; Peña, J.M.; Jiménez-Brenes, F.M.; Csillik, O.; López-Granados, F. An automatic random forest-OBIA algorithm for early weed mapping between and within crop rows using UAV imagery. *Remote Sens.* **2018**, *10*, 285. [CrossRef]
6. Eltner, A.; Kaiser, A.; Castillo, C.; Rock, G.; Neugirg, F.; Abellan, A. Image-based surface reconstruction in geomorphometry—Merits, limits and developments of a promising tool for geoscientists. *Earth Surf. Dyn. Discuss.* **2015**, 1445–1508. [CrossRef]
7. James, M.R.; Robson, S.; d'Oleire-Oltmanns, S.; Niethammer, U. Optimising UAV topographic surveys processed with structure-from-motion: Ground control quality, quantity and bundle adjustment. *Geomorphology* **2017**, *280*, 51–66. [CrossRef]
8. Bemis, S.P.; Micklethwaite, S.; Turner, D.; James, M.R.; Akciz, S.; Thiele, S.T.; Bangash, H.A. Ground-based and UAV-Based photogrammetry: A multi-scale, high-resolution mapping tool for structural geology and paleoseismology. *J. Struct. Geol.* **2014**, *69*, 162–178. [CrossRef]
9. Nikolakopoulos, K.G.; Kozarski, D.; Kogkas, S. Coastal areas mapping using UAV photogrammetry. *Proc. SPIE* **2017**, *10428*, 104280O.
10. Szantoi, Z.; Smith, S.E.; Strona, G.; Koh, L.P.; Wich, S.A. Mapping orangutan habitat and agricultural areas using Landsat OLI imagery augmented with unmanned aircraft system aerial photography. *Int. J. Remote Sens.* **2017**, *38*, 2231–2245. [CrossRef]
11. Adams, S.M.; Friedland, C.J. A Survey of Unmanned Aerial Vehicle (UAV) Usage for Imagery Collection in Disaster Research and Management. In Proceedings of the 9th International Workshop on Remote Sensing for Disaster Response, Stanfort University, Stanford, CA, USA, 15–16 September 2011.
12. Erdelj, M.; Natalizio, E.; Chowdhury, K.R.; Akyildiz, I.F. Help from the Sky: Leveraging UAVs for Disaster Management. *IEEE Pervasive Comput.* **2017**, *16*, 24–32. [CrossRef]
13. Pajares, G. Overview and Current Status of Remote Sensing Applications Based on Unmanned Aerial Vehicles (UAVs). *Photogramm. Eng. Remote Sens.* **2015**, *81*, 281–330. [CrossRef]
14. Stöcker, C.; Bennett, R.; Nex, F.; Gerke, M.; Zevenbergen, J. Review of the current state of UAV regulations. *Remote Sens.* **2017**, *9*, 459. [CrossRef]
15. Manyoky, M.; Theiler, P.; Steudler, D.; Eisenbeiss, H. Unmanned Aerial Vehicle in Cadastral Applications. *ISPRS Int. Arch. Photogramm. Remote Sens. Spat. Inf. Sci.* **2012**, *XXXVIII-1*, 57–62. [CrossRef]
16. Rijsdijk, M.; Van Hinsbergh, W.H.M.; Witteveen, W.; Buuren, G.H.M.; Schakelaar, G.A.; Poppinga, G.; Van Persie, M.; Ladiges, R. Unmanned Aerial Systems in the Process of Juridical Verification of Cadastral Border. *Int. Arch. Photogramm. Remote Sens.* **2013**, *XL*, 4–6. [CrossRef]
17. Barnes, G.; Volkmann, W. High-Resolution Mapping with Unmanned Aerial Systems. *Surv. Land Inf. Sci.* **2015**, *74*, 5–13.
18. Mumbone, M.; Bennett, R.M.; Gerke, M.; Volkmann, W. Innovations in boundary mapping: Namibia, customary lands and UAVs. Proceedigs of the 16th Annual World Bank Conference on Land and Poverty 2015: Linking Land Tenure and Use for Shared Prosperity, Washington, DC, USA, 23–27 March 2015; Volume 1.
19. Ramadhani, S.A.; Bennett, R.M.; Nex, F.C. Exploring UAV in Indonesian cadastral boundary data acquisition. *Earth Sci. Inform.* **2018**, *11*, 129–146. [CrossRef]
20. Stöcker, C.; Koeva, M.; Bennett, R.; Zevenbergen, J. Evaluation of UAV-based technology to capture land rights in Kenya: Displaying stakeholder perspectives through interactive gaming. In Proceedings of the 20th Annual World Bank Conference on Land and Poverty 2019: Catalyzing Innovation, Washington, DC, USA, 25–29 March 2019.
21. Enemark, S.; Bell, K.C.; Lemmen, C.; McLaren, R. *Fit-For-Purpose Land Administration*; World Bank and International Federation of Surveyors: Washington, DC, USA, 2014; ISBN 9788792853103.
22. Zevenbergen, J.; De Vries, W.T.; Bennett, R.M. *Advances in Responsible Land Administration*; CRC Press: Boca Raton, FL, USA, 2015.

23. Zevenbergen, J.; Augustinus, C.; Antonio, D.; Bennett, R. Pro-poor land administration: Principles for recording the land rights of the underrepresented. *Land Use Policy* **2013**, *31*, 595–604. [CrossRef]
24. Ottens, M.; Stubkjær, E. A socio-technical analysis of cadastral systems. In *Real Property Transactions*; Zevenbergen, J., Frank, A., Stubkjær, E., Eds.; IOS Press: Clifton, VA, USA, 2007; p. 143. ISBN 9781586035815.
25. *Fourth Population and Housing Census, 2012 Census Atlas*; National Institute of Statistics Rwanda (NISR): Kigali, Rwanda, 2014.
26. *Rwanda National Human Development Report 2014 (March 2015)*; UNDP and Government of Rwanda 2015: New York, NY, USA, 2015.
27. Gillingham, P.; Buckle, F. *Rwanda Land Tenure Regularisation Case Study*; HTSPE Limited for Evidence on Demand and UK Department for International Development (DFID): Hertfordshire, UK, 2014.
28. Ngoga, T.H. *Rwanda's Land Tenure Reform*; CABI: Wallingford, UK, 2018.
29. Rosen, J. MIT Technology Review. 2017.
30. Delbecq, A.L.; Van de Ven, A.H.; Gustafson, D.H. *Group Techniques for Program Planning: A Guide to Nominal Group and Delphi Processes*; Green Briar Press: Denver, OR, USA, 1975; ISBN 9780673075918.
31. Sink, D.S. Using the nominal group technique effectively. *Natl. Product. Rev.* **1983**, *2*, 173–184. [CrossRef]
32. Witkin, B.R.; Altschuld, J.W. *Planning and Conducting Needs Assessments: A Practical Guide*; SAGE Publications Inc.: Thousand Oaks, CA, USA, 1995.
33. *World Bank Needs Assess. Knowl. Base*; World Bank Institute Evaluation Group Nominal Group Technique: Washington, DC, USA, 2007.
34. Delbecq, A.L.; Van de Ven, A.H. A group process model for problem identification and program planning. *J. Appl. Behav. Sci.* **1971**, *7*, 466–492. [CrossRef]
35. Langford, B.E.; Schoenfeld, G.; Izzo, G. Nominal grouping sessions vs focus groups. *Qual. Mark. Res. Int. J.* **2002**, *5*, 58–70. [CrossRef]
36. Lloyd, S. Applying the nominal group technique to specify the domain of a construct. *Qual. Mark. Res. Int. J.* **2011**, *14*, 105–121. [CrossRef]
37. Cantrill, J.A.; Sibbald, B.; Buetow, S. The Delphi and nominal group techniques in health services research The Delphi technique. *Int. J. Pharm. Pract.* **1996**, *4*, 67–74. [CrossRef]
38. *RCAA Ministerial Regulations No 01/MOS/Trans/016 of 26/04/2016 Relating to Unmanned Civil Aircraft System*; Rwanda Civil Aviation Authority: Kigali, Rwanda, 2016.
39. Stöcker, C.; Ho, S.; Bennett, R.; Koeva, M.; Schmidt, C.; Nkerabigwi, P.; Zevenbergen, J. Towards UAV-based Land Tenure Data Acquisition in Rwanda: Needs Assessment and Technology Response. In Proceedings of the XXVI FIG Congress 2018: Embracing Our Smart World Where the Continents Connect: Enhancing the Geospatial Maturity of Societies, Istanbul, Turkey, 6–11 May 2018.
40. *RCAA Ministerial Regulations No 01/MOS/Trans/2018 of 23/01/2018 Relating to Unmanned Civil Aircraft System*; Rwanda Civil Aviation Authority: Kigali, Rwanda, 2018.
41. Stöcker, C.; Nex, F.; Koeva, M.; Gerke, M. Quality assessment of combined IMU/GNSS data for direct georeferencing in the context of UAV-based mapping. In Proceedings of the ISPRS 2017: International Conference on Unmanned Aerial Vehicles in Geomatics, Bonn, Germany, 4–7 September 2017; Volume 42.
42. Remondino, F.; Barazzetti, L.; Nex, F.; Scaioni, M.; Sarazzi, D. Uav Photogrammetry for Mapping and 3D Modeling—Current Status and Future Perspectives. *Int. Arch. Photogramm.* **2011**, *XXXVIII*, 14–16. [CrossRef]
43. Benassi, F.; Dall'Asta, E.; Diotri, F.; Forlani, G.; di Cella, U.M.; Roncella, R.; Santise, M. Testing accuracy and repeatability of UAV blocks oriented with gnss-supported aerial triangulation. *Remote Sens.* **2017**, *9*, 172. [CrossRef]
44. Gerke, M.; Przybilla, H.J. Accuracy Analysis of Photogrammetric UAV Image Blocks: Influence of Onboard RTK-GNSS and Cross Flight Patterns. *Photogramm. Fernerkundung Geoinf.* **2016**, 17–30. [CrossRef]
45. Agüera-Vega, F.; Carvajal-Ramírez, F.; Martínez-Carricondo, P. Assessment of photogrammetric mapping accuracy based on variation ground control points number using unmanned aerial vehicle. *Meas. J. Int. Meas. Confed.* **2017**, *98*, 221–227. [CrossRef]
46. Enemark, S.; McLaren, R.; Lemmen, C.; Antonio, D.; Gitau, J. Guiding principles for building fit-for-purpose land administration systems in developing countries. In Proceedings of the Land and Poverty Conference 2016: Scaling up Responsible Land Governance, Washington, DC, USA, 14–18 March 2016.

47. Widodo, S.; Safik, A.; Chaerani, M.; Pramudhiarta, N.; Hardiono, M.; Feld, S.; Barthel, K.; Rolfes, L.; Biringkanae, N.; Sugiri, A. Empowering Communities to Mark Boundaries and Map Resources with Geospatial Technology: Early Results of using UAVs in Participatory Village Boundary Setting/Resource Mapping (VBS/RM) activity in Indonesia. In Proceedings of the Land and Poverty Conference 2016: Scaling up Responsible Land Governance, Washington, DC, USA, 14–18 March 2016.
48. *NEPAD Drones on the Horizon—Transforming African's Horizon*; Union African: Addis Ababa, Ethiopia, 2018.
49. Enemark, S. Building Land Information Policies. In Proceedings of the Special Forum on Building Land Information Policies in the Americas, Aguascalientes, Mexico, 26–27 October 2004.

MDPI
St. Alban-Anlage 66
4052 Basel
Switzerland
Tel. +41 61 683 77 34
Fax +41 61 302 89 18
www.mdpi.com

Remote Sensing Editorial Office
E-mail: remotesensing@mdpi.com
www.mdpi.com/journal/remotesensing

www.ingramcontent.com/pod-product-compliance
Lightning Source LLC
Chambersburg PA
CBHW061455210926
51850CB00008B/1720
9783039430543